Amir A. Zadpoor (Ed.)

Mechanics of Biomaterials

MDPI

This book is a reprint of the special issue that appeared in the online open access journal *Materials* (ISSN 1996-1944) in 2015 (available at: http://www.mdpi.com/journal/materials/special_issues/mechanics-biomaterials).

Guest Editor
Amir A. Zadpoor
Delft University of Technology (TU Delft)
The Netherlands

Editorial Office
MDPI AG
Klybeckstrasse 64
Basel, Switzerland

Publisher
Shu-Kun Lin

Managing Editor
Leo Jiang

1. Edition 2015

MDPI • Basel • Beijing • Wuhan

ISBN 978-3-03842-127-6 (PDF)
ISBN 978-3-03842-128-3 (Hbk)

Table of Contents

Chapter 1: Mechanics of Biological Tissues

Chapter 2: Mechanics of Biomaterials

List of Contributors

Seyed Mohammad Ahmadi: Faculty of Mechanical, Maritime and Materials Engineering, Delft University of Technology (TU Delft), Mekelweg 2, 2628 CD Delft, The Netherlands

Mahmoud Baniasadi: Department of Mechanical Engineering, University of Texas at Dallas, 800 W. Campbell Rd, Richardson, TX 75080, USA

Ricardo Bentini: Instituto de Química, Universidade de São Paulo, Av. Prof. Lineu Prestes, 748, 05508-000 São Paulo, SP, Brazil

Bryn Brazile: Department of Agricultural and Biological Engineering, Mississippi State University, Starkville, MS 39762, USA

Mariana C. Burrows: Instituto de Química, Universidade de São Paulo, Av. Prof. Lineu Prestes, 748, 05508-000 São Paulo, SP, Brazil

Ana C. O. Carreira: Instituto de Química, Universidade de São Paulo, Av. Prof. Lineu Prestes, 748, 05508-000 São Paulo, SP, Brazil; Faculdade de Medicina, Núcleo de Terapia Celular e Molecular (NUCEL)—Núcleo de Estudos e Terapia Celular e Molecular (NETCEM), Universidade de São Paulo, Rua Pangaré 100, 05360-130 São Paulo, SP, Brazil

Luiz H. Catalani: Instituto de Química, Universidade de São Paulo, Av. Prof. Lineu Prestes, 748, 05508-000 São Paulo, SP, Brazil

Kai Wang Chan: Department of Physics and Materials Science, City University of Hong Kong, Tat Chee Avenue, Kowloon, Hong Kong

Xuesi Chen: Key Laboratory of Polymer Ecomaterials, Changchun Institute of Applied Chemistry, Chinese Academy of Sciences, Changchun 130022, China

Yun Cui: Beijing Allgens Medical Science and Technology Co., Ltd., No. 1 Disheng East Road, Yizhuang Economic and Technological Development Zone, Beijing 100176, China

Fu-Zhai Cui: School of Materials Science and Engineering, Tsinghua University, Haidian District, Beijing 100084, China

Cristina del Amo: Multiscale in Mechanical and Biological Engineering (M2BE), Department of Mechanical Engineering, Aragon Institute of Engineering Research (I3A), Universidad de Zaragoza, Mariano Esquillor Street, 50018 Zaragoza, Spain

Jan Deprest: Center for Surgical Technologies, Faculty of Medicine, Universitair Ziekenhuis "Gasthuisberg" Leuven, Katholieke Universiteit Leuven, Leuven 3000, Belgium

Jianxun Ding: Key Laboratory of Polymer Ecomaterials, Changchun Institute of Applied Chemistry, Chinese Academy of Sciences, Changchun 130022, China

Nasser Fatouraee: Cardiovascular Engineering Laboratory, Faculty of Biomedical Engineering, Amirkabir University of Technology, 424 Hafez Ave., Tehran 15875-4413, Iran

Andrew Feola: Center for Surgical Technologies, Faculty of Medicine, Universitair Ziekenhuis "Gasthuisberg" Leuven, Katholieke Universiteit Leuven, Leuven 3000, Belgium

José Manuel García-Aznar: Multiscale in Mechanical and Biological Engineering (M2BE), Department of Mechanical Engineering, Aragon Institute of Engineering Research (I3A), Universidad de Zaragoza, Mariano Esquillor Street, 50018 Zaragoza, Spain

Uwe Gbureck: Department for Functional Materials in Medicine and Dentistry, University Hospital Würzburg, Pleicherwall 2, D-97070 Würzburg, Germany

Martha Geffers: Department for Functional Materials in Medicine and Dentistry, University Hospital Würzburg, Pleicherwall 2, D-97070 Würzburg, Germany

Flávia Gonçalves: Instituto de Química, Universidade de São Paulo, Av. Prof. Lineu Prestes, 748, 05508-000 São Paulo, SP, Brazil

Jürgen Groll: Department for Functional Materials in Medicine and Dentistry, University Hospital Würzburg, Pleicherwall 2, D-97070 Würzburg, Germany

Linxia Gu: Nebraska Center for Materials and Nanoscience, Lincoln, NE 68588-0656, USA

Xinyi Guo: Department of Physics, Wake Forest University, 7507 Reynolda Station, Winston-Salem, NC 27109, USA

Martin Guthold: Department of Physics, Wake Forest University, 7507 Reynolda Station, Winston-Salem, NC 27109, USA

Liping Heng: School of Chemistry and Environment, Beihang University, Beijing 100191, China

Hong-Jiang Jiang: Wendeng Orthopaedic Hospital, No. 1 Fengshan Road, Wendeng 264400, Shandong, China

Lei Jiang: School of Chemistry and Environment, Beihang University, Beijing 100191, China

Patricia M. Kossugue: Faculdade de Medicina, Núcleo de Terapia Celular e Molecular (NUCEL)—Núcleo de Estudos e Terapia Celular e Molecular (NETCEM), Universidade de São Paulo, Rua Pangaré 100, 05360-130 São Paulo, SP, Brazil

Shiva Kotha: Biomedical Engineering, Rensselaer Polytechnic Institute, Troy, NY 12180, USA

Baosheng Li: Key Laboratory of Cancer Prevention and Therapy, Tianjin Medical University Cancer Institute and Hospital, Tianjin 300070, China; Department of Radiation Oncology, Shandong Cancer Hospital, Shandong Academy of Medical Sciences, Jinan 250117, China

Dongsong Li: Department of Orthopaedic Surgery, the First Hospital of Jilin University, Changchun 130021, China

Tianyi Li: Orthopedics Dept. 2, Heilongjiang Provincial Corps Hospital of Chinese People's Armed Police Forces, Harbin 150076, China

Wei Li: Department of Physics, Wake Forest University, 7507 Reynolda Station, Winston-Salem, NC 27109, USA

Jun Liao: Department of Agricultural and Biological Engineering, Mississippi State University, Starkville, MS 39762, USA

Shengmao Lin: Department of Mechanical and Materials Engineering, University of Nebraska-Lincoln, Lincoln, NE 68588-0656, USA

Jianguo Liu: Department of Orthopaedic Surgery, the First Hospital of Jilin University, Changchun 130021, China

Tomas Lucioni: Department of Physics, Wake Forest University, 7507 Reynolda Station, Winston-Salem, NC 27109, USA

Xin-Long Ma: Tianjin Hospital, No. 406 Jiefang South Road, Tianjin 300211, China

James Macione: Biomedical Engineering, Rensselaer Polytechnic Institute, Troy, NY 12180, USA

Manfred M. Maurer: Institute of Mechanical Systems, ETH Zurich, Leonhardstrasse 21, Zurich 8092, Switzerland

Edoardo Mazza: Institute of Mechanical Systems, ETH Zurich, Leonhardstrasse 21, Zurich 8092, Switzerland; Empa—Swiss Federal Laboratories for Materials Science and Technology, Überlandstrasse 129, Dübendorf 8600, Switzerland

Johann G. Meier: ITAINNOVA Instituto Tecnológico de Aragón, 7-8 María de Luna Street, 50018 Zaragoza, Spain

Xiangfu Meng: Department of Chemistry, Capital Normal University, Beijing 100048, China

Majid Minary-Jolandan: Department of Mechanical Engineering, University of Texas at Dallas, 800 W. Campbell Rd, Richardson, TX 75080, USA

Bojan Mitevski: Brandenburg University of Technology Cottbus-Senftenberg, Konrad-Wachsmann-Allee 17, Cottbus 03046, Germany

Oihana Moreno-Arotzena: Multiscale in Mechanical and Biological Engineering (M2BE), Department of Mechanical Engineering, Aragon Institute of Engineering Research (I3A), Universidad de Zaragoza, Mariano Esquillor Street, 50018 Zaragoza, Spain

Sterling Nesbitt: Biomedical Engineering, Rensselaer Polytechnic Institute, Troy, NY 12180, USA

Sourav Patnaik: Department of Agricultural and Biological Engineering, Mississippi State University, Starkville, MS 39762, USA

Behdad Pouran: Faculty of Mechanical, Maritime and Materials Engineering, Delft University of Technology (TU Delft), Mekelweg 2, 2628 CD Delft, The Netherlands; Department of Orthopedics and Department of Rheumatology, University Medical Center Utrecht, Heidelberglaan 100, 3584 CX Utrecht, The Netherlands

Raj Prabhu: Department of Agricultural and Biological Engineering, Mississippi State University, Starkville, MS 39762, USA

Zhi-Ye Qiu: School of Materials Science and Engineering, Tsinghua University, Haidian District, Beijing 100084, China; Beijing Allgens Medical Science and Technology Co., Ltd., No. 1 Disheng East Road, Yizhuang Economic and Technological Development Zone, Beijing 100176, China

Barbara Röhrnbauer: Institute of Mechanical Systems, ETH Zurich, Leonhardstrasse 21, Zurich 8092, Switzerland

Mary Rougeau-Browning: Department of Genetics, University of Georgia, Athens, GA 30602, USA

Jan Schrooten : Department of Metallurgy and Materials Engineering, KU Leuven, Kasteelpark Arenberg 44, PB 2450, 3001 Leuven, Belgium

Wentzell Scott: Biomedical Engineering, Rensselaer Polytechnic Institute, Troy, NY 12180, USA

Mohammad B. Shadmehr: Tracheal Diseases Research Center, National Research Institute of Tuberculosis and Lung Diseases (NRITLD), ShahidBeheshti University of Medical Sciences, Tehran 19575-154, Iran

Amanda Smelser: Department of Biochemistry and Molecular Biology, School of Medicine Wake Forest University, Winston-Salem, NC 27157, USA

Mari C. Sogayar: Instituto de Química, Universidade de São Paulo, Av. Prof. Lineu Prestes, 748, 05508-000 São Paulo, SP, Brazil ; Faculdade de Medicina, Núcleo de Terapia Celular e Molecular (NUCEL)—Núcleo de Estudos e Terapia Celular e Molecular (NETCEM), Universidade de São Paulo, Rua Pangaré 100, 05360-130 São Paulo, SP, Brazil

Mohammad Tafazzoli-Shadpour: Cardiovascular Engineering Laboratory, Faculty of Biomedical Engineering, Amirkabir University of Technology, 424 Hafez Ave., Tehran 15875-4413, Iran

Hadi Taghizadeh: Cardiovascular Engineering Laboratory, Faculty of Biomedical Engineering, Amirkabir University of Technology, 424 Hafez Ave., Tehran 15875-4413, Iran

Xun-Xiang Tan: Wendeng Orthopaedic Hospital, No. 1 Fengshan Road, Wendeng 264400, Shandong, China

Yuqi Teng: School of Chemical & Environmental Engineering, China University of Mining & Technology, Beijing 100083, China

Sie Chin Tjong: Department of Physics and Materials Science, City University of Hong Kong, Tat Chee Avenue, Kowloon, Hong Kong

Juan Wang: Key Laboratory of Cancer Prevention and Therapy, Tianjin Medical University Cancer Institute and Hospital, Tianjin 300070, China; Department of Radiation Oncology, Shandong Cancer Hospital, Shandong Academy of Medical Sciences, Jinan 250117, China

Ruebn Wauthle: LayerWise NV, Kapeldreef 60, 3001 Leuven, Belgium

Benjamin Weed: Department of Agricultural and Biological Engineering, Mississippi State University, Starkville, MS 39762, USA

Harrie Weinans: Faculty of Mechanical, Maritime and Materials Engineering, Delft University of Technology (TU Delft), Mekelweg 2, 2628 CD Delft, The Netherlands; Department of Orthopedics and Department of Rheumatology, University Medical Center Utrecht, Heidelberglaan 100, 3584 CX Utrecht, The Netherlands

Sabine Weiss: Brandenburg University of Technology Cottbus-Senftenberg, Konrad-Wachsmann-Allee 17, Cottbus 03046, Germany

Lakiesha N. Williams: Department of Agricultural and Biological Engineering, Mississippi State University, Starkville, MS 39762, USA

Hoi Man Wong: Department of Orthopedics and Traumatology, Li Ka Shing Faculty of Medicine, the University of Hong Kong, Hong Kong

Jin Xu: Kangda College of Nanjing Medical University, No. 8 Chunhui Road, Xinhai District, Lianyungang 222000, Jiangsu, China

Qiaowen Yang: School of Chemical & Environmental Engineering, China University of Mining & Technology, Beijing 100083, China

Saber Amin Yavari: Faculty of Mechanical, Maritime and Materials Engineering, Delft University of Technology (TU Delft), Mekelweg 2, 2628 CD Delft, The Netherlands; Department of Orthopedics and Department of Rheumatology, University Medical Center Utrecht, Heidelberglaan 100, 3584 CX Utrecht, The Netherlands

Kelvin Wai Kwok Yeung: Department of Orthopedics and Traumatology, Li Ka Shing Faculty of Medicine, the University of Hong Kong, Hong Kong

Amir A. Zadpoor: Department of Biomechanical Engineering, Faculty of Mechanical, Maritime, and Materials Engineering, Delft University of Technology (TU Delft), Mekelweg 2, Delft 2628CD, The Netherlands

Yuqi Zhang: College of Chemistry and Chemical Engineering, Yan'an University, Yan'an, Shaanxi 716000, China

Zi-Qiang Zhang: Beijing Allgens Medical Science and Technology Co., Ltd., No. 1 Disheng East Road, Yizhuang Economic and Technological Development Zone, Beijing 100176, China

About the Guest Editor

Amir Zadpoor is Associate Professor and Chair of Biomaterials & Tissue Biomechanics section at the Department of Biomechanical Engineering, Delft University of Technology (TU Delft). He obtained his PhD cum laude from the same university after completing his MSc and BSc studies respectively in Biomedical Engineering and Mechanical Engineering. His research interests are in biofabircation, biomaterials for tissue regeneration, skeletal tissue biomechanics, and mechanobiology. Amir has extensively published in his above-mentioned areas of interest and serves on the editorial boards of several international journals.

Preface

Mechanics of Biological Tissues and Biomaterials: Current Trends

Amir A. Zadpoor

Abstract: Investigation of the mechanical behavior of biological tissues and biomaterials has been an active area of research for several decades. However, in recent years, the enthusiasm in understanding the mechanical behavior of biological tissues and biomaterials has increased significantly due to the development of novel biomaterials for new fields of application, along with the emergence of advanced computational techniques. The current Special Issue is a collection of studies that address various topics within the general theme of "mechanics of biomaterials". This editorial aims to present the context within which the studies of this Special Issue could be better understood. I, therefore, try to identify some of the most important research trends in the study of the mechanical behavior of biological tissues and biomaterials.

Reprinted from *Materials*. Cite as: Zadpoor, A.A. Mechanics of Biological Tissues and Biomaterials: Current Trends. *Materials* **2015**, *8*, 4505-4511.

1. Introduction

The mechanical behavior of biological tissues and biomaterials has been intensively studied for decades, but has recently been receiving increasing attention. The mechanical properties of biological tissues were traditionally studied within the biomechanics community. However, the biomaterials community is becoming interested in this field of research through analyzing the most important predictors of biomaterials suitability, especially their stiffness and strength. During the last decade, our ability to characterize and analyze biological tissues, on the one hand, and design and synthesize "multi-functional biomaterials", on the other hand, has improved substantially. In many cases, these multi-functional biomaterials either replace or enable the regeneration of damaged tissues. There are, therefore, either temporary or permanent interactions between (evolving) tissues and the multi-functional biomaterials that come in contact with them. These interactions take several forms but mechanical interaction is one of the most important types, particularly for load-bearing tissues, such as musculoskeletal tissues. Within the context of these developments, a wider range of researchers have become interested in studying the mechanical interactions between tissues and biomaterials. In many cases, this means the study of the mechanical behavior of both biological tissues and biomaterials, not only to determine the basic mechanical properties, but also to extract the type of data that is needed for advanced constitutive modeling of those materials. Moreover, the multi-functional nature of many biomaterials conveys that their mechanical properties are not only important from the mechanical and load-bearing viewpoints, but also in the way that they influence their other bio-functionalities. There are indeed many examples where the mechanical properties of biomaterials influence and/or regulate their biological performance. For example, it is shown that the physical and mechanical properties of the matrix on which stem cells are cultured could influence the behavior of stem cells [1,2]. Moreover, post-manufacturing treatments of biomaterials, which are usually aimed at improving one or more functionalities of the biomaterials, could influence the mechanical function of biomaterials as well. That is why the different functionalities of biomaterials need to be simultaneously studied. Finally, there is a recent trend in the "rational design" of biomaterials, where materials with specific micro-architectures are designed to achieve specific mechanical and biological properties. Given the recent advances in 3D printing and additive manufacturing, it is now possible to manufacture almost any such design, meaning that an unlimited number of rationally designed biomaterials have become available that need to be studied, among other aspects, from the mechanical viewpoint.

Given all the above-mentioned developments, we felt it is a good time to dedicate a Special Issue of the journal *Materials* to the study of the mechanical behavior of biological tissues and biomaterials. Many authors from all around the world contributed their latest research, which were subsequently reviewed to select the studies that form this Special Issue. This editorial tries to present the context within which these selected studies could be better understood.

2. Mechanics of Biological Tissues

There are several reasons why one may be interested in the mechanical behavior of biological tissues. The relevance of such studies is very clear for skeletal tissues, such as bone, cartilage, and tendon, whose main functions are structural. That is why many of the earliest studies on the mechanical behavior of biological tissues were focused on skeletal tissues. To date, skeletal tissues are among the most intensively studied biological tissues in terms of their mechanical behavior. However, many more types of tissues are now being studied, including brain [3–5], liver [6,7], muscle [8,9], and adipose tissue [10]. In the remainder of this section, I will highlight three of the most important areas where the mechanical properties of tissues are needed.

2.1. Constitutive Modeling of Biological Tissues

Advanced materials models are needed when describing the mechanical response of biological tissues to multi-axial loading. That is partly due to the heterogeneity of the mechanical properties, anisotropy, time-dependency of the mechanical behavior, presence of several phases (fluid, solid, ions, *etc.*), and adaptation of mechanical properties to mechanical loading. That is particularly important when developing computational models of tissues.

Since most biological tissues are strongly hierarchical, it is particularly interesting to relate the microstructure of tissues to their macro-scale mechanical behavior. In the present issue, Taghizadeh *et al.* [11] relate the mechanical behavior of aortic tissue to the lamellar structure. In another study, Li *et al.* [12] have developed a highly stretchable substrate made from fugitive glue that could be used to study "the effects of large strains on biological samples".

2.2. Tissue Regeneration

Regeneration of damaged tissues using tissue engineering and regenerative medicine approaches is an important aim pursued by the biomedical engineering community. In order to regenerate tissues, one needs to provide the proper environment for tissue regeneration including media (*i.e.*, scaffolds, gels) that are mechanically strong enough to support the process of tissue regeneration. At the same time, tissue engineering scaffolds should not be overly stiff, because they might otherwise impede the regeneration of tissues [13,14]. It is therefore natural to ask: "What would are the optimal range of the mechanical properties of tissue engineering scaffolds and gels?" One approach is to characterize the native tissue to gain some insight into the expected range of mechanical properties [13,14]. This approach has some limitations, because the mechanical properties required to optimally support the process of tissue regeneration may not necessarily be the same as those of the native tissue. Despite those limitations, the properties of the native tissue are, in many cases, the best available starting point. Moreover, the mechanical properties of biological tissues could be used for diagnosis of diseases that manifest themselves in terms of changes in the mechanical properties of tissues. In this issue, Nesbitt *et al.* [15] study the mechanical behavior of the skin tissue and how collagen fibrils respond to mechanical loading. This type of information could be potentially useful both for diagnosis of skin diseases and for tissue engineering applications.

2.3. Tissue Damage and Trauma

Mechanical loading of tissues combined with underlying diseases could lead to tissue damage. This includes not only the non-physiological loading that is experienced in traumatic events but also physiological loading of tissues when a chronic disease such as osteoporosis is present. In a chronic disease such as osteoporosis, one is interested in knowing what level of mechanical loading could the bones tolerate without risking osteoporotic fracture. Knowing the answer to that question requires information regarding the mechanical properties of bones. Similarly, studying the changes in the mechanical properties of cartilage is crucial when studying osteoarthritis. Indeed, it has been demonstrated that changes in the mechanical properties of cartilage are one of the first indicators of osteoarthritis onset [16]. As for trauma, one is concerned about how biological tissues respond to non-physiological loading. In this issue, Weed *et al.* [17] report on the mechanical isotropy of porcine lung parenchyma, which is an important property when deciding what kind of constitutive modeling approach should be used for analysis of the mechanical behavior of that tissue. The mechanical behavior of the lung tissue is important, for example, when studying the pulmonary injuries caused by trauma.

3. Mechanics of Biomaterials

3.1. Implants

Implants that are aimed to stay in the human body for a long time were among the first biomaterials. It is important to ensure that the implants do not fail under their service load. Therefore, the mechanical properties of implants, such as static mechanical properties and fatigue behavior, need to be studied. In addition to the implants themselves, the biomaterials that are used for fixation of the implants or filling the cavities inside (hard) tissue need to satisfy certain requirements in terms of their mechanical properties. Finally, the mechanical properties of implants could have consequences for their function even when there is no risk of mechanical failure. Perhaps the most important example is the stress shielding phenomenon [18], where overly stiff implants could cause tissue resorption, implant loosening, and ultimately implant failure. All these concerns have motivated the study of the mechanical behavior of implant systems. In the current issue, Maurer *et al.* [19] study the mechanical behavior of different designs of prosthetic meshes that are used to repair hernia and pelvic organ prolapse. In another study, Weiss and Mitevski [20] report on the microstructure and deformation of different designs of CoCr coronary stents. Bone cements are the subjects of two other studies published in the current Special Issue. In the first study, Geffers *et al.* [21] review the strategies for reinforcement of calcium phosphate cements. In the second study, Jiang *et al.* [22] investigate the effects of adding mineralized collagen on the mechanical properties and cytocompatibility of PMMA (polymethyl methacrylate) bone cements.

3.2. Biomaterials for Tissue Regeneration

As discussed in Section 2.2, the biomaterials that are used to facilitate tissue regeneration need to satisfy a set of requirements concerning their physical, mechanical, and biological properties. Assuming we know the expected range of the mechanical properties of the tissues that need to be regenerated, the next step is to develop biomaterials that exhibit the desired mechanical properties while satisfying other requirements. Characterizing the mechanical properties of tissue engineering scaffolds is therefore an important line of research, as is clear from the large number of related studies appearing in the current issue. Goncalves *et al.* [23] study hybrid membranes of PLLA (poly-l-lactide) and collagen and how their production techniques influence the mechanical properties and osteoinduction ability of the resulting bone tissue engineering scaffolds. In another study, Wang *et al.* [24] used phase separation to fabricate bone tissue engineering scaffolds based on poly (lactide-*co*-glycolide) and tight-coated with gelatin. The effects of gelatin modification on hydrophilicity and mechanical properties of the scaffolds were investigated. Teng *et al.* [25] developed porous films whose wettability and adhesion could be tuned. This technology has potential applications in tissue engineering of various types of tissues. Finally, Chan *et al.* [26] combined polypropylene with boron nitride and nanohydroxyapatite to develop biocomposites aimed for application as bone substitutes. The effects of above-mentioned reinforcements on the mechanical properties and biocompatibility of the resulting biomaterials were studied.

3.3. Biofabrication

The application of advanced manufacturing techniques, such as 3D printing and additive manufacturing, in the fabrication of medical devices and biomaterials is often called biofabrication. Biofabrication techniques have enabled us to manufacture new categories of biomaterials that intimately interact with cells and organs and could have arbitrarily complex micro-architectures. There is a direct relationship between the micro-architecture of biomaterials and their physical, biological, and mechanical properties [27]. The micro-architecture of such biomaterials could therefore be used to create unique combinations of biological, mechanical, and physical properties. Many research groups worldwide are researching the mechanical properties of additively manufactured biomaterials. For example, in the current issue, Ahmadi *et al.* [28] study the relationship between the type of repeating unit cell and the static and morphological properties of selective laser melted porous biomaterials.

3.4. Soft Biomaterials

Soft biomaterials and matrices particularly specific types of (hydro-) gels have been in the center of recent attention of many research groups. Among other applications, these soft biomaterials could provide suitable environments for tissue regeneration. However, the mechanical properties of many types of hydrogels are not high enough to enable them provide enough mechanical support for tissue regeneration. That is why improving the mechanical properties of gels is particularly important and new variants of (hydro-) gels with significantly improved mechanical properties have been developed during the last few years, see, for example, [29]. In addition to improving the mechanical properties of gels, there are other mechanical aspects that need to be fully understood. One example is the swelling behavior of gels in presence of water and ions and development of computational models and constitutive equations that could simulate the swelling behavior.

A number of studies appearing in the current special issue study the various aspects of the mechanical, physical, and biological behavior of soft biomaterials including (hydro-) gels. Baniasadi and Minary-Jolandan [30] report on development and mechanical characterization of a composite hydrogel based on alginate and collagen. Lin and Gu [31] study the effects of crosslink density and stiffness on the mechanical properties of type I collagen gel. In another study, Moreno-Arotzena *et al.* [25] study the properties of collagen and fibrin gels aimed for application in wound healing.

4. Conclusions

Study of the mechanical behavior of biological tissues and biomaterials has been flourishing during the last few years partly due to recent developments in the biomechanics and biomaterials communities. Increasingly, the study of the mechanical behavior is combined with the study of the other aspects of biomaterial functionality. The studies appearing in the current Special Issue contribute towards better understanding of the various aspects of the mechanical behavior of both biological tissues and biomaterials.

Conflicts of Interest

The author declares no conflict of interest.

References

1. Discher, D.E.; Janmey, P.; Wang, Y.-L. Tissue cells feel and respond to the stiffness of their substrate. *Science* **2005**, *310*, 1139–1143.
2. Engler, A.J.; Sen, S.; Sweeney, H.L.; Discher, D.E. Matrix elasticity directs stem cell lineage specification. *Cell* **2006**, *126*, 677–689.
3. Karimi, A.; Navidbakhsh, M.; Yousefi, H.; Haghi, A.M.; Sadati, S.A. Experimental and numerical study on the mechanical behavior of rat brain tissue. *Perfusion* **2014**, *29*, 307–314.
4. Prevost, T.P.; Balakrishnan, A.; Suresh, S.; Socrate, S. Biomechanics of brain tissue. *Acta Biomater.* **2011**, *7*, 83–95.
5. Rashid, B.; Destrade, M.; Gilchrist, M.D. Mechanical characterization of brain tissue in compression at dynamic strain rates. *J. Mech. Behav. Biomed. Mater.* **2012**, *10*, 23–38.
6. Umale, S.; Deck, C.; Bourdet, N.; Dhumane, P.; Soler, L.; Marescaux, J.; Willinger, R. Experimental mechanical characterization of abdominal organs: Liver, kidney & spleen. *J. Mech. Behav. Biomed. Mater.* **2013**, *17*, 22–33.
7. Yarpuzlu, B.; Ayyildiz, M.; Tok, O.E.; Aktas, R.G.; Basdogan, C. Correlation between the mechanical and histological properties of liver tissue. *J. Mech. Behav. Biomed. Mater.* **2014**, *29*, 403–416.
8. Fouré, A.; Nordez, A.; Cornu, C. Effects of eccentric training on mechanical properties of the plantar flexor muscle-tendon complex. *J. Appl. Physiol.* **2013**, *114*, 523–537.
9. Takaza, M.; Moerman, K.M.; Gindre, J.; Lyons, G.; Simms, C.K. The anisotropic mechanical behaviour of passive skeletal muscle tissue subjected to large tensile strain. *J. Mech. Behav. Biomed. Mater.* **2013**, *17*, 209–220.
10. Sommer, G.; Eder, M.; Kovacs, L.; Pathak, H.; Bonitz, L.; Mueller, C.; Regitnig, P.; Holzapfel, G.A. Multiaxial mechanical properties and constitutive modeling of human adipose tissue: A basis for preoperative simulations in plastic and reconstructive surgery. *Acta Biomater.* **2013**, *9*, 9036–9048.

11. Taghizadeh, H.; Tafazzoli-Shadpour, M.; Shadmehr, M.B.; Fatouraee, N. Evaluation of biaxial mechanical properties of aortic media based on the lamellar microstructure. *Materials* **2015**, *8*, 302–316.

12. Li, W.; Lucioni, T.; Guo, X.; Smelser, A.; Guthold, M. Highly stretchable, biocompatible, striated substrate made from fugitive glue. *Materials* **2015**, *8*, 3508–3518.

13. Hollister, S.J. Porous scaffold design for tissue engineering. *Nat. Mater.* **2005**, *4*, 518–524.

14. Hollister, S.J. Scaffold design and manufacturing: From concept to clinic. *Adv. Mater.* **2009**, *21*, 3330–3342.

15. Nesbitt, S.; Scott, W.; Macione, J.; Kotha, S. Collagen fibrils in skin orient in the direction of applied uniaxial load in proportion to stress while exhibiting differential strains around hair follicles. *Materials* **2015**, *8*, 1841–1857.

16. Stolz, M.; Gottardi, R.; Raiteri, R.; Miot, S.; Martin, I.; Imer, R.; Staufer, U.; Raducanu, A.; Düggelin, M.; Baschong, W. Early detection of aging cartilage and osteoarthritis in mice and patient samples using atomic force microscopy. *Nat. Nanotechnol.* **2009**, *4*, 186–192.

17. Weed, B.; Patnaik, S.; Rougeau-Browning, M.; Brazile, B.; Liao, J.; Prabhu, R.; Williams, L.N. Experimental evidence of mechanical isotropy in porcine lung parenchyma. *Materials* **2015**, *8*, 2454–2466.

18. Sumner, D. Long-term implant fixation and stress-shielding in total hip replacement. *J. Biomech.* **2015**, *48*, 797–800.

19. Maurer, M.M.; Röhrnbauer, B.; Feola, A.; Deprest, J.; Mazza, E. Prosthetic meshes for repair of hernia and pelvic organ prolapse: Comparison of biomechanical properties. *Materials* **2015**, *8*, 2794–2808.

20. Weiss, S.; Mitevski, B. Microstructure and deformation of coronary stents from cocr-alloys with different designs. *Materials* **2015**, *8*, 2467–2479.

21. Geffers, M.; Groll, J.; Gbureck, U. Reinforcement strategies for load-bearing calcium phosphate biocements. *Materials* **2015**, *8*, 2700–2717.

22. Jiang, H.-J.; Xu, J.; Qiu, Z.-Y.; Ma, X.-L.; Zhang, Z.-Q.; Tan, X.-X.; Cui, Y.; Cui, F.-Z. Mechanical properties and cytocompatibility improvement of vertebroplasty pmma bone cements by incorporating mineralized collagen. *Materials* **2015**, *8*, 2616–2634.

23. Gonçalves, F.; Bentini, R.; Burrows, M.C.; Carreira, A.C.; Kossugue, P.M.; Sogayar, M.C.; Catalani, L.H. Hybrid membranes of plla/collagen for bone tissue engineering: A comparative study of scaffold production techniques for optimal mechanical properties and osteoinduction ability. *Materials* **2015**, *8*, 408–423.

24. Wang, J.; Li, D.; Li, T.; Ding, J.; Liu, J.; Li, B.; Chen, X. Gelatin tight-coated poly (lactide-co-glycolide) scaffold incorporating rhbmp-2 for bone tissue engineering. *Materials* **2015**, *8*, 1009–1026.

25. Moreno-Arotzena, O.; Meier, J.G.; del Amo, C.; García-Aznar, J.M. Characterization of fibrin and collagen gels for engineering wound healing models. *Materials* **2015**, *8*, 1636–1651.

26. Chan, K.W.; Wong, H.M.; Yeung, K.W.K.; Tjong, S.C. Polypropylene biocomposites with boron nitride and nanohydroxyapatite reinforcements. *Materials* **2015**, *8*, 992–1008.

27. Zadpoor, A.A. Bone tissue regeneration: The role of scaffold geometry. *Biomater. Sci.* **2015**, *3*, 231–245.

28. Ahmadi, S.M.; Yavari, S.A.; Wauthle, R.; Pouran, B.; Schrooten, J.; Weinans, H.; Zadpoor, A.A. Additively manufactured open-cell porous biomaterials made from six different space-filling unit cells: The mechanical and morphological properties. *Materials* **2015**, *8*, 1871–1896.

29. Sun, J.-Y.; Zhao, X.; Illeperuma, W.R.; Chaudhuri, O.; Oh, K.H.; Mooney, D.J.; Vlassak, J.J.; Suo, Z. Highly stretchable and tough hydrogels. *Nature* **2012**, *489*, 133–136.

30. Baniasadi, M.; Minary-Jolandan, M. Alginate-collagen fibril composite hydrogel. *Materials* **2015**, *8*, 799–814.

31. Lin, S.; Gu, L. Influence of crosslink density and stiffness on mechanical properties of type I collagen gel. *Materials* **2015**, *8*, 551–560.

Chapter 1:
Mechanics of Biological Tissues

Experimental Evidence of Mechanical Isotropy in Porcine Lung Parenchyma

Benjamin Weed, Sourav Patnaik, Mary Rougeau-Browning, Bryn Brazile, Jun Liao, Raj Prabhu and Lakiesha N. Williams

Abstract: Pulmonary injuries are a major source of morbidity and mortality associated with trauma. Trauma includes injuries associated with accidents and falls as well as blast injuries caused by explosives. The prevalence and mortality of these injuries has made research of pulmonary injury a major priority. Lungs have a complex structure, with multiple types of tissues necessary to allow successful respiration. The soft, porous parenchyma is the component of the lung which contains the alveoli responsible for gas exchange. Parenchyma is also the portion which is most susceptible to traumatic injury. Finite element simulations are an important tool for studying traumatic injury to the human body. These simulations rely on material properties to accurately recreate real world mechanical behaviors. Previous studies have explored the mechanical properties of lung tissues, specifically parenchyma. These studies have assumed material isotropy but, to our knowledge, no study has thoroughly tested and quantified this assumption. This study presents a novel methodology for assessing isotropy in a tissue, and applies these methods to porcine lung parenchyma. Briefly, lung parenchyma samples were dissected so as to be aligned with one of the three anatomical planes, sagittal, frontal, and transverse, and then subjected to compressive mechanical testing. Stress-strain curves from these tests were statistically compared by a novel method for differences in stresses and strains at percentages of the curve. Histological samples aligned with the anatomical planes were also examined by qualitative and quantitative methods to determine any differences in the microstructural morphology. Our study showed significant evidence to support the hypothesis that lung parenchyma behaves isotropically.

Reprinted from *Materials*. Cite as: Weed, B.; Patnaik, S.; Rougeau-Browning, M.; Brazile, B.; Liao, J.; Prabhu, R.; Williams, L.N. Experimental Evidence of Mechanical Isotropy in Porcine Lung Parenchyma. *Materials* **2015**, *8*, 2454-2466.

1. Introduction

Pulmonary injury, including pulmonary contusion and pulmonary laceration, is a serious source of morbidity and mortality following blunt or blast trauma [1–7]. Common causes for civilians include motor vehicle accidents, falls, assaults, and sports injuries [8–10]. Additionally, explosions and other sources of blast trauma can cause severe pulmonary injuries, and are commonly seen in military or civilian victims of such events [1–3,11,12]. Blast lung injury is the primary cause of death in those who initially survive an explosion [12]. The prevalence of these types of injuries has made pulmonary injury a crucial area of trauma research.

Lungs have a complex structure, which can be summarized as a network of stiff airway tubes, bronchi and bronchioles, embedded in a soft and porous tissue, lung parenchyma. The lungs are enclosed by the thin pleural membrane and immersed in pleural fluid. The parenchyma contains the gas-exchanging alveoli which are very soft and highly susceptible to damage. Damage to the parenchyma leads to bleeding, edema, and collapse of the microstructure, which prevents gas exchange [13]. In serious cases this ultimately leads to the inability of the lungs to adequately oxygenate the blood and even medically supplemented oxygen may not be sufficient to prevent death [4]. The degree of lung damage a trauma victim has sustained is difficult to diagnose in a clinical setting. Physical experiments which could create a predictive index for these injuries involving human cadavers or live animals are costly and logistically difficult to perform. Finite element methods are a promising option for performing controlled experiments of different forms and severities of traumatic events.

Finite element methods are commonly used in research for simulating different mechanisms of injury [4,7,11,14,15]. These simulations require constitutive relationships to represent the mechanical properties of the materials. Previous studies have explored the mechanical responses of lung parenchyma under different loading conditions [16–18]. These studies have all assumed that lung parenchyma behaves isotropically, but to our knowledge, no study has verified this assumption. Moreover, reports that have addressed isotropy in other materials have used the comparison of values derived from stress-strain curves such as modulus, extensibility, or stress and strain at failure [19–22]. These methods of comparisons offer insight into the isotropy in a material, but do not compare the entire stress-strain relationships of different loading directions along quantifiable histological data, which may be useful for certain materials or loading multiaxial conditions.

In this study we evaluated lung tissue in compression with three experimental groups comprising specimens aligned with the three anatomical planes: sagittal, frontal, and coronal. We subjected these groups to uniaxial compression, with each group corresponding to one of the anatomical planes. Stress-strain curves of each group were generated from the test data, and the groups were compared to determine the differences among loading directions in the stress and strain values at 10% intervals of the curve. These differences were evaluated for statistical significance between direction groups. Furthermore, histological investigations (Movat, Massons) and analysis of the histological micrographs were used to determine if any morphological differences were apparent at the microstructure level.

2. Materials and Methods

2.1. Tissue Procurement and Preparation

Porcine lungs were obtained from a local abattoir, in accordance with Mississippi State University Institutional Animal Care and Use Committee regulations. All specimens were obtained from market ready pigs of approximately 12 months of age weighing 225 pounds. Lungs were obtained immediately following sacrifice, placed in sealed plastic bags in a cooler on ice and transferred within thirty minutes to the laboratories in the Department of Agricultural and Biological Engineering at Mississippi State University for mechanical testing (Figure 1a). To alleviate the effects of post mortem tissue changes, all mechanical tests were performed within twelve h of sacrifice. Lungs were dissected of the pleural membranes and a section of the lung, approximately 60 mm by 100 mm rectangular sections were isolated from the center area (Figure 1b). Twenty mm cylindrical test specimens were dissected from the larger section of lung. These cylindrical samples were 10 mm thick (Figure 2c). The cylindrical samples were prepared such that they were aligned with one of the three anatomical planes; sagittal, frontal, and coronal. The three alignments were used as our three comparison groups.

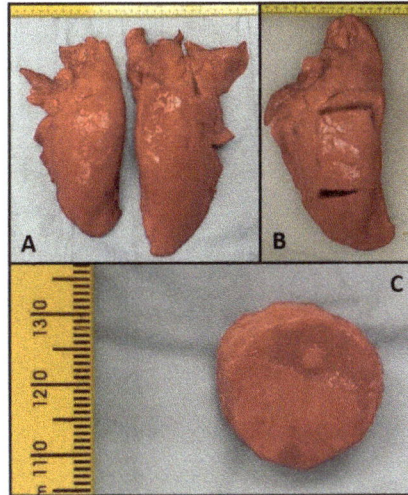

Figure 1. (**A**) Freshly obtained pig lung; (**B**) 60 mm × 100 mm rectangular sections were removed; (**C**) 10 mm thick cylindrical samples were prepared.

2.2. Histological Preparation and Light Microscopy

Three specimens, one from each of the aforementioned anatomical planes, were prepared for histological examination. Specimens were fixed using 10% neutral buffered formalin, paraffin embedded, sectioned to 5 μm thickness, and mounted to glass slides. One set of slides was stained with Movat's Pentachrome to observe the composition of structural proteins, and a second set of slides was stained with Masson's Trichrome to observe the blood vessels and fibrous tissue. Prepared specimens were analyzed via light microscopy using a Leica DM2500 light microscope (Leica Microsystems, Wetzlar, Germany) at 100× magnification to observe any qualitative differences in microstructure.

Specimens were quantitatively analyzed for differences in microstructure using ImageJ (National Institute of Health, Bethesda, MD, USA). For each orientation, a series of twenty five overlapping light microscope images were obtained. These micrographs were compiled using the MosaicJ plugin (Ecole Polytechnique Federale De Lausanne, Lausanne, Switzerland) for ImageJ. Briefly, overlapping micrographs were positioned within the MosaicJ window and the software was used to create a single cohesive image based on overlapping portions of separate micrographs. This allowed for larger regions of the slide-mounted sample to be imaged without omitting portions of larger alveoli. The cohesive images were then cropped to create the largest rectangle available. Figure 2 illustrates how individual micrographs contributed to the composites, and how the uneven edges were cropped. The final composites for each direction were thresholded, inverted, and analyzed using the "Analyze Particles" feature in ImageJ. In this tissue the most relevant, and most readily observed by this method, feature is the alveolar spaces (white) of the lung tissue. "Analyze Particles" was configured to filter out particles with an area of less than 2000 pixels, approximately 2839 square microns, to eliminate extremely small particles or artifacts. Edge particles were also filtered. Roundness, Circularity, Solidity, and Aspect Ratio of the Fit Ellipse were plotted as histograms and analyzed to discern the 1st, 2nd and 3rd quartiles of the histograms to compare any differences among the alignment planes. These parameters describe particle shape and provide insight into particle elongation and tortuosity. Formulas for these parameters are shown in Appendix.

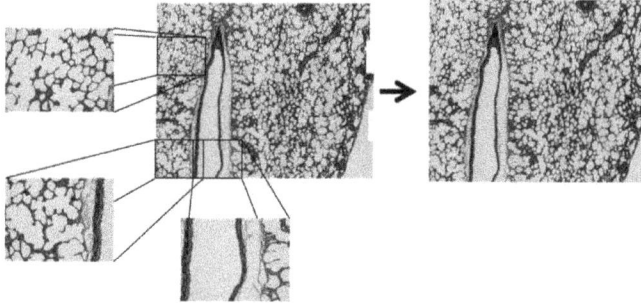

Figure 2. Composite micrograph composed of individual micrographs.

2.3. Compression Testing

Specimens were subjected to uniaxial compressive loading using the Mach1 Micromechanical Testing System® (Biomomentum, Laval, QC, Canada). A flat, disc-shaped metal platen was used for compression testing. The sample was attached with a small amount of high-viscosity cyanoacrylate ester adhesive (Permabond LLC, Pottstown, PA, USA) to prevent slipping of the sample from below the compression head, without introducing confinement effects that would compromise the uniform stress state assumption. Samples were submerged in Phosphate buffer saline (PBS) to simulate physiological conditions and prevent samples from drying during the test. Figure 3 shows the testing apparatus. The samples were pre-loaded to two grams-force, and pre-conditioned by cyclic loading to ten percent strain for ten cycles. We chose to precondition because lung is under constant cyclic loading during normal physiological conditions, and the preconditioning mimics this cycling. Specimens were then loaded to ten kilograms-force at a rate of ten percent strain (engineering) per second, with the loading velocity adjusted accordingly for each specimen. Load and displacement data were collected by the Mach1 software for further analysis.

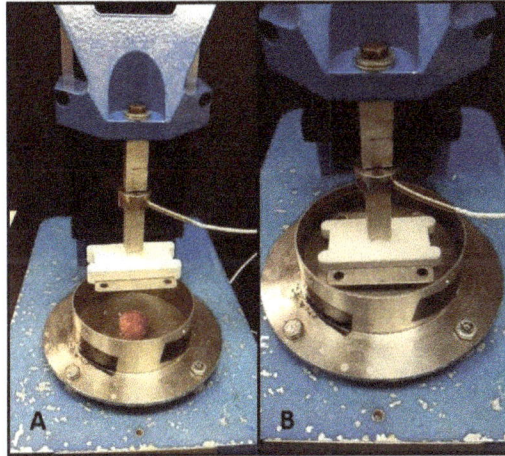

Figure 3. Mach1 Micromechanical Testing System®. (**A**) Sample in dish with compression platen removed; (**B**) Compression platen lowered onto sample in testing dish.

3. Data Analysis

3.1. True Stress versus True Strain Conversions

Data from our mechanical tests were processed in Microsoft Excel (Microsoft, Redmond, WA, USA) to create true stress-true strain curves. Data from each test were converted to engineering stress and engineering strain using the following Equations (1) and (2) (Appendix). Data were further processed to true stress and true strain using Equations (3) and (4) (Appendix). These true stress-strain conversions assume a uniform stress state and incompressibility. These assumptions are discussed further in the limitations section of this manuscript.

3.2. Interpolation of True Stress-True Strain Curves

We observed the true stress and true strain at equivalent percentages of the stress-strain curve in order to compare tests within each anatomical plane and between anatomical planes. Because the sampling rate during testing yields results that do not coincide with even percentages, an Excel-based interpolation script was used to create an equivalent curve with 10 data points at multiples of 10% (10%, 20%, …, 100%) for each experimental data set. Briefly and for a given set of test data, this interpolation script determined the maximum true stress and true strain values and normalized all true stress and true strain values relative to the maximum, respectively. The distance was then determined between each recorded data point by using the Pythagorean distance formula. For each data point, the accumulated distance was divided by the total distance for the data set to determine the percentage of the total for that data point. Using these newly assigned percentages for each data point, the true stress-true strain data points closest to a 10% strain were then selected. A linear interpolation was done between these two points to create a new data point, which represented the 10% point of the experimentally recorded curves. This process was repeated for each multiple of 10% until 90% was reached. The final experimentally recorded data point was used as the 100% data point. This script was run for each experimental data set to generate a bank of curves that would allow for precise analysis.

3.3. Comparison of Recorded Data among Anatomical Planes

The interpolated data sets from each experimental test were compiled in Excel and the data at increments of 10% strain within each anatomical plane were averaged to create a characteristic curve for each anatomical plane. The average of the characteristic curves were plotted with standard deviation (Figure 4).

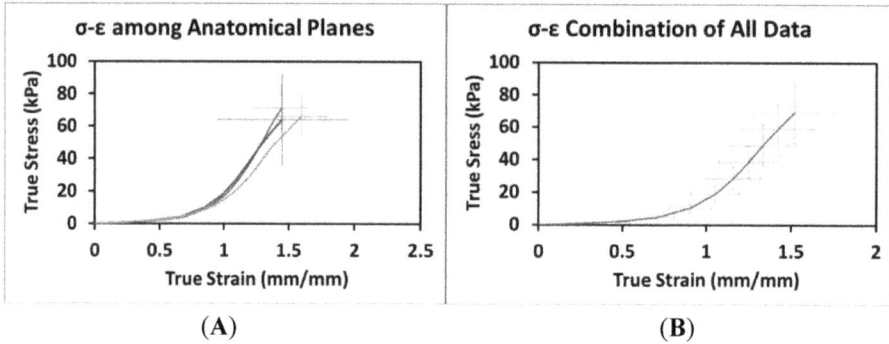

Figure 4. (**A**) Stress-Strain relationship across anatomical planes (Frontal (blue): $n = 18$, Sagittal (red): $n = 13$, Transverse: $n = 12$); (**B**) Combination of all Stress-Strain data ($n = 43$). Error bars indicate +/− one standard deviation.

Statistical analysis was performed to compare the differences in stress and strain in each of the three anatomical planes. At 10% strain increments, the stress-strain data were tested using a single-factor ANOVA with a 95% confidence interval to determine if there was a significant difference among each of the three anatomical planes. Statistics were performed using in Microsoft Excel.

4. Results

Micrographs for each anatomical plane are presented in Figure 5. Figure 5A-B, C-D, and E-F show Movat's Pentachrome for the sagittal, frontal, and transverse planes, respectively. These micrographs do not indicate any apparent qualitative differences in morphology or alignment of microstructural features among the three anatomical planes. Composite Micrographs created with MosaicJ are shown in Figure 6; Figure 6A is sagittal, Figure 6B is frontal, Figure 6C is transverse. The corresponding histograms for a particle analysis, particles being alveolar spaces, are shown in Figure 7. Histograms for Roundness (Figure 7A), Circularity (Figure 7B), and Solidity (Figure 7C), were prepared with bins from 0 to 1, incremented by 0.05, where the columns indicate the percentage of particles that fell into each bin. The aspect ratio (Figure 7D), was prepared with bins from 1.0 to 3.4, incremented by 0.1, and columns denoting the percentage of particles for each bin. It should be noted that some aspect ratio data was not displayed in this histogram in the interest of more effectively highlighting the majority of the data; 3.4 was chosen as the end of the axis because this was the final value at which more than two particles occurred within a single bin for any of the three alignments. Table 1 shows the 1st quartile, median, and 3rd quartiles for the histograms. Histograms of particle analysis parameters had little difference among the different orientation planes.

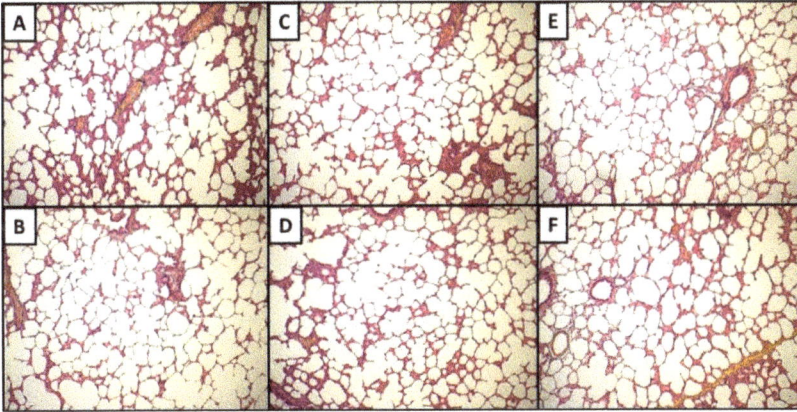

Figure 5. Movat's Pentachrome for the sagittal (**A,B**), frontal (**C,D**) and transverse (**E,F**) planes.

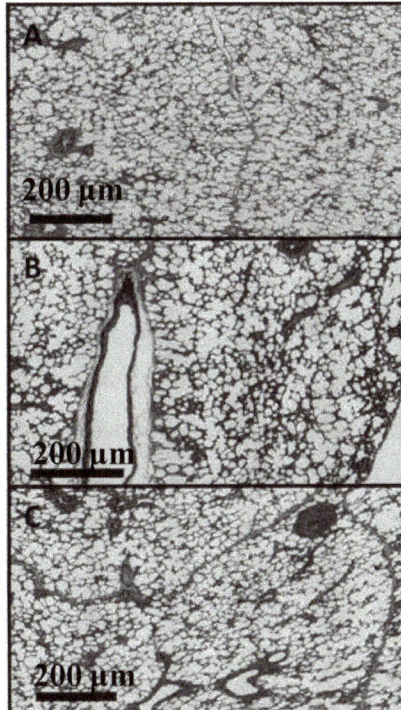

Figure 6. Composite micrographs of (**A**) saggital, (**B**) frontal, (**C**) transverse planes (200 μm scale).

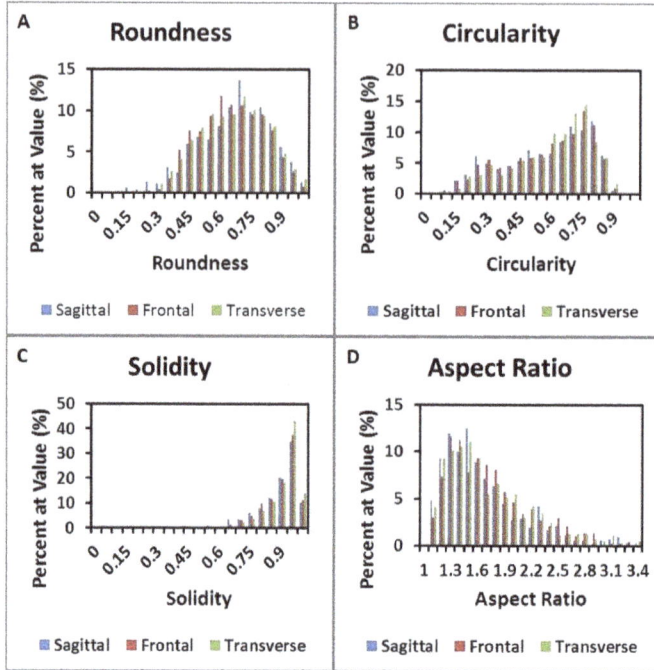

Figure 7. Histograms for (**A**) Roundness; (**B**) Circularity; and (**C**) Solidity; (**D**) shows the aspect ratio of samples.

Table 1. The 1st quartile, median, and 3rd quartiles for the histograms.

Roundness	Sagittal	Frontal	Transverse
1st Quartile	0.53375	0.512	0.512
Median	0.66	0.626	0.647
3rd Quartile	0.77225	0.75	0.761
Circularity	**Sagittal**	**Frontal**	**Transverse**
1st Quartile	0.385	0.411	0.451
Median	0.5855	0.6	0.618
3rd Quartile	0.72475	0.722	0.723
Solidi	**Sagittal**	**Frontal**	**Transverse**
1st Quartile	0.812	0.824	0.853
Median	0.889	0.898	0.912
3rd Quartile	0.93525	0.936	0.94
Aspect Ratio	**Sagittal**	**Frontal**	**Transverse**
1st Quartile	1.29475	1.333	1.314
Median	1.515	1.598	1.546
3rd Quartile	1.87475	1.953	1.955

True stress-true strain curves for each anatomical plane are shown in Figure 4A. The composite curve for the average of the three planes is shown in Figure 4B. The results of the statistical comparisons for each percentage are presented in Tables 2 and 3. There were no significant differences in either stress or strain among the three anatomical planes for any percentage level.

Table 2. Results of statistical comparisons for the samples.

Path %	Frontal Avg. True σ	Sagittal Avg. True σ	Transverse Avg. True σ	F	p-value	Frontal Avg. True ε	Sagittal Avg. True ε	Transverse Avg. True ε	F	P-value
10	619.39	729.75	957.42	2.6893	0.0793	0.2293	0.2251	0.2477	0.0470	0.9542
20	1,588.29	1,934.04	2,390.74	2.5329	0.0912	0.4580	0.4488	0.4937	0.0468	0.9544
30	3,984.76	4,568.52	5,304.14	1.4654	0.2423	0.6817	0.6642	0.7301	0.0462	0.9549
40	9,563.28	9,819.38	10,977.00	0.3470	0.7087	0.8778	0.8506	0.9318	0.0453	0.9557
50	18,377.42	17,719.62	19,178.35	0.0115	0.9886	1.0198	0.9872	1.0789	0.0448	0.9562
60	28,466.14	26,711.31	28,380.77	0.0223	0.9780	1.1261	1.0896	1.1895	0.0444	0.9566
70	39,023.73	36,054.37	37,928.60	0.0771	0.9259	1.2130	1.1766	1.2817	0.0442	0.9568
80	49,760.69	45,514.93	47,479.43	0.1367	0.8726	1.2907	1.2584	1.3727	0.0443	0.9567
90	60,601.53	54,822.69	56,676.20	0.2193	0.8040	1.3625	1.3489	1.4820	0.0453	0.9558
100	71,323.68	64,005.78	65,778.25	0.2841	0.7541	1.4406	1.4472	1.5977	0.0461	0.9550

Table 3. Standard deviations for statistical comparisons of the samples. Dev.: Deviations.

Part %	Frontal Strain Dev.	Frontal Stress Dev.	Sagittal Strain Dev.	Sagittal Stress Dev.	Transverse Strain Dev.	Transverse Stress Dev.
10	0.030915423	127.2781798	0.075739326	240.5005433	0.027138761	331.5140315
20	0.061629963	347.8942716	0.150734704	648.4193448	0.054122752	817.186667
30	0.090535682	829.7268617	0.221484316	1556.705447	0.078778275	1624.193247
40	0.110923627	1890.534598	0.280026781	3438.735718	0.097867071	2602.47132
50	0.123904198	3539.070152	0.32436637	6711.154692	0.113630437	3650.112918
60	0.137258653	5585.462839	0.359870173	10716.82231	0.127062902	5247.527794
70	0.150673246	7795.522934	0.392218749	14986.77743	0.139112415	7176.190338
80	0.164835636	10072.46936	0.424828203	19386.53237	0.151445696	9218.119181
90	0.180556847	12424.48858	0.46166626	23807.36975	0.168660868	11333.30934
100	0.208370407	14937.09112	0.504503056	28290.23423	0.198543701	13672.60307

5. Discussion

True stress-true strain curves for each anatomical plane appear to be similar, and the error bars indicate the variation among specimens is much greater than the variation between the averaged curves for each anatomical plane. It is well known that biological tissues exhibit a large degree of inter-specimen variation, and the degree to which the plane curves vary appears to be within one standard deviation for each other curve, respectively. This observation, while qualitative, strongly supports our hypothesis of isotropy. Statistical comparison of the stress and strain values for each multiple of 10% indicates no significant differences between any of the three planes or the average of the three planes. This quantitative observation also supports our hypothesis of isotropy.

The interpolation method presented was formulated with the interest of comparing curves effectively. The method for determining the percentages of the curve was chosen as a means of giving equivalent weight to both stress and strain. We believed this approach to be preferable to basing percentages on only one parameter because the characteristic curves for biological materials, which are often described as hyperelastic, are commonly observed to have two distinct regions of tissue behavior. The first region is described as the toe-in region and is dominated by the realignment of tissue structures to resist loading. The second region is described as the linear region and is dominated by significant resistance to loading as the tissue builds towards failure. A comparison based on strain percentages would capture more of the toe-in region and less of the linear region, essentially hiding a large amount of tissue behavior in the higher strain percentages. Likewise, a percentage comparison based on stress would capture more of the linear region and less of the toe-in, hiding a large amount of tissue behavior in the lower stress values. The combined weight of the two parameters allows the entire stress-strain path to be considered equally.

Quantitative analysis of micrographs did not indicate apparent differences in the morphology of porcine lungs (Figure 7 and Table 1). The quantitative particle analysis, which showed the morphological parameters had very similar values and distributions among the orientation planes, complimented the qualitative observations. Overall, the histological examinations supported the hypothesis that porcine lung is an isotropic material. Analysis of microstructural features is an important part of understanding the hierarchical organization of biological tissues. The change in organization, shape, and size of some features under deformation lends insight to possible physiological changes in the biological system. Microstructural changes during loading can be used to understand the sub structural evolution of lung, or other biological tissues, under loading and thus aid in modeling complex tissue behaviors. The data presented here strongly supports the widely held belief that lung parenchyma is an isotropic material. Despite this belief, we are not aware of a thorough analysis of this assumption. Beyond the implications for lung tissue, we believe the methods presented here represent a strong system for assessing isotropy in other tissues. This is valuable for future research as biological tissues may be isotropic or anisotropic, and a confident understanding of a given tissue's behavior is crucial for effective constitutive modeling. Additionally, verification of isotropy allows the analysis of more complex material properties such as viscoelasticity, strain rate dependency, and stress state dependency without considering the complexity of anisotropy. The ultimate result is a more efficient path towards a complete understanding of the material properties of lung, and other biological tissues of interest to the research community.

6. Conclusions

This study presents a novel method for testing mechanical isotropy in biological tissues. This method was used to demonstrate the isotropic nature of porcine lung parenchyma. This was a widely used assumption in the body of literature, but had not been explicitly analyzed or proven. This is significant for soft-tissue biomechanics in general, as it presents a means for testing isotropy in other tissues. It is significant for lung biomechanics research because it allows for future research to be conducted without concern for overlooking anisotropy in more complex material properties.

Future studies will address important behaviors such as viscoelasticity, strain rate dependency, stress-state dependency, and microstructural evolution of lung parenchyma.

7. Limitations

The conversion formulas for true stress and true strain rely on assumptions of uniform stress state and incompressibility. Our mechanical testing setup is intended to provide the most uniform stress state possible, but there may be some uneven distributions that limit this aspect of the assumption. Moreover, the assumption of incompressibility is inexact, as the lung contains solid and gas components, which have a high probability of being compressible. Considering that our test procedure does not actively track the changing cross-sectional area it is necessary to use some form of conversion to achieve true stress, as this method was used in our previous publication [23].

The interpolation method described uses a linear interpolation process, which slightly overestimates the interpolated values given that our curves are concave upward. Future studies will explore the incorporation of Newton-Raphson methodology to alleviate this limitation.

Acknowledgments

This material is based upon work supported by the National Nuclear Security Administration, Department of Energy, under award number (DE-FC26-06NT42755). The authors would also like to thank Sansing Meat Service, Maben, MS, USA, for their generous provision of porcine lung tissue.

Author Contributions

Benjamin Weed was a graduate student who led and directed the mechanical and microstructural testing and analysis of this study. Sourav Patnaik and Bryn Brazile are graduate students who worked with Benjamin on obtaining the tissues, as well as equipment design and calibration for mechanical testing. Mary Rougeau-Browning was an undergraduate student working in the laboratory under the mentorship of the corresponding author. She helped in obtaining tissues, tissue dissection, and mechanical testing. Jun Liao and Raj Prabhu assisted with data analysis. Lakiesha N. Williams assisted with the study design and was a lead in writing the manuscript. All authors contributed to an aspect of writing the manuscript.

Appendix

Formulas for Stress-Strain Calculations

1. $\sigma_{engineering} = \dfrac{force}{area_{undeformed}}$

2. $\varepsilon_{engineering} = \dfrac{length_{undeformed} - displacement}{length_{undeformed}}$

3. $\sigma_{true} = \sigma_{engineering} * \left(1 + \varepsilon_{engineering}\right)$

4. $\varepsilon_{true} = \ln(1 + \varepsilon_{engineering})$

Formulas for Shape Descriptors Used in Particle Analysis

1. $Roudness = 4 \times \left(\frac{Area}{\pi \times (Major\ Axis\ of\ Fit\ Ellipse)^2} \right)$

2. $Circularity = 4\pi \times \left(\frac{Area}{Perimeter^2} \right)$

3. $Solidity = \frac{Area}{Convex\ Area}$

4. $Aspect\ Ratio = \frac{Major\ Axis\ of\ Fit\ Ellipse}{Minor\ Axis\ of\ Fit\ Ellipse}$

Conflicts of Interest

The authors declare no conflict of interest.

References

1. Champion, H.R.; Holcomb, J.B.; Young, L.A. Injuries from explosions: Physics, biophysics, pathology, and required research focus. *J. Trauma Acute Care Surg.* **2009**, *66*, 1468–1477.

2. Wolf, S.J.; Bebarta, V.S.; Bonnett, C.J.; Pons, P.T.; Cantrill, S.V. Blast injuries. *Lancet* **2009**, *374*, 405–415.

3. Mayorga, M.A. The pathology of primary blast overpressure injury. *Toxicology* **1997**, *121*, 17–28.

4. Stuhmiller, J.H. Biological response to blast overpressure: A summary of modeling. *Toxicology* **1997**, *121*, 91–103.

5. Stuhmiller, J.H.; Ho, K.H.; Vander Vorst, M.J.; Dodd, K.T.; Fitzpatrick, T.; Mayorga, M. A model of blast overpressure injury to the lung. *J. Biomech.* **1996**, *29*, 227–234.

6. Cooper, G.J.; Townend, D.J.; Cater, S.R.; Pearce, B.P. The role of stress waves in thoracic visceral injury from blast loading: Modification of stress transmission by foams and high-density materials. *J. Biomech.* **1991**, *24*, 273–285.

7. Stuhmiller, J.H.; Chuong, C.J.; Phillips, Y.Y.; Dodd, K.T. Computer modeling of thoracic response to blast. *J. Trauma Acute Care Surg.* **1988**, *28*, 132–139.

8. Dehghan, N.; de Mestral, C.; McKee, M.D.; Schemitsch, E.H.; Nathens, A. Flail chest injuries: A review of outcomes and treatment practices from the national trauma data bank. *J. Trauma Acute Care Surg.* **2014**, *76*, 462–468.

9. Pauzé, D.R.; Pauzé, D.K. Emergency management of blunt chest trauma in children: An evidence-based approach. *Pediatr. Emerg. Med. Pract.* **2013**, *10*, 1–22.

10. Idriz, S.; Abbas, A.; Sadigh, S.; Padley, S. Pulmonary laceration secondary to a traumatic soccer injury: A case report and review of the literature. *Am. J. Emerg. Med.* **2013**, *31*, 1625.

11. D'Yachenko, A.I.; Manyuhina, O.V. Modeling of weak blast wave propagation in the lung. *J. Biomech.* **2006**, *39*, 2113–2122.

12. Born, C.T. Blast trauma: The fourth weapon of mass destruction. *Scand. J. Surg.* **2005**, *94*, 279–285.

13. Sasser, S.M.; Sattin, R.W.; Hunt, R.C.; Krohmer, J. Blast lung injury. *Prehosp. Emerg. Care* **2006**, *10*, 165–172.

14. Gayzik, F.S.; Hoth, J.J.; Stitzel, J. Finite element–based injury metrics for pulmonary contusion via concurrent model optimization. *Biomech. Model. Mechanobiol.* **2011**, *10*, 505–520.

15. Vawter, D.L. A finite element model for macroscopic deformation of the lung. *J. Biomech. Eng.* **1980**, *102*, 1–7.

16. Suki, B.; Bates, J.H.T. Lung tissue mechanics as an emergent phenomenon. *J. Appl. Physiol.* **2011**, *110*, 1111–1118.

17. Gao, J.; Huang, W.; Yen, R.T. Mechanical properties of human lung parenchyma. *Biomed. Sci. Instrum.* **2006**, *42*, 172–180.

18. Freed, A.D.; Einstein, D.R. An implicit elastic theory for lung parenchyma. *Int. J. Eng. Sci.* **2013**, *62*, 31–47.

19. Feng, Y.; Okamoto, R.J.; Namani, R.; Genin, G.M.; Bayly, P.V. Measurements of mechanical anisotropy in brain tissue and implications for transversely isotropic material models of white matter. *J. Mech. Behav. Biomed. Mater.* **2013**, *23*, 117–132.

20. Lillie, M.A.; Shadwick, R.E.; Gosline, J.M. Mechanical anisotropy of inflated elastic tissue from the pig aorta. *J. Biomech.* **2010**, *43*, 2070–2078.

21. Sacks, M.S.; Sun, W. Multiaxial mechanical behavior of biological materials. *Annu. Rev. Biomed. Eng.* **2003**, *5*, 251–284.

22. Sacks, M. Biaxial mechanical evaluation of planar biological materials. *J. Elast.* **2000**, *61*, 199–246.

23. Weed, B.C.; Borazjani, A.; Patnaik, S.S.; Prabhu, R.; Horsteemyer, M.F.; Ryan, P.L.; Franz, T.; Williams, L.N.; Liao, J. Stress state and strain rate dependence of human placenta. *Ann. Biomed. Eng.* **2012**, *40*, 2255–2265.

Collagen Fibrils in Skin Orient in the Direction of Applied Uniaxial Load in Proportion to Stress while Exhibiting Differential Strains around Hair Follicles

Sterling Nesbitt, Wentzell Scott, James Macione and Shiva Kotha

Abstract: We determined inhomogeneity of strains around discontinuities as well as changes in orientation of collagen fibrils under applied load in skin. Second Harmonic Generation (SHG) images of collagen fibrils were obtained at different strain magnitudes. Changes in collagen orientation were analyzed using Fast Fourier Transforms (FFT) while strain inhomogeneity was determined at different distances from hair follicles using Digital Image Correlation (DIC). A parameter, defined as the Collagen Orientation Index (COI), is introduced that accounts for the increasingly ellipsoidal nature of the FFT amplitude images upon loading. We show that the COI demonstrates two distinct mechanical regimes, one at low strains (0%, 2.5%, 5% strain) in which randomly oriented collagen fibrils align in the direction of applied deformation. In the second regime, beginning at 5% strain, collagen fibrils elongate in response to applied deformation. Furthermore, the COI is also found to be linearly correlated with the applied stress indicating that collagen fibrils orient to take the applied load. DIC results indicated that major principal strains were found to increase with increased load at all locations. In contrast, minimum principal strain was dependent on distance from hair follicles. These findings are significant because global and local changes in collagen deformations are expected to be changed by disease, and could affect stem cell populations surrounding hair follicles, including mesenchymal stem cells within the outer root sheath.

Reprinted from *Materials*. Cite as: Nesbitt, S.; Scott, W.; Macione, J.; Kotha, S. Collagen Fibrils in Skin Orient in the Direction of Applied Uniaxial Load in Proportion to Stress while Exhibiting Differential Strains around Hair Follicles. *Materials* **2015**, *8*, 1841-1857.

1. Introduction

The biomechanics of skin is complex, typically demonstrating three deformation regimes in response to stress (Figure 1). Initially, the collagen is oriented in a random pattern (Figure 1a) [1]. While the collagen fibrils are in this random pattern, it orients readily to the applied load, hence demonstrating low elastic modulus [1,2]. Within this region, the elastic behavior of skin is not determined by collagen, but rather, an elastin mesh [1]. As the skin is stretched further, the collagen fibrils straighten and realign parallel to one another, thus, requiring more load to induce further elongation (Figure 1b) [1–3]. This process can continue until the fibrils are mostly aligned in the direction of the applied load (Figure 1b). Subsequently, failure occurs as fibrils begin to slide past one other, or by a multitude of other mechanisms (Figure 1c).

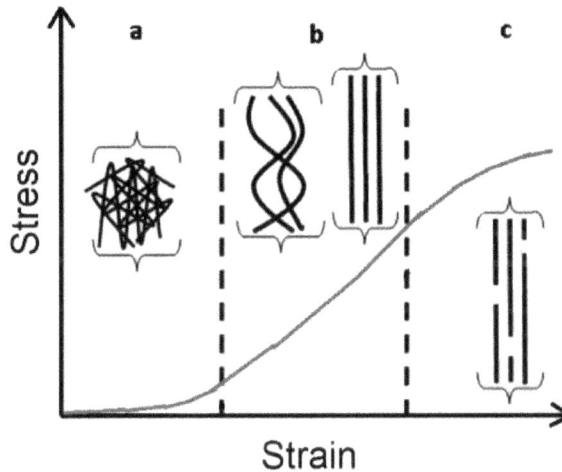

Figure 1. The stress-strain relationship of the skin is associated with collagen orientation. (**a**) The collagen fibrils are randomly aligned and readily orient with the application of small amounts of stress (low elastic modulus); (**b**) The collagen fibrils begin to straighten and gradually align in the direction of the applied stress. At the end, almost all collagen fibrils are aligned in the direction of loading; (**c**) Damage begins to accumulate due to sliding of fibrils past one other, among many other mechanisms. Figure adapted from Brinckmann [1] and Daly [3].

To determine the mechanical behavior of skin, most studies have used the initial orientation of skin collagen fibrils. The initial orientations of the collagen fibrils were obtained by histological analysis of excised skin. Upon application of load, the orientation of collagen fibrils is assumed to change, leading to increased load bearing by collagen fibrils. However, the change in collagen fibril orientation under different stress magnitudes has yet to be characterized experimentally. Experimental characterization of collagen fibrillar orientation is now possible, because of advances in multiphoton microscopy and because collagen fibrils emit second harmonic generation (SHG) due to their inherent non-centrosymmetry [4]. Within the last few years, the periodic structure of collagen fibrils in skin has been analyzed using fast fourier transforms (FFT) of SHG images. The amplitude component in the FFT of SHG images indicates the frequency of spacing between collagen fibrils. This parameter has been demonstrated as a diagnostic potential for medical disorders, including, distinguishing between healthy and diseased dermis [5–7], evaluating skin cancer [8], accessing photoaging [9], and scarring [10], among a diverse array of other applications.

Mechanical loading may contribute towards skin biomechanical behavior by regulating stem cells residing around hair follicles, including mesenchymal stem cells in the outer root sheath [11,12]. Therefore, an understanding of the deformation of collagen fibrils around these discontinuities could lead to an understanding of how the skin cells regulate its collagen content, as a means to achieve a desired mechanical environment, as well as during tissue regeneration. Strains in and adjacent to discontinuities have been investigated using digital image correlation (DIC) [13,14]. DIC has been used to determine inhomogeneity of strains on the surface of soft [15,16] and inside hard tissues [17], but, strains in tissues with pigmentation, such as skin, have not been evaluated.

The goal of this paper is to determine the stress-collagen fibril orientation relationship by utilizing the symmetry of the FFT of SHG images. A further goal of this manuscript is to determine the inhomogeneity of strains, especially, as it relates to the presence of discontinuities, such as hair follicles. Since the deformation of collagen fibrils is primarily responsible for soft tissue deformation and subsequent mechanobiological response, and since collagen fibril orientation at various depths can be measured non-invasively, this has the potential to add to our understanding of its contribution in various disease states.

2. Results and Discussion

2.1. FFT Analysis of Global Deformation

Skin stretching causes collagen fibrils to change orientation towards the direction of applied load (Figure 2a). SHG images of skin taken at low strain (0%, 2.5% strain) indicate that the collagen fibrils (white lines) appear to be coiled and wavy. At 5% and 10% applied strain, the collagen fibrils appear more organized along the axis in which the strain is applied and some of the waviness disappears. At the largest strain (15%), the collagen fibrils appear aligned along the axis of induced strain.

The shape of each FFT (Figure 2b) can be visually compared to the spatial frequency of the collagen fibrils (Figure 2a). Note that the arrows indicate the direction of applied load (direction of arrows is different in Figure 2a,b). As strain increases, the collagen fibrils orient along the direction of applied load and become more densely packed, *i.e.*, periodicity decreases, in the perpendicular direction. As a result, there is an increase in the higher frequency components in the FFT amplitude images perpendicular to the direction of applied load (Figure 2b). Thus, the FFT images appear to become more ellipsoidal, which is characterized by a parameter termed COI (Equation (2)–see Methods).

Applied strains, stresses and COI were compared to one other. ANOVA followed by Tukey *post-hoc* analysis, was performed using R (R 3.01, http://www.r-project.org/). Analysis of COI *vs.* strain relationships (Figure 3a) indicated that the COI's at 0% and 2.5% are different from COI's at 10% and 15% ($p < 0.05$ and power > 0.8). Also, the COI at 5% were found to be different from COI's at 15% ($p < 0.05$ and power > 0.8). The COI is observed to be non-linear with respect to applied strain, with two distinct regimes observed above and below the 5% strain threshold, that were fit linearly (Figure 3a). At less than 5% strain, there is a slope of approximately 0.0066 (first regime, $R^2 = 0.9557$), which, at greater than 5% strain, increases to 0.0159 (second regime, $R^2 = 0.9973$), an increase of 2.4 times ($p < 0.05$, power > 0.8). Analysis of the stress *vs.* strain relationship (Figure 3b) indicated that stresses at 0% and 2% were different from stresses at 10% and 15%. The stress is also observed to be non-linear with respect to strain (Figure 3b), exhibiting two distinct regimes, occurring below and above 5% strain. A linear regression performed on both regions, $R^2 = 0.9438$ and 0.9994 below and above 5% strain, respectively; indicates that the slopes increases by 2.5 times from 0.0142 to 0.0351 ($p < 0.05$, power > 0.8). Analysis of the COI *vs.* stress relationship (Figure 3c) indicated that the COI's at low stresses (that correspond to strains of 0%, 2.5% and 5%) were different from COI's at higher stresses (that correspond to strains of 10% and the 15%) ($p < 0.05$, power > 0.8). Moreover, COI's at stress that corresponds to 10% strain was different from the stress that corresponds to 15% strain ($p < 0.05$, power > 0.8). In contrast to the non-linear relationship observed between stress *vs.* strain and COI *vs.* strain, the COI and applied stress relationship (Figure 3c) demonstrates a strong linear relationship between the orientation of collagen and the strain being applied to the skin ($R^2 = 0.9969$).

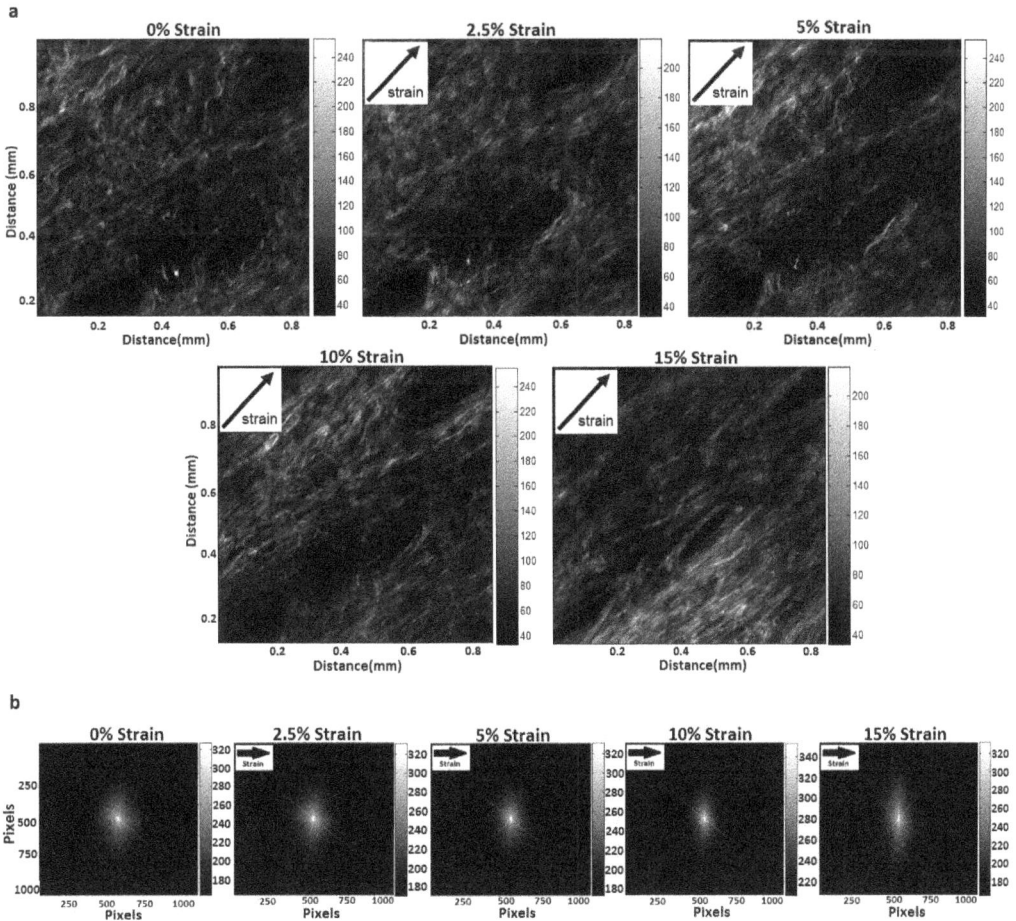

Figure 2. Representative Second Harmonic Generation (SHG) images and the respective fast Fourier transforms (FFT) for a random sample. (**a**) SHG images of skin at different strain magnitudes. Strain was applied at $45°$ to the x-y scans. Each image is single frame from a 70–100 z-stack images obtained at each strain level. As the strain increases, a greater number of fibrils align in that direction; (**b**) FFT amplitude images reoriented in the direction of applied strain (horizontal direction). Arrows indicate the direction of applied strain.

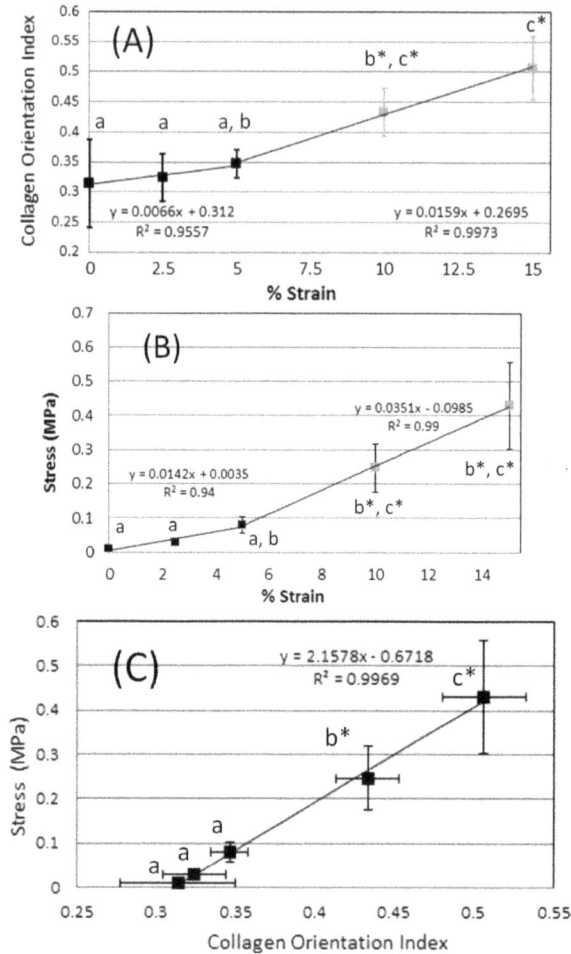

Figure 3. The collagen orientation index (COI) and stress plotted as functions of induced strain. (**a**) The COI plotted as a function of induced strain. Linear regression is performed on two different regions above and below 5% strain; (**b**) Stress *vs.* strain demonstrates similar properties to COI *vs.* strain; (**c**) Stress (used to induce strain) *vs.* COI demonstrates high correlation. Error bars indicate standard deviation and $n = 4$. Similar letters indicate that the groups are statistically similar. The symbol * indicates power > 0.8.

2.2. DIC Analysis of Local Deformation by Strain Maps

To determine inhomogeneity of local strains, strains in three regions were investigated based on distance from the hair follicle, namely, region with primarily collagenous matrix, region surrounding the hair follicle, and a thin region immediately adjacent to the hair follicle (labeled as regions A, B, and C in Figure 4).

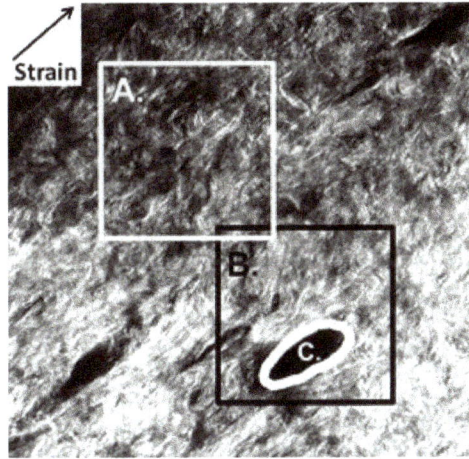

Figure 4. Maximum and minimal principal strains were averaged over three different regions; collagen matrix region (white box denoted by **A**), surrounding hair follicle collagen (black box denoted by **B**), and hair follicle collagen, (which is the area covered by the white ellipse **C**). The direction of applied deformation is indicated in the upper left hand corner.

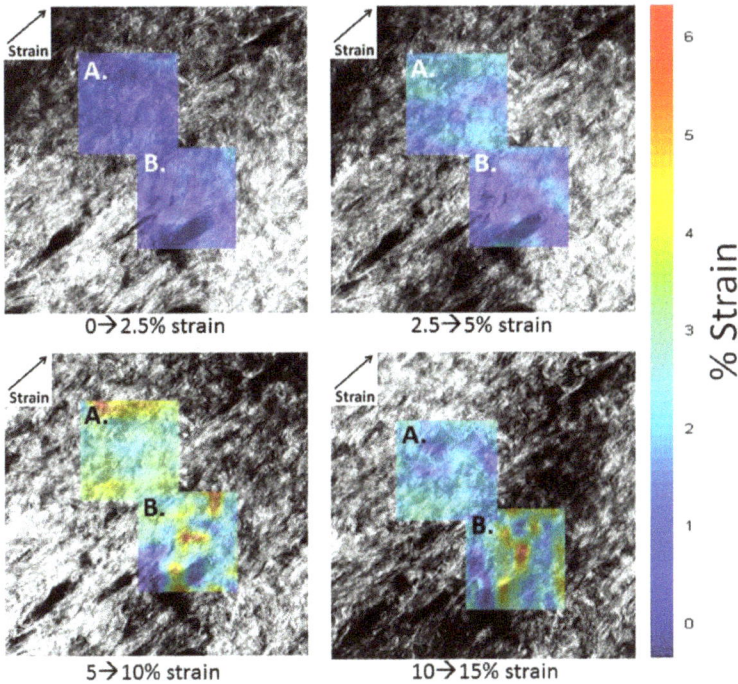

Figure 5. Maximum principal strain maps for matrix region (**A**) and hair follicle region (**B**). Percent strain is given as the difference between images as indicated. Strain direction is indicated in the upper left corner.

Maximum principal strain was found to be positive over all collagen regions (Figure 5 and supplementary Figure S1). Figure 5 shows maximum principal strains scaled to the color map on the right for percent strain. The color map is then overlaid onto gray scale SHG images to show underlying structure of collagen fibrils as a function of stretch. Averaged strain values (Figure 6) are lower than the overall applied strain, which was determined from the separation between grips as measured using a Vernier caliper. This is expected as there are typically higher strain values immediately around the grips, making strain levels in the center of the sample lower. It was found that maximum principal strain increased for all three areas (collagen matrix, collagen surrounding the hair follicle, and hair follicle) with respect to increasing applied strain, as determined by linear regression ($p < 0.001$). Results from ANOVA indicated that there was also no interaction between applied strain and location ($p = 0.93$).

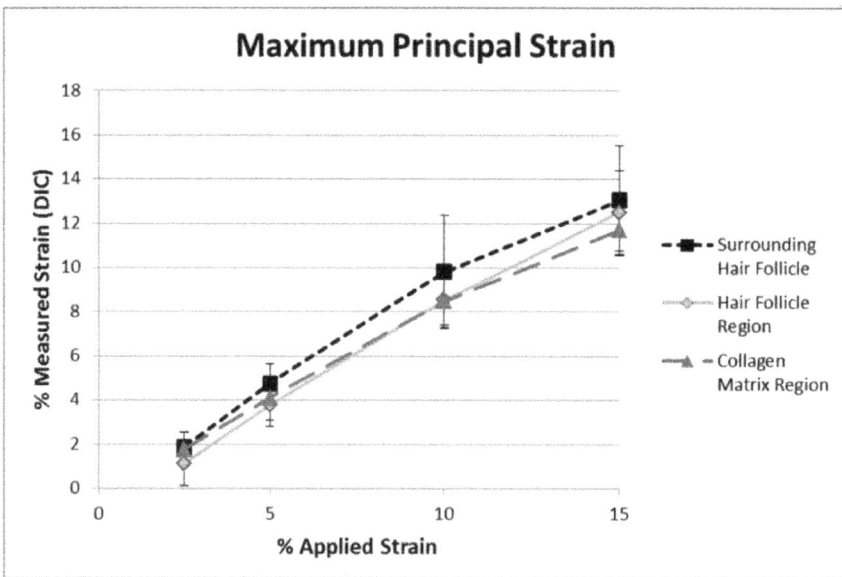

Figure 6. Maximum principal strains averaged over the three regions indicated 4, *versus* applied strain. All three regions of collagen (*i.e.*, hair follicle, surrounding hair follicle, and matrix) increase with applied strain ($p < 0.001$). There is no statistical significance between locations. Error bars represent ±1 Standard deviation.

There was some non-linearity in averaged strains in the 10%–15% strain regime. This difference may be because of relative slippage of fibrils in skin as it was stretched beyond 10% strains, as observed in the change in direction of principal strains at some locations (Figure S1, bottom left).

Minimum principal strains were generally negative (Figure 7), as would be expected because of the Poisson's ratio effect, perpendicular to the direction of stretch. In contrast to maximum principal strains, minimum principal strains were affected by the collagen's location, *i.e.*, proximity with respect to hair follicles ($p = 0.02$). Minimum principal strains in predominantly collagen rich regions (Figure 8) consistently become more negative with higher applied strains of up to 5%, and only gradually increased thereafter (Figure 8). Averaged minimum principal strains determined from collagen fibril deformation surrounding the hair follicle were slightly negative, whereas, they were slightly positive in the hair follicle region. This indicates that the hair follicles are expanding in the direction perpendicular to applied strain. Minimum principal strain was shown to only be significantly different between the hair follicle and collagen matrix regions ($p < 0.01$), and moderately different between surrounding hair follicle and collagen matrix regions ($p = 0.10$). Furthermore, minimum principal strains at all three locations were independent of applied strain ($p = 0.60$).

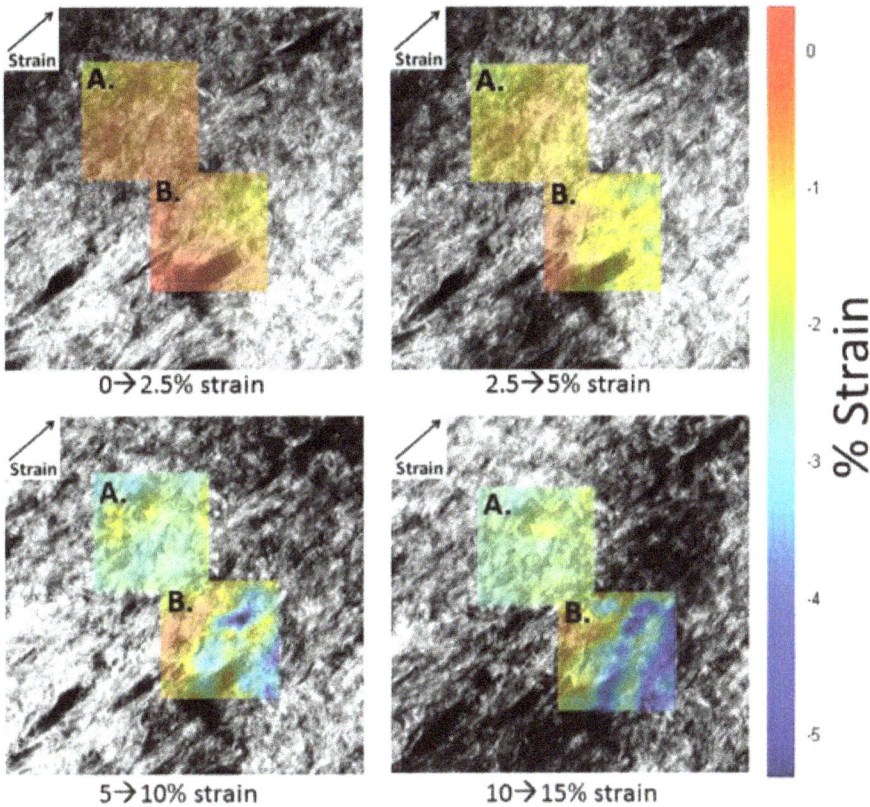

Figure 7. Minimum principal strain maps for matrix region (**A**) and hair follicle region (**B**). Percent strain is given as the difference between images as indicated. Strain direction is indicated in the upper left corner.

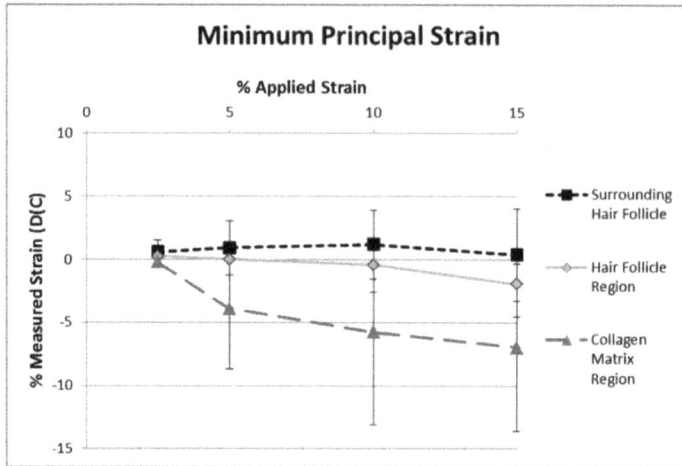

Figure 8. Averaged minimum principal strain, as measured by DIC, *versus* applied strain. There was a statistically significant difference ($p = 0.02$) between the different locations. Minimum principal strains immediately surrounding the hair follicle were positive, indicating an expansion of collagen in this zone. In contrast, minimum principal strains in primarily collagenous zones were negative, indicating a Poisson's effect. Error bars represent ±1 Standard deviation.

2.3. Analysis and Limitations

Reorientation of collagen fibrils was assessed by FFT of SHG images under different strain magnitudes (global changes) while the inhomogeneity of strains were determined from DIC of SHG images (local changes). SHG images were obtained after complete stress relaxation had occurred. As the skin is stretched, collagen fibrils re-orient towards the direction of applied strain (Figure 2a). This change in fibril orientation leads to a change in the shape of the FFT (Figure 2b) which is quantified with the COI (Figure 3a). The COI demonstrates two regions of skin stretching, when the collagen starts to reorient (0%, 2.5%, and 5% strain) and when the collagen fibrils start to elongate (5%, 10%, and 15% strain). These two regions compare well with the known stress-strain analysis of skin (Figure 1) and also with the measured stress-strain relationship (Figure 3b). In contrast to the two linear regions observed for the COI-strain and stress-strain relationships, there is only one linear relationship for the stress *vs.* COI for the strain ranges investigated (Figure 3c). This indicates that the proportion of collagen fibrils re-orienting is proportional to the applied load.

DIC of SHG images allowed us to measure the maximum and minimum principal strain based on deformation of collagen fibrils, especially in proximity to discontinuities such as hair follicles. It was found that maximum principal strain was highly affected by applied uniaxial strain and was unchanged by the presence of discontinuities. Minimum principal strain was affected by the location of the collagen matrix with respect to the discontinuity (*i.e.*, the hair follicle). Minimum principal strain in the predominantly collagen matrix region (region A in Figure 4) demonstrated that the collagen fibrils are closer in the direction perpendicular to direction in which it was stretched (Poisson's effect). However, collagen surrounding hair follicles exhibited slightly negative minimum principal strains and collagen in the hair follicle region has a positive minimal principal strain, indicating it is expanding in the direction perpendicular to the direction of stretching. This has important implications for the mechanobiological response of skin, because of the presence of stem cells in the hair follicle region.

Most studies of tissues that have utilized SHG to evaluate changes in organization to collagen fibrils have evaluated damage or changes to the collagen fibrils after applying loading of different types, typically, after inducing some type of damage—not, when the tissue is being subjected to mechanical loading [18–21]`. Recent studies have begun to characterize changes in orientations of bundles of collagen fibrils in soft tissues (e.g., heart valves) in response to external load, by measuring 3-D changes in angle of bundles of collagen fibrils using a fit to the FFT of the SHG images [22]. One angle was used to characterize the averaged orientation of the fibrils in a 2-D plane, which is equivalent to the use of the major axis of the ellipsoid, in this study. As this axis does not change orientation during loading, we used changes in the ratio of the two axes (defined as COI) to characterize soft tissue mechanical behavior. It is noted that there are minimal confounding effects when studying the heart valve (e.g., there is no hair, or pigmentation).

In complex tissue such as skin, experimental studies of skin deformation are mostly limited to measurements of homogenized deformation measurements on skin surfaces [23] or to evaluation of deformation after removing layers [24]. It is noted that studies have been performed using small-angle x-ray scattering (SAXS) to relate collagen orientation to tear strength, but, not to changes in orientation while loading of skin [25,26]. As a result, the mechanical contribution of collagen fibrils in skin while it is undergoing loading has been mostly studied using computational and analytical models [27–30]. In these models, almost the entire load bearing capacity of skin was ascribed to the behavior of collagen fibrils under load. In these models, and as observed in our experiments, collagen fibrils change orientation at low loads and they are more densely packed at high loads (in the direction perpendicular to loading).

These studies are important because determining local and global changes to the main load bearing constituent of skin (*i.e.*, collagen fibrils), can provide an enhanced understanding of the mechanobiological responses of soft tissues [22], and for regeneration of soft tissues [31]. This technique could potentially provide earlier detection of disease or improve the robustness of the phenotypes detected using SHG techniques which can currently provide gross pathological changes, but, are relatively insensitive to subtle and progressive pathological changes to the dermis [5–7,9,32], for example, through changes in the slope of the COI-stress relationship or in the local strains around structures where stem cells may be present.

One major limitation of our study is that load bearing at the individual collagen fibril scale was not assessed. Complementary studies using polarization resolved SHG to evaluate anisotropy factor during loading [33] may enable one to relate changes in the network to those at the individual fibril. Another limitation of this experiment is that SHG was performed at 5 discrete strains levels and loading was limited to strains of 15%. The duration of imaging at each strain (30 min for stress relaxation, and 30 min to obtain 100 z-stacks) means that it took over 6 h (including microscope start-up time) to image one sample. These long time-periods precluded us from obtaining SHG images at intermediate strain values. Strain values were limited to 15% in order to ensure that damage induced movement of collagen fibrils were not occurring during loading (zone III in Figure 1). Movement of collagen fibrils during imaging causes motion artifacts that would have caused distortions in the FFT patterns (as was observed in some of the samples that were loaded in this study). Another limitation was that z-direction strains were not accounted for. We used 3-D FFT to determine the three axes of the ellipsoid, but, differences were not found in the third dimension. Note that the third axis was extremely truncated as there were only 30 pixels in that direction as opposed to 2048×2048 pixels in the X-Y axis. Furthermore, qualitatively, most of the fiber re-orientation was observed to take place along the long-axis. Another limitation was that the bi-axial behavior of skin was not measured, primarily because of the time required to conduct imaging. Another limitation is that images were obtained half an hour after stretching, a time-scale over which relaxation was complete, which would amplify the changes in orientation. Newer techniques, which can enable imaging of collagen fibril deformation quasi-static loading [34], might provide a more complete description of the physiological contribution of collagen fibrils to mechanical loading. A final limitation is that only a small region of the skin was assessed in few samples, and the results may not be representative.

3. Experimental Section

3.1. Sample Preparation

Skin was excised from nearly identical locations (the mid-central portion of the backs) of female, TOPGAL mice (roughly C57B6 background) of approximately 3 months of age. Before excision, the skin was shaved rigorously to remove hair, in order to remove spurious motion artifacts related to motion of hair, when stretching the skin. Each specimen was made into a strip that was approximately 10 mm × 50 mm and stored in tris-buffered saline (TBS) containing 0.05 wt% sodium azide. All experiments were performed within a week of excision. Individual specimens were about 2 mm thick, inclusive of the fat pad. Loading was performed in excised skin oriented along the cranial-caudal direction.

3.2. Imaging

Imaging was conducted using a Carl Zeiss LSM 5 multiphoton and confocal microscope in reflection mode consisting of a Zeiss Axioskop 2 FS upright scope, with a mode locked Ti:Sapphire laser (Chameleon Ultra II pulsed laser, Coherent Inc., Santa Clara, CA, USA) operating at 80 MHz and 0.7 W (20% of a full-scale average output power of 3.5 W). The incident laser wavelength was 800 nm, which was low-pass filtered at 680 nm, allowing second harmonics to pass. The reflected second harmonic was band-pass filtered (405/30 nm; Chroma) and intensity detected using an NDD PMT detector at 12-bit resolution and a gain setting of 346. Image stacks were obtained by scanning parallel to the skin surface using a Zeiss 20× water immersion objective (Zeiss W Plan Apo with 1.0 NA and WD of 1.8 mm). A stack of 70–100 images were obtained separated by 1 μm along the z-axis. Each image was 2048 × 2048 pixels, with a 0.5 μm/pixel resolution in the x-y plane. In these images, only a set of roughly 30 images contained SHG signal. The relevant 30 image stacks were isolated at each level of strain (0%, 2.5%, 5%, 10%, 15% strain) for 4 different samples. Note that multiple skin samples were imaged (n = 7) but, because of sample movement during an imaging period of 30 min per strain level, data at all strain values were obtained for only 4 samples (in 30 z-stack images) for FFT and in only 3 samples for DIC. So, only these samples were analyzed.

3.3. Stretching

In order to preserve the elastic properties of skin, it was kept wet using a tris-buffered saline drip throughout the imaging procedure. To generate controlled strain in the skin, the sample was placed between two grips which were below the microscope objective (Figure 9). Both ends of the grips were connected to a stepper motor through a gearing mechanism, which was able to generate displacements with a resolution of 0.5 μm. To insure that samples did not slip at the grips, grips were designed with pins and clamped. The initial distance between grips was 40 mm as measured by a caliper (0.01 mm resolution) and strain was applied to the samples by displacing the stepper motor through defined distances at a rate of 0.5 mm/min. After each level of strain was applied, the skin was given 30 min to relax. Note that if the collagen fibrils in the samples moved during an imaging period of 30 min, it was immediately discernible in the images. During this time, the stage was moved under the objective in order to obtain images from similar locations. Strains were applied in incremental order, starting with 0% and working to 15% strain (0%, 2.5%, 5%, 10%, 15%). The force generated was recorded with a 22.2 N load cell with a customized amplifier with ±0.05 N resolution [35]. The force at the end of 30 min of relaxation was used in stress calculations. Loads and displacements were recorded at 10 Hz. It is noted that the direction of loading was at 45° to the x-y scans performed by the servo motors on the objective as verified using images from the top of the sample stage.

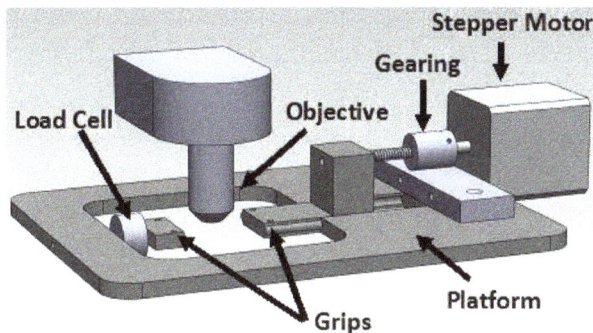

Figure 9. Schematic of a loading machine used to generate stress on the skin sample was made which allowed the skin to be stretched while under the objective of the multi-photon microscope.

3.4. Fourier Transform Power Spectra Analysis (FFT)

Each image stack ($di(x,y)$), which represents a set of 30 relevant images though the skin at a particular strain value, was imported into MATLAB for processing. Images were smoothed by filtering with a 9 × 9 array of ones to reduce high frequency noise in the FFT images. The 2-dimensional FFT of each image in the stack was shifted (fftshift) to move its low frequency components to the center of each image. The absolute value and then log of FFT was taken to reduce the dynamic range. The resultant sets of log compressed frequency transformed images were averaged into a two dimensional array, which can be expressed as the following:

$$I\_FFT_{X,Y} = \frac{1}{n} \sum_{i=1}^{n} [\log_{10}(abs|FFT2(d_i(x,y))|)] \tag{1}$$

This averaging scheme is viable because the collagen fibrils in murine skin form a layered mesh network which oriented in the direction of applied load. As the x-y scans were performed at 45° to the loading axis, the FFT images were then reoriented by rotation to align with the axis of applied strain (the resulting FFT image is shown in Figure 10a). The symmetry of the frequency response (FFT) was evaluated by determining the length of two profiles across and normal to the axis of applied strain [7,9,10,36,37]. Different thresholds were chosen to determine the long- and short- axis of the ellipsoid in the FFT amplitude images. It is noted that results were not affected when using other thresholds, −3 dB to −8 dB values. Therefore, results are presented for the −6 dB value (which is full-width at half maximum or FWHM), as it is halfway between the maximum and background, in other words (maximum + background)/2 (Figure 10b). The FWHM was determined as the sum of pixels from the center of the image to the first point which was below the FWHM value (on each side) as shown in Figure 10b.

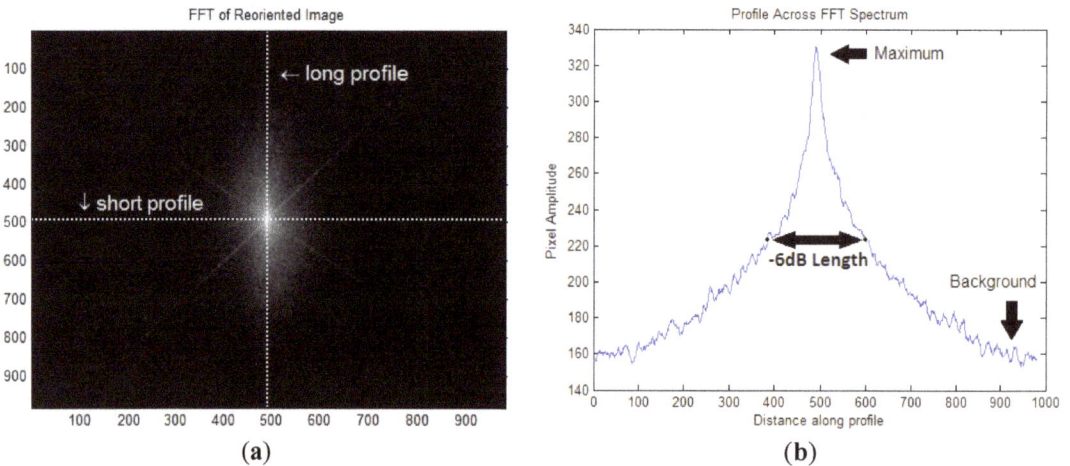

(a)   (b)

Figure 10. Measuring the short and long profiles (**a**) A FFT of an image of skin in which the lowest frequencies have been shifted to the center of the image. Two profiles (short and long) can be defined with respect to the induced strain; (**b**) When the short profile is examined, the FWHM (or –6 dB length) is a pixel distance between values which are half the amplitude between the maximum and background.

The FWHM lengths (short and long) across each of the dimensions would be used to compute the collagen orientation index (COI) [7,9,10,36]. The COI has been used to demonstrate the periodicity of collagen fibrils along and perpendicular to the direction of loading. The COI has been defined as,

$$COI = 1 - (short\ profile/long\ profile) \qquad (2)$$

Thus, a tissue with random collagen orientation would have a COI of 0. As collagen micro-fibrils orient along the direction of applied load, they are squeezed in the perpendicular direction. In the FFT amplitude images, this change in periodicity is identified by increases in the frequency component perpendicular to the direction of applied load. Therefore, the long profile, which is perpendicular to the applied load, is expected to increase, resulting in increasing COI values. The short profile is expected to remain constant and serves as a normalizing parameter.

3.5. Digital Image Correlation (DIC)

DIC was performed to find maximum and minimum principal strain every twenty pixels apart and for 32 × 32 iterations along the x- and y-planes [17]. DIC created data sets consist of displacement (V and U), strain (VX and UY), and shear (VY and UX) values obtained every twenty pixels apart using a 100 × 100 pixel window. For each set of images, two DIC displacement data sets (32 × 32 iterations in the X- and Y- axis) were obtained based on proximity to hair follicles. The displacements, U and V, were corrected for rotation and filtered, following which they were interpolated to the size of the image (601 × 601 pixels). Strain values were determined using a Savitsky-Golay filter, from which maximum and minimum principal strains were calculated. Principal strain values were then averaged over three different areas of the image; the collagen matrix far away from hair follicles (601 × 601 pixels or 300 × 300 μm² this conversion needs to be modified), the hair follicle (601 × 601 pixels), and the collagen immediately surrounding the hair follicle (10 pixels or 5 μm—adjacent to the hair follicle, where the collagen was found to be oriented along its perimeter) (Figure 6).

DIC was conducted between images that were close in applied strain (*i.e.*, between 0% and 2.5%, between 2.5% and 5%, *etc.*) resulting in 4 deformation maps for each sample. These averages were then summed such that all strain values are referenced from 0% strain (Equation (3)).

$$\begin{aligned}
\varepsilon\ 0\% &\rightarrow 2.5\% = \varepsilon\ 0\% \rightarrow 2.5\% \\
\varepsilon\ 0\% &\rightarrow 5\% = \varepsilon\ 0\% \rightarrow 2.5\% + \varepsilon\ 2.5\% \rightarrow 5\% \\
\varepsilon\ 0\% &\rightarrow 10\% = \varepsilon\ 0\% \rightarrow 2.5\% + \varepsilon\ 2.5\% \rightarrow 5\% + \varepsilon\ 5\% \rightarrow 10\% \\
\varepsilon\ 0\% &\rightarrow 15\% = \varepsilon\ 0\% \rightarrow 2.5\% + \varepsilon\ 2.5\% \rightarrow 5\% + \varepsilon\ 5\% \rightarrow 10\% + \varepsilon\ 10\% \rightarrow 15\%
\end{aligned} \qquad (3)$$

3.6. Statistics

DIC was performed to determine collagen fibril deformation adjacent to and far away from hair follicles. The images were divided into regions where hair follicles were absent (shown as white box in Figure 6), where hair follicles were present (shown as black box), and immediately next to the hair follicle (10 pixels along hair follicle perimeter–shown as white thick outline).

Two-way ANOVA was performed followed by Tukey's significant difference test. Applied strain and the region of collagen were the independent variables, while averaged maximum and minimum principal strain values for skin collagen were the dependent variables. Mice were considered to come from the same population and were binned along with the dependent variables.

4. Conclusions

Analysis of SHG images of murine skin to discern local and global changes indicates that minimum principal strains are not-homogenous (especially being positive around hair follicles) as well as that closer packing of collagen fibrils perpendicular to the loading direction is correlated to stress. The global change is collagen fibril orientation with applied load is captured by the collagen orientation index (COI), which is a measure of the ratio of the FFT in the periodicity in the collagen fibrils along and perpendicular to the direction of applied load. The COI-strain and stress-strain relationships exhibit bi-linear behavior while the COI-stress relationship was linear over the range of strains studied. This data combined with previous computational models suggests that collagen fibrils in skin are primarily responsible in contributing to its mechanical behavior. Further studies are necessary to determine whether parameters, such as COI and strain inhomogeneity, can be used for sensitive detection of changes to the extracellular matrix as a consequence of disease.

Supplementary Materials

Supplementary materials can be accessed at: http://www.mdpi.com/1996-1944/8/4/1841/s1.

Acknowledgments

Funding for this work was provided by NSF grant number 0846869 and Rensselaer Polytechnic Institute start up grant. We thank Robert Knapp for his contribution in the initial development of the digital image correlation code used in this study.

Author Contributions

S.N. built the loading machine, conducted the experiments and analyzed DIC results, J.M. conducted the FFT analysis, S.W. expanded on the DIC code and conducted DIC analysis of skin, S.K. wrote the manuscript and supervised the experiments.

Conflicts of Interest

The authors declare no conflict of interest.

References

1. Brinckmann, P.; Frobin, W.; Leivseth, G. *Musculoskeletal Biomechanics*; Thieme: Stuttgart, NY, USA, 2002; p. 243.
2. Silver, F.H.; Freeman, J.W.; de Vore, D. Viscoelastic properties of human skin and processed dermis. *Skin Res. Technol.* **2001**, *7*, 18–23.
3. Daly, C.H. Biomechanical properties of dermis. *J. Invest. Dermatol.* **1982**, *79*, 17s–20s.
4. Williams, R.M.; Zipfel, W.R.; Webb, W.W. Interpreting second-harmonic generation images of collagen i fibrils. *Biophys. J.* **2005**, *88*, 1377–1386.

5. Auada, M.P.; Adam, R.L.; Leite, N.J.; Puzzi, M.B.; Cintra, M.L.; Rizzo, W.B.; Metze, K. Texture analysis of the epidermis based on fast fourier transformation in sjogren-larsson syndrome. *Anal. Quant. Cytol. Histol.* **2006**, *28*, 219–227.

6. Cicchi, R.; Kapsokalyvas, D.; de Giorgi, V.; Maio, V.; van Wiechen, A.; Massi, D.; Lotti, T.; Pavone, F.S. Scoring of collagen organization in healthy and diseased human dermis by multiphoton microscopy. *J. Biophoton.* **2010**, *3*, 34–43.

7. Lu, K.; Chen, J.; Zhuo, S.; Zheng, L.; Jiang, X.; Zhu, X.; Zhao, J. Multiphoton laser scanning microscopy of localized scleroderma. *Skin Res. Technol.* **2009**, *15*, 489–495.

8. Levitt, J.M.; McLaughlin-Drubin, M.E.; Munger, K.; Georgakoudi, I. Automated biochemical, morphological, and organizational assessment of precancerous changes from endogenous two-photon fluorescence images. *PLoS One* **2011**, *6*, e24765.

9. Wu, S.; Li, H.; Yang, H.; Zhang, X.; Li, Z.; Xu, S. Quantitative analysis on collagen morphology in aging skin based on multiphoton microscopy. *J. Biomed. Opt.* **2011**, *16*, 040502.

10. Zhu, X.; Zhuo, S.; Zheng, L.; Lu, K.; Jiang, X.; Chen, J.; Lin, B. Quantified characterization of human cutaneous normal scar using multiphoton microscopy. *J. Biophoton.* **2010**, *3*, 108–116.

11. Greco, V.; Chen, T.; Rendl, M.; Schober, M.; Pasolli, H.A.; Stokes, N.; Dela Cruz-Racelis, J.; Fuchs, E. A two-step mechanism for stem cell activation during hair regeneration. *Cell Stem Cell* **2009**, *4*, 155–169.

12. Tiede, S.; Kloepper, J.E.; Bodo, E.; Tiwari, S.; Kruse, C.; Paus, R. Hair follicle stem cells: Walking the maze. *Eur. J. Cell Biol.* **2007**, *86*, 355–376.

13. Bruck, H.A.; McNeill, S.R.; Sutton, M.A.; Peters, W.H., III. Digital image correlation using newton-raphson method of partial differential correction. *Exp. Mech.* **1989**, *29*, 261–267.

14. Chu, T.C.; Ranson, W.F.; Sutton, M.A. Applications of digital-image-correlation techniques to experimental mechanics. *Exp. Mech.* **1985**, *25*, 232–244.

15. Ni Annaidh, A.; Bruyere, K.; Destrade, M.; Gilchrist, M.D.; Ottenio, M. Characterization of the anisotropic mechanical properties of excised human skin. *J. Mech. Behav. Biomed. Mater.* **2012**, *5*, 139–148.

16. Staloff, I.A.; Guan, E.; Katz, S.; Rafailovitch, M.; Sokolov, A.; Sokolov, S. An *in vivo* study of the mechanical properties of facial skin and influence of aging using digital image speckle correlation. *Skin Res. Technol.* **2008**, *14*, 127–134.

17. Wentzell, S.; Sterling Nesbitt, R.; Macione, J.; Kotha, S. Measuring strain using digital image correlation of second harmonic generation images. *J. Biomech.* **2013**, *46*, 2032–2038.

18. Abraham, T.; Fong, G.; Scott, A. Second harmonic generation analysis of early achilles tendinosis in response to *in vivo* mechanical loading. *BMC Musculoskel Dis.* **2011**, *12*, 26.

19. Frisch, K.E.; Duenwald-Kuehl, S.E.; Kobayashi, H.; Chamberlain, C.S.; Lakes, R.S.; Vanderby, R. Quantification of collagen organization using fractal dimensions and fourier transforms. *Acta Histochem.* **2012**, *114*, 140–144.

20. Fung, D.T.; Sereysky, J.B.; Basta-Pljakic, J.; Laudier, D.M.; Huq, R.; Jepsen, K.J.; Schaffler, M.B.; Flatow, E.L. Second harmonic generation imaging and fourier transform

spectral analysis reveal damage in fatigue-loaded tendons. *Ann. Biomed. Eng.* **2010**, *38*, 1741–1751.

21. Sinclair, E.B.; Andarawis-Puri, N.; Ros, S.J.; Laudier, D.M.; Jepsen, K.J.; Hausman, M.R. Relating applied strain to the type and severity of structural damage in the rat median nerve using second harmonic generation microscopy. *Muscle Nerve* **2012**, *46*, 899–907.

22. Alavi, S.H.; Ruiz, V.; Krasieva, T.; Botvinick, E.L.; Kheradvar, A. Characterizing the collagen fiber orientation in pericardial leaflets under mechanical loading conditions. *Ann. Biomed. Eng.* **2013**, *41*, 547–561.

23. Jor, J.W.; Nash, M.P.; Nielsen, P.M.; Hunter, P.J. Estimating material parameters of a structurally based constitutive relation for skin mechanics. *Biomech. Model. Mechanobiol.* **2011**, *10*, 767–778.

24. Brown, I.A. A scanning electron microscope study of the effects of uniaxial tension on human skin. *Br. J. Dermatol.* **1973**, *89*, 383–393.

25. Basil-Jones, M.M.; Edmonds, R.L.; Cooper, S.M.; Kirby, N.; Hawley, A.; Haverkamp, R.G. Collagen fibril orientation and tear strength across ovine skins. *J. Agric. Food Chem.* **2013**, *61*, 12327–12332.

26. Basil-Jones, M.M.; Edmonds, R.L.; Cooper, S.M.; Haverkamp, R.G. Collagen fibril orientation in ovine and bovine leather affects strength: A small angle x-ray scattering (saxs) study. *J. Agric. Food Chem.* **2011**, *59*, 9972–9979.

27. Annaidh, A.N.; Bruyere, K.; Destrade, M.; Gilchrist, M.D.; Maurini, C.; Ottenio, M.; Saccomandi, G. Automated estimation of collagen fibre dispersion in the dermis and its contribution to the anisotropic behaviour of skin. *Ann. Biomed. Eng.* **2012**, *40*, 1666–1678.

28. Chen, H.; Liu, Y.; Zhao, X.F.; Lanir, Y.; Kassab, G.S. A micromechanics finite-strain constitutive model of fibrous tissue. *J. Mech. Phys. Solids* **2011**, *59*, 1823–1837.

29. Jor, J.W.Y.; Parker, M.D.; Taberner, A.J.; Nash, M.P.; Nielsen, P.M.F. Computational and experimental characterization of skin mechanics: Identifying current challenges and future directions. *Wires Syst. Biol. Med.* **2013**, *5*, 539–556.

30. Lokshin, O.; Lanir, Y. Micro and macro rheology of planar tissues. *Biomaterials* **2009**, *30*, 3118–3127.

31. Raub, C.B.; Suresh, V.; Krasieva, T.; Lyubovitsky, J.; Mih, J.D.; Putnam, A.J.; Tromberg, B.J.; George, S.C. Noninvasive assessment of collagen gel microstructure and mechanics using multiphoton microscopy. *Biophys. J.* **2007**, *92*, 2212–2222.

32. Balbir-Gurman, A.; Denton, C.P.; Nichols, B.; Knight, C.J.; Nahir, A.M.; Martin, G.; Black, C.M. Non-invasive measurement of biomechanical skin properties in systemic sclerosis. *Ann. Rheum. Dis.* **2002**, *61*, 237–241.

33. Gusachenko, I.; Tran, V.; Houssen, Y.G.; Allain, J.M.; Schanne-Klein, M.C. Polarization-resolved second-harmonic generation in tendon upon mechanical stretching. *Biophys. J.* **2012**, *102*, 2220–2229.

34. Smith, D.R.; Winters, D.G.; Bartels, R.A. Submillisecond second harmonic holographic imaging of biological specimens in three dimensions. *Proc. Natl. Acad. Sci. USA* **2013**, *110*, 18391–18396.

35. Macione, J.; Nesbitt, S.; Pandit, V.; Kotha, S. Design and analysis of a novel mechanical loading machine for dynamic *in vivo* axial loading. *Rev. Sci. Instrum.* **2012**, *83*, 025113.

36. Van Zuijlen, P.P.; Ruurda, J.J.; van Veen, H.A.; van Marle, J.; van Trier, A.J.; Groenevelt, F.; Kreis, R.W.; Middelkoop, E. Collagen morphology in human skin and scar tissue: No adaptations in response to mechanical loading at joints. *Burns* **2003**, *29*, 423–431.

37. Xia, D.; Zhang, S.; Hjortdal, J.O.; Li, Q.; Thomsen, K.; Chevallier, J.; Besenbacher, F.; Dong, M. Hydrated human corneal stroma revealed by quantitative dynamic atomic force microscopy at nanoscale. *ACS Nano* **2014**, *8*, 6873–6882.

Evaluation of Biaxial Mechanical Properties of Aortic Media Based on the Lamellar Microstructure

Hadi Taghizadeh, Mohammad Tafazzoli-Shadpour, Mohammad B. Shadmehr and Nasser Fatouraee

Abstract: Evaluation of the mechanical properties of arterial wall components is necessary for establishing a precise mechanical model applicable in various physiological and pathological conditions, such as remodeling. In this contribution, a new approach for the evaluation of the mechanical properties of aortic media accounting for the lamellar structure is proposed. We assumed aortic media to be composed of two sets of concentric layers, namely sheets of elastin (Layer I) and interstitial layers composed of mostly collagen bundles, fine elastic fibers and smooth muscle cells (Layer II). Biaxial mechanical tests were carried out on human thoracic aortic samples, and histological staining was performed to distinguish wall lamellae for determining the dimensions of the layers. A neo-Hookean strain energy function (SEF) for Layer I and a four-parameter exponential SEF for Layer II were allocated. Nonlinear regression was used to find the material parameters of the proposed microstructural model based on experimental data. The non-linear behavior of media layers confirmed the higher contribution of elastic tissue in lower strains and the gradual engagement of collagen fibers. The resulting model determines the nonlinear anisotropic behavior of aortic media through the lamellar microstructure and can be assistive in the study of wall remodeling due to alterations in lamellar structure during pathological conditions and aging.

Reprinted from *Materials*. Cite as: Taghizadeh, H.; Tafazzoli-Shadpour, M.; Shadmehr, M.B.; Fatouraee, N. Evaluation of Biaxial Mechanical Properties of Aortic Media Based on the Lamellar Microstructure. *Materials* **2015**, *8*, 302-316.

1. Introduction

The arterial wall is functionally and structurally a complicated tissue. Successful biomechanical investigations of the aortic wall go back to only recent decades due to these complications [1]. The significance of such studies becomes multifold considering that pathological conditions of the cardiovascular system, such as coronary heart disease and hypertension, are among the leading causes of world mortality [2].

Since the mechanical properties of soft biological tissues highly depend on their microstructure, proposing a reliable mechanical model for these tissues, including the arterial wall, depends on the level of microstructure integration attained in the constitutive model. The early proposed models assumed the arterial wall to be a single continuous medium and suggested different forms of the strain energy function (SEF) to characterize its mechanical behavior [3,4]. These models contributed by providing primary insights into the mechanical features of arteries, but more realistic models with a focus on the microstructure were vital to fully understand the mechanisms involved in the mechanical behavior of arteries. As a consequence, a new category of arterial tissue models, called "microstructural models", was adopted.

Microstructural modeling led to significant refinements, particularly in cardiovascular tissue modeling and recognizing how physiological and pathological states affect the state of arterial tissue [5,6]. In the case of the arterial wall, such models are classified into two categories. In the first category, arterial tissue is regarded to be composed of three main layers, *i.e.*, intima, media and adventitia [7,8]. The second category incorporates proposing different SEFs for the main wall fibrous constituents, such as elastin and collagen [9,10]. Since the mechanical behavior of the arterial wall highly depends on its fibrous lamellar structure, multiscale models have been recruited to relate the microstructural features of the arterial wall to its bulk mechanical behavior [11,12]. Homogenization techniques used in a multiscale approach are a powerful tool in modeling complex microstructures, such as collagen bundles and crosslinks within the arterial wall [13]. It is believed that elastic tissue contributes dominantly at low strains, while at higher strain ranges, collagen fibers become gradually engaged and their contribution ascends [14]. Although the active behavior of smooth muscle cells (SMCs) also contributes to the mechanical response of the tissue [15], their passive mechanical properties are negligible compared to those of elastin and collagen [16]. Since collagen fibers are three orders of magnitude stiffer than elastin [17], uncrimping of collagen fibers drastically stiffens the mechanical response of the tissue, leading to a nonlinear incremental stress-strain relationship for the entire tissue.

It is well known that the mechanical behavior of the arterial wall depends mainly on its composite-like lamellar structure and the mechanical properties of the media [18–20]. However, it should be noted that in supraphysiological loads, adventitia also contributes to the mechanical behavior of the arterial wall [5,14]. The lamellar organization of the media and its main building fractions have been delineated previously [21–23]. Wolinski and Glagov first uncovered the well-organized lamellar structure of arteries among different mammalian species. They observed alternating dark and light layers in stained slides of media and named them the "lamellar unit" of the media. The relationship between the numbers of these lamellar units (from six units in rat to 70 units in pig) and the physiologic pressure for the examined species revealed almost an equal force per lamellar unit among mammalian species [18]. Dingemans *et al.* provided a neat schematic representation of the lamellar unit of aortic media and depicted the organization of the extracellular matrix and the connection to smooth muscle cells (SMCs) [22]. O'Connel *et al.* provided a 3D schematic of the structure combining confocal and electron microscopy imaging, in which elastin sheets as elastic lamellae were distinguished from stiffer layers of collagen bundles and fine elastic fibers accompanied by SMCs [24]. Concentric elastin sheets are almost identical in terms of the thickness and also possess a nearly isotropic composition [18,25]. Within the interlamellar space, SMCs are located as the main organic constituent of media, with collagen and fine elastic fibers running between SMCs [22,24].

The incorporation of the histological and structural data of the arterial wall into mechanical models is crucial to inspect ongoing changes of the tissue from the healthy to diseased state, such as hypertension and also age-related wall remodeling. In this way, the material parameters of the model can be better correlated with the physical features of the tissue, and the alterations in these parameters define the respective alterations in the mechanics of the tissue.

Combining mechanical and histological data, it is convenient to inspect how the microstructure contributes to the mechanical characteristics of the wall among major arteries. Hence, the wall media can be considered as two sets of contiguous and concentric lamellae: sheets of elastin (Layer I) and interstitial layers consisting of collagen bundles and fine elastic fibers containing SMCs (Layer II).

To the best of our knowledge, the mechanical analysis of the lamellar structure of the media and its contribution to the whole wall mechanics have not been considered before; as a result, we have proposed a new approach to integrate the microstructure of the aortic media in a mechanical model. In our study, the pair of Layers I and II defines the "lamellar unit" of the media. Approximately 60 lamellar units form the aortic media in the thoracic region of an adult human aorta [18,23]. Utilizing biaxial mechanical data and geometric measurements of stained tissue rings, the material parameters of the proposed microstructural model are calculated, and the contributions of Layers I and II to the overall mechanical behavior of the media are explored.

2. Materials and Methods

2.1. Surgical Procedures

In the thoracic region of the descending aorta, the diameter changes are minimal, and the microstructural parameters, including the thickness and number of lamellae, do not vary notably. Hence, samples of human descending thoracic aorta were used. Arterial samples were provided from brain-dead patients after organ donation according to the Ethical Committee instructions of Masih Daneshvari Hospital, the main site of organ donation and transplantation in Iran. Aortic samples from three male donors, aged 25 (M25), 28 (M28) and 42 (M42) years, were used with special attention to the medical history of the donors. None of the subjects had shown a cardiovascular disorder and disease background. This is of particular relevance, since the mechanical properties of mature and healthy aortic media are addressed. Tubes of 5–6 cm in length were cut just above the diaphragm in the descending region of the aorta. Small branches were cropped carefully to achieve undamaged and intact tissues (Figure 1a). Sections from these samples were prepared for biaxial tests, and some adjacent blocks were extracted for histological staining.

2.2. Tissue Preparation

Aortic samples were preserved in phosphate-buffered saline (PBS) immediately after harvest and transferred to our lab. Before tests, the adventitial and loose connective tissues were carefully removed using a surgical scalpel (Figure 1). The remaining media-intima was submerged in PBS and then refrigerated at 4 °C for the test, preferably within the same day. Prior to tests, samples were allowed to reach room temperature.

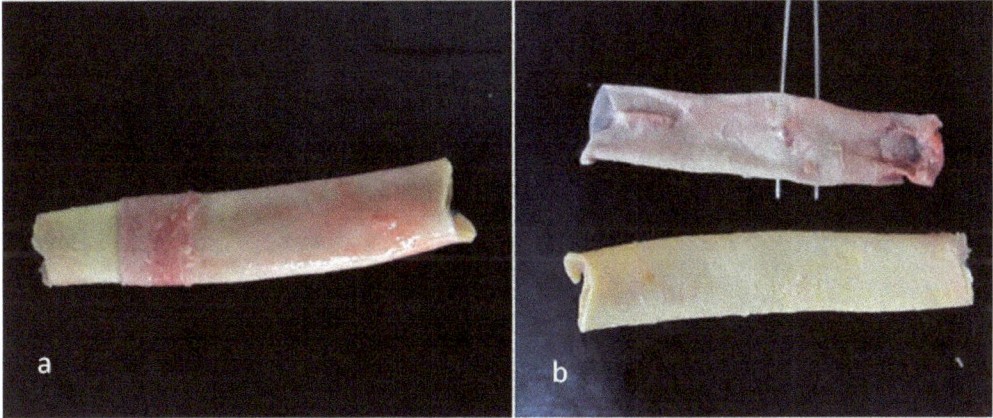

Figure 1. (**a**) Removal of adventitial tissue; (**b**) resulting intima-media composite and loose adventitial layer.

Cylinders with a height of 11 mm were cut, and 11 mm × 11 mm squares were extracted for biaxial tests. Simultaneously, adjacent rings of 2–3 mm in height were cut to be used in histological staining.

2.3. Biaxial Testing

Biaxial testing of soft biological tissues is difficult, yet necessary, to fully comprehend the mechanical characteristics due to the complex microstructure and resulting anisotropic nature of the tissue. Such tests require a precise experimental setup, including the testing machine and attachment of the tissue to the jaws of the testing apparatus. Some researchers used uniaxial test data to characterize the mechanical properties [20,26]; others utilized a variety of biaxial tests, including planar biaxial tests [27] or extension-inflation tests [8]. In the present study, we carried out planar biaxial tests using a custom-made biaxial test apparatus consisting of four stepping motors to stretch the samples, while two load-cells recorded the load magnitude during extensions (Figure 2a). We used hooks to mount samples to the test machine. Such attachments are crucial to avoid unwanted stress concentrations. To record displacements, four markers were used on the central part of the square specimens, and a CCD camera was used to track the marker coordinates during the tests. The tissue thickness was measured several times with a caliper, and the average thickness was used in subsequent stress calculations. Considering homogeneous strain in the central region of the specimen, the resulting force-displacement data for both of the test axes were converted to second Piola stress and Green–Lagrange strain. The details of these conversions are provided in subsequent sections.

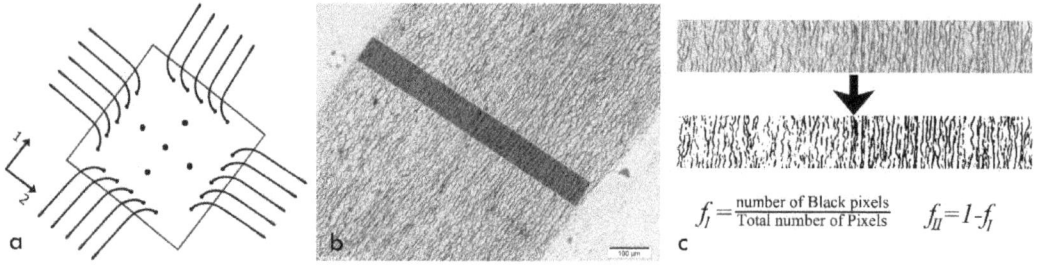

$$f_I = \frac{\text{number of Black pixels}}{\text{Total number of Pixels}} \qquad f_{II} = 1 - f_I$$

Figure 2. (**a**) A schematic of the sample under biaxial stretch; (**b**) Stained aortic media. Dark regions in this tissue section denote elastic fibers (Layer I), and light colored regions denote interlamellar zones, including collagen bundles, fine elastic fibers and smooth muscle cells (SMCs) (Layer II); (**c**) Volume fraction computation steps, including radial strip extraction, conversion to black and white and, finally, counting the number of pixels of the dark (Layer I) and light (Layer II) areas, as well as finding their proportion regarding the total number of pixels.

2.4. Histological Staining

In addition to biaxial test data, microstructural information, including dimensions and volume fractions of lamellae, is required in constitutive modeling. The volume fraction of each type of layer is based on the overall non-liquid phase, which contributes to the mechanical properties of the whole structure. These features were extracted by histological staining of extracted samples from human aorta. Sliced rings of aortic samples were first fixed in formalin, then paraffin-embedded and, finally, cut into micron-height tubes using a microtome [27]. The resulting slices were stained and investigated under light microscopy (Figure 2b). We used Verhoeff Van-Gieson (VVG) to stain elastin sheets (Layer I) to be distinguished from adjacent layers [27]. After staining, slides were photographed under a microscope, and the resultant images were processed by a MATLAB code to find the volume fractions. To do so, images were converted to black and white, and after removal of the artifacts, the percentage of the black (Layer I) and white (Layer II) zones were computed as the volume fractions. To minimize location dependency, we measured and averaged the volume fraction into six equally apart radial sites. Representative volume fraction computation steps are depicted in Figure 2c.

2.5. Constitutive Equations and Parameter Estimation

Experimental force-displacement data were converted to stress-strain. The resulting stress–strain response characterizes the mechanical behavior of the whole media. On the other hand, we assigned strain energy functions (SEFs) to Layers I and II and proposed the SEF of the media based on the SEFs of the layers. It should be noted that the SEFs of the layers are furnished with unknown material parameters, which should be identified. The SEF of the media was differentiated to give computational stress in terms of the mentioned material parameters. To evaluate these parameters, we compared experimental and computational stresses and minimized their differences by optimization of the material parameters through a nonlinear regression algorithm.

2.5.1. Theoretical Framework

Describing the nonlinear and anisotropic response of the arterial wall requires utilization of continuum mechanics and hyperplastic models, due to the large deformations that arteries experience in the physiological environment [28]. Arteries are assumed to be incompressible, because of the high water content [29]. In a recent study, the degree of compressibility of different arteries has been shown to be small, especially for elastic arteries [30]. The SEF for such materials depends only on the deformation gradient F. Assuming that X and x denote the coordinates of the point in reference and deformed configurations, respectively, the deformation gradient is given by:

$$F = \frac{\partial x}{\partial X} \tag{1}$$

Considering arteries as cylindrical tubes, it has been well described that axial, circumferential and radial directions coincide with the principal axes of arteries [31]. In the principal directions, F is represented by a diagonal matrix with components that are stretch ratios in the respective principal directions, Equation (2). Axial (λ_1) and circumferential (λ_2) stretch ratios are calculated from planar biaxial tests, and the radial component (λ_3) is given by the incompressibility constraint ($J = \det(F) = \lambda_1 \lambda_2 \lambda_3 = 1$):

$$F = \begin{bmatrix} \lambda_1 & 0 & 0 \\ 0 & \lambda_2 & 0 \\ 0 & 0 & \frac{1}{\lambda_1 \lambda_2} \end{bmatrix} \tag{2}$$

SEFs are usually formulated as functions of left or right Green–Cauchy strain tensor ($B = FF^T$ and $C = F^T F$, respectively) invariants (I_1, I_2 and J) [32] or a Green–Lagrange strain tensor ($E = \frac{1}{2}(C - I)$) [31], which are interrelated functions of the deformation gradient. The second Piola stress for an incompressible material can be represented as [33]:

$$S = -pC^{-1} + 2\frac{\partial W}{\partial C} = -pC^{-1} + \frac{\partial W}{\partial E} \tag{3}$$

where p is a Lagrange multiplier term to enforce incompressibility and W describes SEF. The Lagrange multiplier parameter is determined from a boundary or pre-defined condition. In our case, the out of plane component of stress is zero (plane stress assumption), and the term p is obtained accordingly. After deriving the second Piola stress, other stress measures can be calculated and utilized accordingly. Cauchy stress is obtained by transforming the second Piola stress into a deformed configuration, which is given by:

$$\sigma = \frac{1}{J} FSF^T \tag{4}$$

Experimental marker coordinates in biaxial tests were used to calculate respective stretch ratios, and subsequently, the deformation gradient tensor was formed. Then, the Green–Lagrange strain tensor ($E_i = \frac{1}{2}(\lambda_i^2 - 1)$, E_i is used instead of E_{ii} to denote diagonal elements of the Green–Lagrange strain tensor for simplicity) and the first invariant of right Green–Cauchy strain tensor ($I_1 = \lambda_1^2 + \lambda_2^2 + \frac{1}{\lambda_1^2 \lambda_2^2}$) were calculated for the time increments during the tests. Furthermore, the measured force per unit of the initial orthogonal cross-section of the tissue was calculated as the first Piola stress (P); then, it is converted to the second Piola stress:

$$S = F^{-1}P \tag{5}$$

In the following sections, the deformation variables (Green–Lagrange strain and the first invariant of the right Green–Cauchy strain tensor) denote the experimental deformations. Parameter S_{exp} will denote the stress obtained from the experiments.

2.5.2. Strain Energy Function

Appropriate types of SEF should be used for the media layers to evaluate the overall mechanical behavior of the tissue. To present proper forms of SEF for media layers, a phenomenological approach was adopted. It has been reported that the arterial elastin component is almost isotropic; hence, a neo-Hookean constitutive model can adequately elucidate its mechanical behavior [25,34,35]. Since, in this study, Layer I represents elastic lamellae, which are made almost entirely of elastin, neo-Hookean-type SEF was assigned to represent their mechanical behavior. Layer II, which is mechanically dominated by collagen fibers, behaves nonlinearly, due to the non-homogeneous distribution and gradual uncrimping of collagen fibers interwoven by fine elastic fibers and ground substances [36]. The four-parameter exponential SEF, which was proposed by Fung et al. [37], was used to simulate the mechanical performance of Layer II of the media. This function is capable of representing the nonlinear, anisotropic and stiffening behavior of the collagen-embedded biological materials and is described as "the most concise potential for biotissues" [38]. We assumed the superimposed contribution of Layers I and II, i.e., Layers I and II act in parallel and contribute proportional to their volume fractions of the media. The superimposed contribution of components is widely used in arterial models [5,39]. Assigned forms of SEF to Layer I, Equation (6), Layer II, Equation (7), and the entire media, Equation (8), are discussed as follows:

$$W_I = c_1(I_1 - 3) \tag{6}$$

$$W_{II} = c_2[\exp(a_1 E_1^2 + a_2 E_2^2 + 2a_3 E_1 E_2) - 1] \tag{7}$$

$$W_{media} = f_I W_I + f_{II} W_{II} \tag{8}$$

in which f_I and f_{II} represent volume fractions of Layers I and II of the overall non-liquid phase, respectively, and W denotes the SEF for Layer I (W_I), Layer II (W_{II}) and the whole media (W_{media}). Additionally, E_i stands for the Green–Lagrange strain and I_1 for the first invariant of the right Green–Cauchy strain tensor. Furthermore, c_1, c_2, a_1, a_2, a_3 are unknown material parameters.

Considering the concentric and parallel configurations of two layers and the fact that layers do not detach during stretch within the physiological range, it is assumed that the deformations of Layers I and II are equal to the deformation of the whole tissue, which is recorded during biaxial tests.

Bidirectional computational stresses for the whole media (W_{media}) can be obtained in terms of layer stresses using Equations (3) and (8). Then, the following set of equations can be written for the computational axial (S^a) and circumferential (S^c) stresses of layers and the whole wall:

$$\begin{cases} S_{comp}^a = f_I S_I^a + f_{II} S_{II}^a \\ S_{comp}^c = f_I S_I^c + f_{II} S_{II}^c \end{cases} \tag{9}$$

The mechanical behavior of the media is expressed as a function of unknown material parameters proposed for the layers, Equations (6) and (7). The volume fractions of the layers were determined from image processing of the histological analysis, as described previously.

2.5.3. Parameter Estimation

To find the unknown material parameters, the obtained experimental data were approximated with the proposed SEF in both circumferential and axial directions. Unknown material parameters accommodating computational stresses were estimated, such that computational stresses could best follow experimental stresses for the range of deformations (E_i, I_1). A MATLAB code was developed to search for the optimized set of unknown material parameters that can simultaneously best interpolate experimental data in circumferential and axial directions. Nonlinear regression was utilized to evaluate the error function as the squared difference of computational (S_{comp}) and experimental (S_{exp}) stresses in a full range of experimental deformations, Equation (10). The material parameters were updated in each iteration, till the minimum difference criterion was met.

$$e(c_1, c_2, a_1, a_2, a_3) = \sum_{n=1}^{k} \left[(S_{comp}^c - S_{exp}^c)_n^2 + (S_{comp}^a - S_{exp}^a)_n^2 \right] \tag{10}$$

Parameter k is used to summate the difference over the full range of deformations.

3. Results

The nonlinear stress–strain responses calculated from the load-displacement data of human thoracic aorta obtained from biaxial tensile tests are depicted in Figure 3 for the axial and circumferential directions of all three test subjects. Comparing the respective strains, stiffer circumferential behavior is observed. The results indicate that in the initial part of the curves, the stress is proportionally increasing with the exerted strain up to a specific region, after which stress increases faster.

These responses were fitted with the proposed lamellar model, and the resultant material parameters are delineated in Table 1. Goodness of fits were investigated for each case using the coefficient of determination (R^2) and are reported in the same Table. The computational stress predicted by our model is plotted together with respective experimental data for one of the cases (M25) in Figure 4 to visualize the ability of the model to describe the anisotropic behavior of the aortic media. The same trend was reported for other cases. In addition, Figure 4 illustrates the contribution of Layers I and II on the stress-strain response of the media. In lower strains, Layer I bears higher stresses. By further straining, Layer II exceeds Layer I in stress. This pattern is observed in both the axial and circumferential directions. The intersection point of layer stresses denotes the equal contributions of them to the mechanical behavior of the whole media.

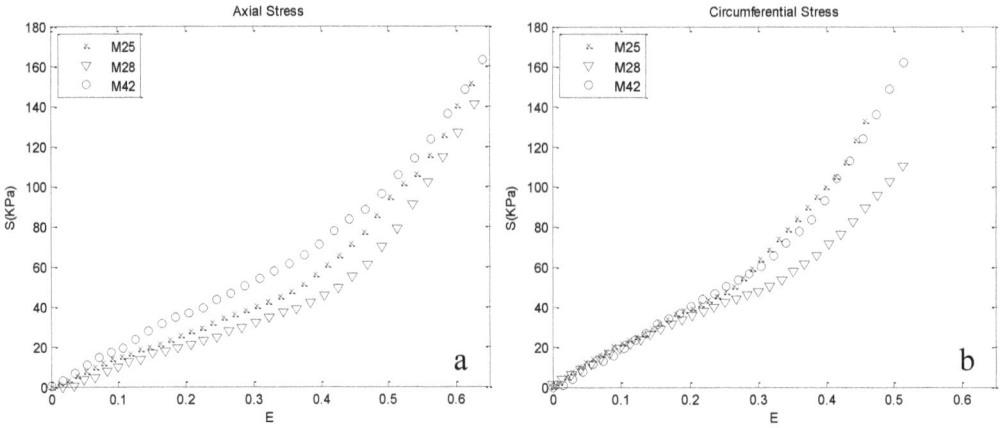

Figure 3. Stress-strain response obtained by biaxial tests: (**a**) axial and (**b**) circumferential responses. At low strain ranges, similar responses of the three cases in the axial and circumferential directions indicate that components leading to anisotropy do not contribute significantly in this region.

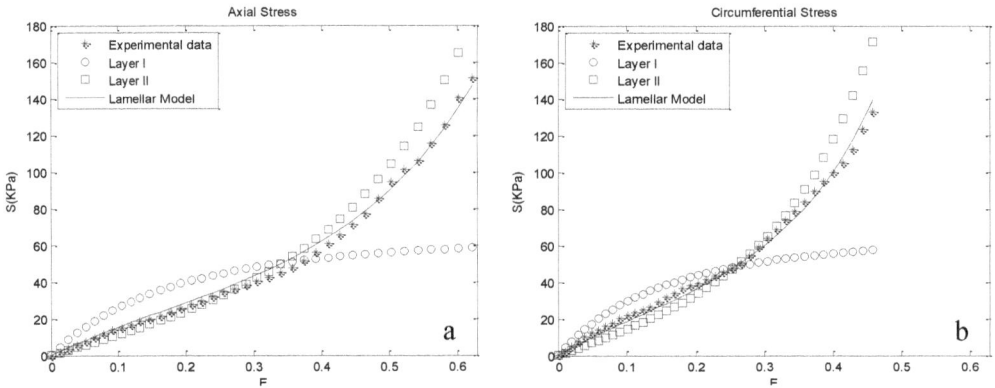

Figure 4. Experimental stresses plotted together with computational stresses for M25 (male donor, aged 25), based on the evaluated material parameters. The contribution of Layers I and II on the mechanical behavior of the media is depicted, as well, for the range of experimental strains: (**a**) axial direction; (**b**) circumferential direction.

The computed volume fractions for Layers I and II utilizing image-processing are depicted in Table 1. For M25 and M28, the same range of volume fractions are found. However, for the third case (M42), representing a middle-aged subject, significant changes in the volume fractions are observed. For this case, the volume fraction of Layer II is drastically (~20%) elevated compared to younger subjects. This is in accordance with the effects of aging on the remodeling of the arterial wall due to new collagen synthesis, which is inherent within Layer II [40]. Moreover, it has been shown that aging is associated with decreased fibrillar crosslinks and changes in collagen fiber orientations in addition to altered collagen content [41,42]. Nevertheless, with such considerable changes in the microstructure, the stress levels of all three cases are similar.

Table 1. Computed volume fractions and obtained material parameters for the three cases tested.

Parameters	Cases		
	M25	M28	M42
f_I (-)	0.2803 ± 0.0113	0.2711 ± 0.0100	0.2172 ± 0.0067
f_{II} (-)	0.7197 ± 0.0113	0.7289 ± 0.0100	0.7828 ± 0.0067
c_1 (kPa)	33.894	43.795	43.428
c_2 (kPa)	35.877	15.692	46.627
a_1 (-)	1.4655	1.9984	1.1188
a_2 (-)	1.85712	2.2043	1.2865
a_3 (-)	0.0473	0.0080	0.2265
R^2 (-)	0.9946	0.9761	0.9959

4. Discussion

Distinct phases can be distinguished in the axial and circumferential stress-strain curves of the arterial wall tissue [20,43] (Figure 3). At low strains, the stress–strain curve is nearly linear and similar in both the axial and circumferential directions, indicating that components leading to anisotropic behavior of the artery wall do not contribute significantly in this range. The results of our lamellar model are in agreement with this fact, because at low strains, Layer I, which is assigned to arterial elastin, bears higher stresses compared to Layer II (Figure 4). The obtained material parameter for Layer I is consistent with recent investigations of the mechanical behavior of arterial elastin [25,44]. In the initial part of the stress–strain curve, collagen fibers behave in a wavy and crimped manner, and they remain almost inactive until further stretching, leading to their load-bearing engagement [45]. Along with further strain, collagen fibers gradually uncrimp and become engaged in the mechanical response, presumably described by the intersection of the stress-strain curves of Layers I and II (Figure 4). Further straining results in their activation, which is shown in the stress–strain curve of the media by a transition from a linear to a nonlinear response. Increasing stiffness of Layer II is owed to the gradual recruitment of collagen fibers. In this range of strain, Layers I and II contribute similarly to the mechanical behavior of the wall. The intersection point of the stress curves of the layers in Figure 4 denotes the equal contributions of Layers I and II. An interesting fact that can be inferred from Figure 4 is that the strain corresponding to the intersection point in the circumferential direction is lower compared to the axial direction, showing a stiffer circumferential response. This can indicate that more collagen fibers are aligned in the circumferential direction compared to the axial direction. In recent studies, the preferred direction of collagen fibers was inquired, and the same dominancy in circumferential direction has been reported [13,24,46].

Physiologic ranges of strains in aorta are reported to be maximally 36% in the axial and 21.5% in the circumferential direction [47]. Interestingly, these strains almost coincide with the intersection points in Figure 4, indicating almost equal layer stresses in physiologic strains. In the last phase, most of the collagen fibers are recruited [48], their behavior becomes dominant and the stress-strain response of the whole media follows the contribution of Layer II.

The measured volume fractions, besides providing geometrical data for the current lamellar model, can be used as an indicator of the structural changes in the lamellar structure of the media with age progression. It should be noted that for a comprehensive judgment, more samples of differently aged subjects should be investigated. Considering the drastic decrease observed in the volume fraction of Layer I in M42, together with the similar mechanical responses at low strains for all three cases, one can conclude that the elastin content of the media is mainly constant and increases in volume for Layer II, and the inherent deviations in the orientation and density of collagen fibers [42] are responsible for these changes. Unchanged elastin content and its decreased concentration in aging is well published [49].

The proposed microstructural model integrates the isotropic nature of arterial elastin, the anisotropic behavior of interstitial collagenous layers and their microstructural features simultaneously. Such models give more appropriate estimates of the mechanical response of the aortic media compared to conventional SEFs, such as the exponential function proposed by Fung *et al.* [37], since the proposed model in this study incorporates a separate term (the neo-Hookean function that is assigned to Layer I), which makes it more flexible to follow the nearly linear initial part of the stress-strain curve together with the exponential and anisotropic behavior of the tissue after the onset of collagen fiber activation (referred to as "biphasic behavior" in the literature [38]).

The lamellar model presented here is useful for establishing the roles of the micro-constituents of aortic media on the macro-behavior; this model is capable of following arterial mechanical behavior in its functional phases. The intended model provides some novel insights into the contributions of elastin and collagen to the mechanical behavior of the whole media. Further investigation is required in terms of the experiments and structural elements taken into account in the model. The mechanical properties of elastin sheets, collagen bundles, SMCs and also the interaction of these components are not fully understood, and novel experimental protocols are required. The accurate mechanical properties of these components along with realistic geometry will lead to more accurate models and new aspects of arterial mechanics.

5. Limitations

It was assumed that Layers I or II remain the same along the thickness of the media with the same mechanical properties. This assumption is adequate for describing the mechanical response of elastin sheets. However, as shown by advanced imaging techniques, such as second harmonic generation and multi-photon microscopy, collagen fibers change their orientation and possibly diameter as we move from the intimal side of the media towards the adventitia [22,50]. Since collagen is one of the main contributors to the mechanical properties of the media, a more accurate model that accounts for these alterations will lead to a better description of the mechanical properties. However, the proposed approach, which focuses on the lamellar microstructure of the media, is potentially capable of clarifying the underlying mechanics. Such models can describe the biomechanics of the arterial wall in health and disease, as well as the remodeling of the arterial wall due to aging or hypertension, based on the changes of the structural components.

6. Conclusions

In this study, a lamellar model of the arterial media based on both mechanical testing and microstructural information is proposed, and related material parameters are obtained. Samples of human thoracic aorta were examined biaxially to provide the mechanical data. Other required data, such as the volume fractions of layers, were computed from images of stained aortic rings using image processing techniques. Utilizing this new approach, some novel insights into the contribution of the media's microstructure to the mechanical response are provided, which can be summarized as:

- Visualizing the contributions of Layers I and II for the range of physiologic and supraphysiological deformations.
- Providing an appropriate fit to biaxial test data of the aortic samples in both the axial and circumferential directions for different functional phases, *i.e.*, initial crimp and gradual activation of collagen fibers. The model predictions were in good agreement with the experimental data in both the circumferential and axial directions.

Compared to common microstructural models, the proposed approach in this study further investigates the mechanical contribution of lamellar units to the bulk of aortic media and determines how wall lamellae share functional loads. The proposed model in Equation (8) with the values of the material parameters reported in Table 1 can be regarded as a new framework of arterial models to investigate the physiological and pathological conditions of the arteries, such as aging.

Author Contributions

Hadi Taghizadeh contributed to the whole process of the preparation of tissues, staining, mechanical tests, interpretation of the results and preparation of the manuscript. Mohammad Tafazzoli-Shadpour supervised the whole procedure and also contributed to the data interpretation and preparation of the manuscript. Mohammad B. Shadmehr contributed to the preparation of the tissues, staining and imaging of stained tissue rings. Nasser Fatouraee contributed to mechanical tests and data extraction.

Conflicts of Interest

The authors declare no conflict of interest.

References

1. Humphrey, J.D. Mechanics of the arterial wall: Review and directions. *Crit. Rev. Biomed. Eng.* **1995**, *23*, 1–162.

2. Nichols, M.; Townsend, N.; Luengo-Fernandez, R.; Leal, J.; Gray, A.; Scarborough, P.; Rayner, M. *European Cardiovascular Disease Statistics 2012*; European Society of Cardiology: Brussels, Belgium, 2012.

3. Takamizawa, K.; Hayashi, K. Strain energy density function and uniform strain hypothesis for arterial mechanics. *J. Biomech.* **1987**, *20*, 7–17.

4. Schulze-Bauer, C.A.; Holzapfel, G.A. Determination of constitutive equations for human arteries from clinical data. *J. Biomech.* **2003**, *36*, 165–169.

5. Gasser, T.C.; Ogden, R.W.; Holzapfel, G.A. Hyperelastic modelling of arterial layers with distributed collagen fibre orientations. *J. R. Soc. Interface* **2006**, *3*, 15–35.

6. Haskett, D.; Johnson, G.; Zhou, A.; Utzinger, U.; Vande Geest, J. Microstructural and biomechanical alterations of the human aorta as a function of age and location. *Biomech. Model. Mechanobiol.* **2010**, *9*, 725–736.

7. Demiray, H.; Vito, R.P. A layered cylindrical–shell model for an aorta. *Int. J. Eng. Sci.* **1991**, *29*, 47–54.

8. Sommer, G.; Regitnig, P.; Koltringer, L.; Holzapfel, G.A. Biaxial mechanical properties of intact and layer-dissected human carotid arteries at physiological and supraphysiological loadings. *Am. J. Physiol. Heart Circ. Physiol.* **2010**, *298*, H898–H912.

9. Martufi, G.; Gasser, T.C. A constitutive model for vascular tissue that integrates fibril, fiber and continuum levels with application to the isotropic and passive properties of the infrarenal aorta. *J. Biomech.* **2011**, *44*, 2544–2550.

10. Zulliger, M.A.; Fridez, P.; Hayashi, K.; Stergiopulos, N. A strain energy function for arteries accounting for wall composition and structure. *J. Biomech.* **2004**, *37*, 989–1000.

11. Maceri, F.; Marino, M.; Vairo, G. A unified multiscale mechanical model for soft collagenous tissues with regular fiber arrangement. *J. Biomech.* **2010**, *43*, 355–363.

12. Tang, H.; Buehler, M.J.; Moran, B. A constitutive model of soft tissue: From nanoscale collagen to tissue continuum. *Ann. Biomed. Eng.* **2009**, *37*, 1117–1130.

13. Marino, M.; Vairo, G. Multiscale elastic models of collagen bio-structures: From cross-linked molecules to soft tissues. In *Multiscale Computer Modeling in Biomechanics and Biomedical Engineering*; Springer: Berlin, Germany, 2013; pp. 73–102.

14. Wagenseil, J.E.; Mecham, R.P. Vascular extracellular matrix and arterial mechanics. *Physiol. Rev.* **2009**, *89*, 957–989.

15. Qiu, H.; Zhu, Y.; Sun, Z.; Trzeciakowski, J.P.; Gansner, M.; Depre, C.; Resuello, R.R.G.; Natividad, F.F.; Hunter, W.C.; Genin, G.M.; *et al.* Short communication: Vascular smooth muscle cell stiffness as a mechanism for increased aortic stiffness with aging. *Circ. Res.* **2010**, *107*, 615–619.

16. Chen, H.; Zhao, X.; Lu, X.; Kassab, G. Non-linear micromechanics of soft tissues. *Int. J. Non-Linear Mech.* **2013**, *58*, 79–85.

17. Fung, Y.C. Bio-viscoelastic solids. In *Biomechanics*; Springer: New York, NY, USA, 1981; pp. 196–214.

18. Wolinsky, H.; Glagov, S. A lamellar unit of aortic medial structure and function in mammals. *Circ. Res.* **1967**, *20*, 99–111.

19. Avolio, A.; Jones, D.; Tafazzoli-Shadpour, M. Quantification of alterations in structure and function of elastin in the arterial media. *Hypertension* **1998**, *32*, 170–175.

20. Sokolis, D.P.; Kefaloyannis, E.M.; Kouloukoussa, M.; Marinos, E.; Boudoulas, H.; Karayannacos, P.E. A structural basis for the aortic stress–strain relation in uniaxial tension. *J. Biomech.* **2006**, *39*, 1651–1662.

21. Clark, J.M.; Glagov, S. Transmural organization of the arterial media. The lamellar unit revisited. *Arteriosclerosis* **1985**, *5*, 19–34.

22. Dingemans, K.P.; Teeling, P.; Lagendijk, J.H.; Becker, A.E. Extracellular matrix of the human aortic media: An ultrastructural histochemical and immunohistochemical study of the adult aortic media. *Anat. Rec.* **2000**, *258*, 1–14.

23. Rhodin, J.A.G. Architecture of the vessel wall. In *Handbook of Physiology, the Cardiovascular System, Vascular Smooth Muscle*; American Physiological Society: Bethesda, MD, USA, 1980; Volume 2, pp. 1–31.

24. O'Connell, M.K.; Murthy, S.; Phan, S.; Xu, C.; Buchanan, J.; Spilker, R.; Dalman, R.L.; Zarins, C.K.; Denk, W.; Taylor, C.A.; *et al.* The three-dimensional micro- and nanostructure of the aortic medial lamellar unit measured using 3D confocal and electron microscopy imaging. *Matrix Biol.* **2008**, *27*, 171–181.

25. Weisbecker, H.; Viertler, C.; Pierce, D.M.; Holzapfel, G.A. The role of elastin and collagen in the softening behavior of the human thoracic aortic media. *J. Biomech.* **2013**, *46*, 1859–1865.

26. Karimi, A.; Navidbakhsh, M.; Shojaei, A.; Faghihi, S. Measurement of the uniaxial mechanical properties of healthy and atherosclerotic human coronary arteries. *Mater. Sci. Eng. C* **2013**, *33*, 2550–2554.

27. Okamoto, R.J.; Wagenseil, J.E.; DeLong, W.R.; Peterson, S.J.; Kouchoukos, N.T.; Sundt, T.M., 3rd. Mechanical properties of dilated human ascending aorta. *Ann. Biomed. Eng.* **2002**, *30*, 624–635.

28. Limbert, G. A mesostructurally-based anisotropic continuum model for biological soft tissues—Decoupled invariant formulation. *J. Mech. Behav. Biomed. Mater.* **2011**, *4*, 1637–1657.

29. Wolinsky, H.; Glagov, S. Comparison of abdominal and thoracic aortic medial structure in mammals. Deviation of man from the usual pattern. *Circ. Res.* **1969**, *25*, 677–686.

30. Yosibash, Z.; Manor, I.; Gilad, I.; Willentz, U. Experimental evidence of the compressibility of arteries. *J. Mech. Behav. Biomed. Mater.* **2014**, *39*, 339–354.

31. Chuong, C.J.; Fung, Y.C. Three-dimensional stress distribution in arteries. *J. Biomech. Eng.* **1983**, *105*, 268–274.

32. Delfino, A.; Stergiopulos, N.; Moore, J.E., Jr.; Meister, J.J. Residual strain effects on the stress field in a thick wall finite element model of the human carotid bifurcation. *J. Biomech.* **1997**, *30*, 777–786.

33. Holzapfel, G.A. Nonlinear solid mechanics: A continuum approach for engineering science. *Meccanica* **2002**, *37*, 489–490.

34. Agrawal, V.; Kollimada, S.A.; Byju, A.G.; Gundiah, N. Regional variations in the nonlinearity and anisotropy of bovine aortic elastin. *Biomech. Model. Mechanobiol.* **2013**, *12*, 1181–1194.

35. Hollander, Y.; Durban, D.; Lu, X.; Kassab, G.S.; Lanir, Y. Constitutive modeling of coronary arterial media—Comparison of three model classes. *J. Biomech. Eng.* **2011**, *133*, doi:10.1115/1.4004249.

36. Driessen, N.J.; Bouten, C.V.; Baaijens, F.P. A structural constitutive model for collagenous cardiovascular tissues incorporating the angular fiber distribution. *J. Biomech. Eng.* **2005**, *127*, 494–503.

37. Fung, Y.C.; Fronek, K.; Patitucci, P. Pseudoelasticity of arteries and the choice of its mathematical expression. *Am. J. Physiol.* **1979**, *237*, H620–H631.

38. Holzapfel, G.A.; Weizsacker, H.W. Biomechanical behavior of the arterial wall and its numerical characterization. *Comput. Biol. Med.* **1998**, *28*, 377–392.

39. Roy, S.; Boss, C.; Rezakhaniha, R.; Stergiopulos, N. Experimental characterization of the distribution of collagen fiber recruitment. *J. Biorheol.* **2011**, *24*, 84–93.

40. Fleenor, B.S. Large elastic artery stiffness with aging: Novel translational mechanisms and interventions. *Aging Dis.* **2013**, *4*, 76–83.

41. Sherratt, M.J. Tissue elasticity and the ageing elastic fibre. *Age* **2009**, *31*, 305–325.

42. Maceri, F.; Marino, M.; Vairo, G. Age-dependent arterial mechanics via a multiscale elastic approach. *Int. J. Comput. Methods Eng. Sci. Mech.* **2013**, *14*, 141–151.

43. Roach, M.R.; Burton, A.C. The reason for the shape of the distensibility curves of arteries. *Can. J. Biochem. Physiol.* **1957**, *35*, 681–690.

44. Lillie, M.A.; Shadwick, R.E.; Gosline, J.M. Mechanical anisotropy of inflated elastic tissue from the pig aorta. *J. Biomech.* **2010**, *43*, 2070–2078.

45. Martufi, G.; Gasser, T.C. Turnover of fibrillar collagen in soft biological tissue with application to the expansion of abdominal aortic aneurysms. *J. R. Soc. Interface* **2012**, *9*, 3366–3377.

46. Schriefl, A.J.; Zeindlinger, G.; Pierce, D.M.; Regitnig, P.; Holzapfel, G.A. Determination of the layer-specific distributed collagen fibre orientations in human thoracic and abdominal aortas and common iliac arteries. *J. R. Soc. Interface* **2012**, *9*, 1275–1286.

47. Debes, J.C.; Fung, Y.C. Biaxial mechanics of excised canine pulmonary arteries. *Am. J. Physiol.* **1995**, *269*, H433–H442.

48. Hill, M.R.; Duan, X.; Gibson, G.A.; Watkins, S.; Robertson, A.M. A theoretical and non-destructive experimental approach for direct inclusion of measured collagen orientation and recruitment into mechanical models of the artery wall. *J. Biomech.* **2012**, *45*, 762–771.

49. Tsamis, A.; Krawiec, J.T.; Vorp, D.A. Elastin and collagen fibre microstructure of the human aorta in ageing and disease: A review. *J. R. Soc. Interface* **2013**, *10*, doi:10.1098/rsif.2012.1004.

50. Schriefl, A.J.; Wolinski, H.; Regitnig, P.; Kohlwein, S.D.; Holzapfel, G.A. An automated approach for three-dimensional quantification of fibrillar structures in optically cleared soft biological tissues. *J. R. Soc. Interface* **2013**, *10*, doi:10.1098/rsif.2012.0760.

Chapter 2:
Mechanics of Biomaterials

Highly Stretchable, Biocompatible, Striated Substrate Made from Fugitive Glue

Wei Li, Tomas Lucioni, Xinyi Guo, Amanda Smelser and Martin Guthold

Abstract: We developed a novel substrate made from fugitive glue (styrenic block copolymer) that can be used to analyze the effects of large strains on biological samples. The substrate has the following attributes: (1) It is easy to make from inexpensive components; (2) It is transparent and can be used in optical microscopy; (3) It is extremely stretchable as it can be stretched up to 700% strain; (4) It can be micro-molded, for example we created micro-ridges that are 6 μm high and 13 μm wide; (5) It is adhesive to biological fibers (we tested fibrin fibers), and can be used to uniformly stretch those fibers; (6) It is non-toxic to cells (we tested human mammary epithelial cells); (7) It can tolerate various salt concentrations up to 5 M NaCl and low (pH 0) and high (pH 14) pH values. Stretching of this extraordinary stretchable substrate is relatively uniform and thus, can be used to test multiple cells or fibers in parallel under the same conditions.

Reprinted from *Materials*. Cite as: Li, W.; Lucioni, T.; Guo, X.; Smelser, A.; Guthold, M. Highly Stretchable, Biocompatible, Striated Substrate Made from Fugitive Glue. *Materials* **2015**, *8*, 3508-3518.

1. Introduction

Cells and other biological samples are often exposed to stresses and strains in their natural environment, and these stresses and strains can have a significant biological effect. For example, the growth of smooth muscle cells and the import of nuclear proteins are stimulated by strains [1], and bone formation is stimulated in the presence of mechanical stimuli [2]. In the last few years, strong evidence has emerged that cells are sensitive to the mechanical properties (e.g., stiffness) of their environment. In a seminal paper, Engler *et al.* showed that the differentiation of stem cells is influenced by substrate mechanical properties [3]. These researchers observed that stem cells will differentiate into bone-like cells when grown on stiff substrates or into neuron-like cells when grown on soft substrates. Moreover, mechanical strain plays an important role in stem cell differentiation and function: global gene expression changes (for example, smooth muscle markers increase, cartilage matrix decrease) when cells are aligned parallel to the strain axis [4]. Furthermore, intracellular calcium oscillation in human mesenchymal stem cells is governed by mechanical tension [5] and mesenchymal stem cell differentiation into vascular smooth muscle cells may be promoted by uniaxial strain [6].

Similarly, biological fibers in the body are often exposed to stresses and strains. For example, blood flow exerts stress on fibrin fibers during blood coagulation, and blood flow can affect the structure of blood clots [7,8] and the interaction of fibrin fibers with platelets [9]. Elastin fibers [10], which are found in the skin and lungs, experience strain during respiration and are very extensible;

collagen fibers, found in cartilage and connective tissue, experience strain during movement, but are not as extensible [11].

Therefore, when investigating the behavior of biological samples in the lab, a stretchable substrate is required in many situations. There are numerous devices and techniques to apply and test the effect of stress and strains on cells. One example is traction force microscopy, which uses a stretchable substrate. The migration of normal 3T3 (3-day transfer, inoculum 3×10^5 cells) cells can be detected by using traction force microscopy, where the cell body was pulled forward by the dynamic traction forces at the leading edge [12]. Shao *et al.* described a homemade cell stretching device to demonstrate that external mechanical stretch plays a key role in regulating subcellular molecular dynamics with the F-actin cytoskeleton [13]. Moore's group used their stretching device to show that phenotype modulation (alignment and altered mRNA expression) can be induced by stretching 10T1/2 (from Embryonic mesenchymal cell line) cells [14].

Nano- and microfiber properties are often measured by suspending fibers over microridges in a substrate and then the fibers are manipulated with an Atomic Force Microscope (AFM) [15]. The mechanical properties of different nanofibers have been determined by this method, such as fibrin fibers [15,16], electrospun collagen fibers [17] and electrospun fibrinogen fibers [18]. This sophisticated AFM technique allows for precise mechanical manipulations of nanofibers, and numerous mechanical properties can be extracted with this technique. However, individual fibers are pulled one at a time, which is tedious, time-consuming and not efficient. Varju *et al.* showed that a strained blood clot lyses slower than an unstrained blood clot [19]. So, to investigate the effect of strain on single fibrin fibers, an AFM could be used, but a technique that allows the investigation of multiple fibers in parallel would be more efficient. With our novel, highly stretchable substrate, it is possible to manipulate an array of single fibers simultaneously, instead of only one single fiber. This facilitates investigations of the effect of strain on single biological nanofibers, such as collagen and fibrin, and other nanofibers, such as electrospun nanofibers [18], that are used in biomedical engineering applications.

Besides investigating biological samples, stretchable substrates may also be used in flexible electronics. Stretchable materials can serve as substrates, onto which circuits and electronics are engineered [20,21]. Other times, conducting materials are injected into the substrate [22,23]. Often these materials have a rather low stretch limit, just a little above 100% [24]. However, some applications may require significantly higher elongations. Another drawback of current stretchable devices, like bio-Microelectromechanical systems (bioMEMS) [25,26] and traction force microscopy [12], is the high cost of these devices.

In this paper, we report tests on fugitive glue (a styrenic block copolymer), which is extraordinarily stretchable (up to a 750% strain). It also is very inexpensive, easy to obtain and easy to handle. Moreover, it can be molded into microstructures, and presumably many other shapes. Furthermore, it is transparent (for use in optical microscopy), it can withstand extreme environments, like strong acids/bases (between pH 0 and 14), it is compatible with salt solutions and biological samples (fibrin fibers), and it is non-toxic to cells.

2. Results

2.1. Stability

2.1.1. Mechanical Stability and Stretchability

We found the substrate to be mechanically stable (it kept its original shape and length) even when stretched to about 250% (detailed experimental conditions are given in Section 4. In our 24-h test, it kept its original shape in air at room temperature. Therefore, we kept the stretching percentage at around 250% in the following experiments. In the most extreme case we tested, the substrate could be stretched to around 750%, but it was only mechanically stable for 10 min. These experiments demonstrate that substrates formed from fugitive glue are extremely extensible and hold their shapes for long enough time periods to do many biological and other experiments. The extensibility far exceeds that of other common materials used in recently described stretching devices, such as Polydimethylsiloxane (PDMS) [20], Poly-ethylene-terephthalate (PET) [27], Polyimide (PI) [28] and silicone [29]. A concentrated strain of 107% has been reported on a soft, thin PDMS film area in microsupercapacitor arrays [20]. PET substrates coated with an acrylic primer can be stretched to over 70% without breaking [27]. None of these materials can be stretched to several times their original length. In separate experiments, we also tested PDMS (Sylgard 184 Silicone Elastomer Kit, 10:1 mix ratio, Dow Corning, Midland, MI, USA), Norland optical adhesive 81 (Norland Products Incorporated, Cranbury, NJ, USA) and silicone (GE Silicone II, Kitchen and Bath Caulk, clear color, purchased at Home Depot, Atlanta, GA, USA); they were all significantly stiffer than fugitive glue and significantly less extensible.

A familiar example of fugitive glue is its common use to attach credit cards to paper (credit card glue), and it can be easily stretched to several times its original length.

Fugitive glue is a styrenic block copolymer, a type of thermoplastic elastomer. Their mechanical properties, which are similar to rubber, stem from their microstructure. Microscopically, styrenic block copolymers consist of two hard polystyrene end blocks that are connected by a soft, elastomeric midblock (linker), typically made of polybutadiene or polyisoprene.

2.1.2. pH Tolerance Test

Some experiments require extreme pH values, so the ability of a substrate to tolerate highly acidic and basic surroundings can become important. We, therefore, also tested the pH tolerance of our fugitive glue substrate from pH 0 to pH 14. Some polymers may degrade at these extreme environments, e.g., polyanhydrides degrades at high pH values [30], poly(dl-lactide-co-glycolic acid)-methoxypoly (ethyleneglycol) (PLGA-mPEG) microparticles show degradation in strong acid and base [31]. However, our stretchable substrate made from fugitive glue maintained its mechanical and chemical stability under both highly acidic and basic environments for up to at least 1.5 h. There was no discernable degradation (by visual inspection under a microscope with a 40× objective lens) and the shape of the grooves and ridges was not affected (Figure 1).

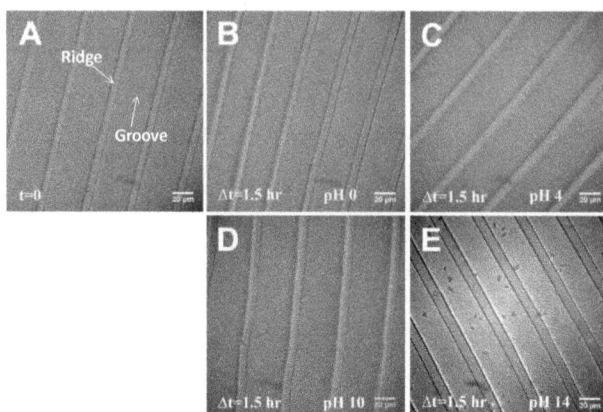

Figure 1. pH Tolerance Test. (**A**) Initial image right after adding pH. (**B**), (**C**), (**D**), (**E**) Images taken after 1.5 h incubation at pH 0, 4, 10, 14. The images before and after adding solutions are at different locations of the same sample. Some crystals formed in the pH 14 solution, probably due to the high Na^+ concentration, but the substrate appears unaffected by the solution.

2.1.3. Salt Solution Tolerance Test

For biological and non-biological experiments, different salt solutions may be applied to the substrate. So, it is also important to test if this material can withstand different salt solutions. We selected two commonly used salts, NaCl and $MgCl_2$, at very high concentrations (near their solubility), reasoning that if fugitive glue can withstand such extremely high salt concentrations, it should also be able to withstand lower salt concentrations. In this experiment, a 5 M NaCl solution and a (2.5 M NaCl + 2.5 M $MgCl_2$) solution were applied to the stretchable substrate. After 4 h, no salt deposits and no deformation or degradation of the ridges and grooves were observed (Figure 2).

In summary, the stretchable substrate made from fugitive glue can accommodate a broad range of solution conditions that may be found in many experiments.

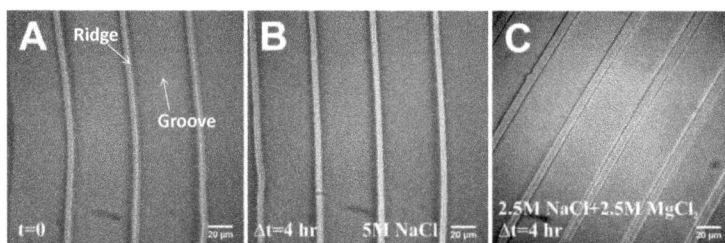

Figure 2. Salt Solution Tolerance Test. (**A**) Initial image that was taken right after adding salt solution to the substrate. (**B**) and (**C**) Images of the substrate after a 4 h-incubation with 5 M NaCl and (2.5M NaCl + 2.5M $MgCl_2$) solutions. The images before and after adding salt solution are at different locations of the same sample.

2.2. Bio-Compatibility

2.2.1. Cell Growth on Fugitive Glue Substrate

Since many biological samples are exposed to stress and strains in their environment, we tested if our stretchable fugitive glue substrate is suitable for biological samples. We tested cells and biological fibers. First, we tested if this material is toxic to cells. Human Mammary Epithelial Cells (HMECs) were grown on fugitive glue for 48 h (Figure 3A,B). Cells were well attached and grew well on the fugitive glue substrate. Thus, it appears that fugitive glue is suitable as a cell substrate. About 85% of cells are alive (stained green), whereas about 15% of cells are dead (stained red).

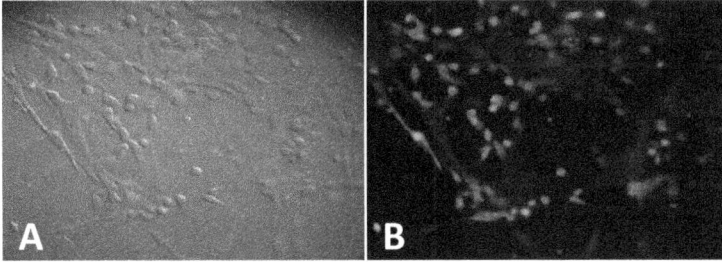

Figure 3. Human mammary epithelial cells grown on fugitive glue substrate. (**A**) Differential Interference Contrast (DIC) images of cells; (**B**) Fluorescence images (20× objective lens) of live cells (stained green) and dead cells (stained red) of same field of view as in (**A**).

2.2.2. Fibrin Fiber Formed on Fugitive Glue Substrate

Besides cells, we also tested if our fugitive glue substrate is compatible with biological fibers. Fibrin fibers are the major structural and mechanical component of a blood clot. They have an average diameter of about 130 nm. Fibrin fibers form from fibrin monomers, the activated form of the blood protein, fibrinogen. Fibrinogen gets converted to fibrin by thrombin in the last step of the coagulation cascade. Fibrin fibers can be easily formed in the lab, by adding thrombin to fibrinogen. In previous work, we have determined various mechanical properties of single fibrin fibers, such as their stiffness, extensibility and elasticity [15,16]. As shown in Figure 4A, fibrin fibers form well on this substrate, and they strongly adhere to the substrate. We did not observe any slipping or detachment, even at over two-fold extensions (Figure 4B). Since fibrin fibers experience stress during blood circulation [7,8], there is a strong interest in investigating fibrin fiber mechanical properties. Our stretchable substrate provides a novel approach for these investigations.

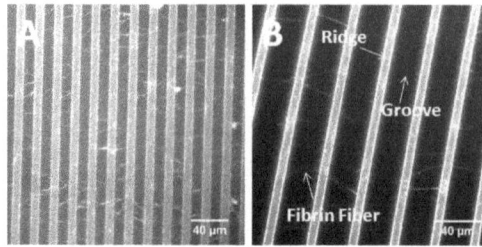

Figure 4. Fibrin fibers on the unstretched substrate (**A**), and stretched substrate (268%) (**B**). The substrate has imprinted ridges and grooves. The width of the groove is 13.5 μm (before stretching), and 36.2 μm (after stretching).

3. Discussion

We have described and tested a highly stretchable substrate made from fugitive glue, a styrenic block copolymer. It is moldable, transparent to visible light (usable as a substrate in optical microscopy), tolerant to high and low pH values and salt concentrations, and compatible with biological cells and fiber samples. Compared to other devices that are used to apply strain to biological samples on the microscale, it is among the least expensive and easiest to handle and manufacture. Other devices include bio-Microelectromechanical systems (BioMEMS) and some home-made stretching devices.

BioMEMS are MEMS devices for biological applications, which are manufactured using similar microfabrication techniques as those used to create integrated circuits. They are usually used in biosensors, pacemakers, immunoisolation capsules, and drug delivery systems [32]. BioMEMS have been used to apply strain to adherent fibroblasts and detect the de-adhesion force [25], and to test cell force responses: strongly linear, reversible, and repeatable under large stretches [26].

Some home-made devices have also been used to apply external strain to biological systems. For example, Heo *et al.* applied input pressure (air input) from underneath to a PDMS layer, so that the cells on the layer can be stretched [33]. Another novel stretching device is based on the movement of computer-controlled, piezoelectrically actuated pins of a refreshable Braille display underneath a sample to generate strain on a elastometric PDMS membrane's top surface. The Braille pins could provide 20%–25% maximal strain in the radial direction [34]. Yang's group used a force sensor probe coated with biomolecules to stretch cells [35]. Wipff *et al.* used PDMS as the elastic membrane, mixed with tracking particles to monitor the degree of substrate expansion under stretch [36].

All these are examples of well-suited devices for biological stretch experiments on a micrometer scale. Many have a limited stretching range (around 20%–30%), since they use PDMS. Our stretchable substrate is a good substrate choice when large strains are required, since it is extremely extensible (about 750%). It is also less stiff than PDMS, and can be stretched manually.

4. Experimental Section

4.1. Stretchable Substrate Preparation

The stretchable substrate was made from fugitive pressure sensitive adhesive (Surebonder AT-10154 Hot Melt, Hotmelt.com, Edina, MN 55439, www.hotmelt.com); this type of glue is also called hot melt pressure sensitive adhesive or, colloquially, credit card glue. Chemically, this adhesive is a styrenic block copolymer. A drop of hot fugitive glue was placed onto the surface of a microscope cover glass slide (No. 1.5, 24 mm × 60 mm) (Fisherbrand, Pittsburgh, PA, USA) from a Surebonder PRO100 Hot Melt Gun (Hotmelt.com, Edina, MN 55439, www.hotmelt.com). Immediately afterward, a rectangular PDMS (Polydimethylsiloxane) (6 mm × 8 mm) stamp with imprinted grooves and ridges was pressed into the glue. After it cooled down and dried (4 min), the PDMS stamp was peeled off, leaving ridges and grooves in the fugitive glue (width and height of the ridges was 6.5 μm, width of the grooves was 13.5 μm, measured by scanning electron microscope (SEM, Amray 1810, AMRAY, Bedford, MA, USA) in a previous publications [18]. Next, the imprinted fugitive glue substrate was manually stretched to the desired length, as follows. The imprinted substrate was carefully peeled off the glass cover slide, then manually stretched to a specific length, and anchored back down again. For anchoring we used Adhesive Squares ($\frac{1}{2}$ inch × $\frac{1}{2}$ inch Adhesive Squares™ RS Industrial, Inc., Buford, GA, USA) as follows. The squares, which are about 1 cm × 1 cm where stretched into a string of about 15 cm. This string was then used to tie down the two sides by wrapping them around the cover glass as shown in Figure 5.

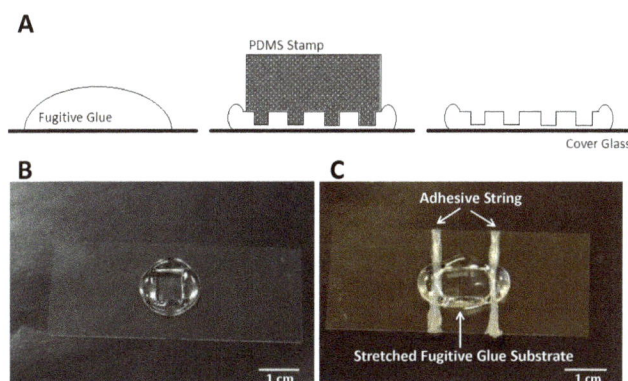

Figure 5. Setup of stretchable substrate. (**A**) Schematic of forming fugitive glue with ridges and grooves; (**B**) Photograph of fugitive glue substrate with ridges and grooves; (**C**) Photograph of stretched substrate.

4.2. pH and Salt Solution Tolerance Test

400 μL solutions with different pH values and salt concentrations were deposited onto the surface of the stretchable substrate and left for 1.5 h at room temperature. The following solutions were used (all solutions from Fisher Scientific, Pittsburgh, PA, USA): 1 N HCl (pH 0), pH 4 buffer solution

(pH-meter calibration solution, Potassium Acid Phthalate), pH 10 buffer solution (pH-meter calibration solution, Boric Acid-Potassium Chloride-Sodium Hydroxide buffer), 1 N NaOH (pH 14), 5 M NaCl (Sigma-Aldrich, St. Louis, MO, USA), 5 M $MgCl_2$ (Sigma-Aldrich). Microscope images were taken before and after the tests with an inverted optical microscope (Axio Observer D1, Zeiss, Thornwood, NY, USA) with a 40× objective lens.

4.3. Fibrin Fibers

An 18 µL solution of purified human fibrinogen (Enzyme Research Laboratories, South Bend, IN, final concentration 1 mg/mL) was placed onto the surface of the stretchable substrate. Then 2 µL of thrombin (Enzyme Research Laboratories, South Bend, IN, final concentration 0.1 NIH (National Institute of Health) units/mL) were added into it and kept in a wet environment at room temperature for 1 h. After that, a skin (part of the clot) on this solution was peeled off with a pipette tip to reduce the density of the clot before imaging. Fibrin fibers on the stretchable substrate were kept continuously in fibrin buffer (140 mM NaCl, 10 mM, Hepes, 5 mM $CaCl_2$, pH 7.4).

4.4. Cell Growth

For the cell growth experiments, we used a flat (instead of a striated) substrate made from fugitive glue. Glass bottom dishes (Willcowells, Amsterdam, The Netherlands) of size 35 mm × 2 mm were purchased and assembled in the lab. A drop of hot fugitive glue was placed onto the surface of a petri dish, then a PDMS stamp with flat surface was pressed into the glue and removed after 4 min.

Human mammary epithelial cells (HMECs) were purchased from Lonza (Lonza Group Ltd, Walkersville, MD, USA) and used within 6 passages from their original state from Lonza. HMECs were cultured in Mammary Epithelial Cell Growth Medium–MEGM (Lonza) with 0.4% bovine pituitary extract (BPE) (Lonza), according to the distributor's recommendations. Cells were cultured and maintained in a culture incubator at 37 °C with 5% CO_2. Pictures were taken 48 h after the cells were seeded (Figure 3).

4.5. Cell Viability Assay

The cell viability assay was done by using LIVE/DEAD Viability/Cytotoxicity Kit (Life Technologies, Grand Island, NY, USA). 400 µL of 1 µM Calcein AM (acetomethoxy derivate of calcein) (used to stain live cells, ex/em~495 nm/515 nm) and 400 µL of 10 µM Ethidium homodimer-1 (used to stain dead cells, ex/em~495 nm/635 nm) were added into the petri dish with cells prepared as described above and incubated for 15–20 min. Green fluorescence indicated the activity of intracellular esterases present in live cells, and red fluorescence indicated the loss of cell membrane integrity in dead cells. Fluorescence images of live and dead cells and Differential Interference Contrast (DIC) images of cells were taken with a Nikon Eclipse Ti, 20× objective, NA 0.75.

Acknowledgments

We thank Ching-Wan Yip for taking pictures of our stretchable substrate. This work was supported by grants from the National Science Foundation (CMMI-1106105) and the Wake Forest University Translational Science Center (CG0006-U01078, CG0006-U01508), and the Wake Forest University Center for Molecular Communication and Signaling (CG0005-U01057).

Author Contributions

Wei Li and Tomas Lucioni performed all experiments, except the cell growth experiments, and wrote the manuscript. Xinyi Guo, Amanda Smelser and Wei Li performed the cell experiments and cell viability assay. Martin Guthold supervised the experiments and edited the manuscript.

Conflicts of Interest

The authors declare no conflict of interest.

References

1. Richard, M.N.; Deniset, J.F.; Kneesh, A.L.; Blackwood, D.; Pierce, G.N. Mechanical stretching stimulates smooth muscle cell growth, nuclear protein import, and nuclear pore expression through mitogen-activated protein kinase activation. *J. Biol. Chem.* **2007**, *282*, 23081–23088.
2. Klein-Nulend, J.; Bacabac, R.G.; Bakker, A.D. Mechanical Loading and How it Affects Bone Cells: The Role of the Osteocyte Cytoskeleton in Maintaining Our Skeleton. *Eur. Cells Mater.* **2012**, *24*, 278–291.
3. Engler, A.J.; Sen, S.; Sweeney, H.L.; Discher, D.E. Matrix elasticity directs stem cell lineage specification. *Cell* **2006**, *126*, 677–689.
4. Kurpinski, K.; Chu, J.; Hashi, C.; Li, S. Anisotropic mechanosensing by mesenchymal stem cells. *Proc. Natl. Acad. Sci. USA* **2006**, *103*, 16095–16100.
5. Kim, T.J.; Sun, J.; Lu, S.Y.; Qi, Y.X.; Wang, Y.X. Prolonged Mechanical Stretch Initiates Intracellular Calcium Oscillations in Human Mesenchymal Stem Cells. *PLoS ONE* **2014**, *9*, e109378.
6. Park, J.S.; Chu, J.S.F.; Cheng, C.; Chen, F.Q.; Chen, D.; Li, S. Differential effects of equiaxial and uniaxial strain on mesenchymal stem cells. *Biotechnol. Bioeng.* **2004**, *88*, 359–368.
7. Gersh, K.C.; Edmondson, K.E.; Weisel, J.W. Flow rate and fibrin fiber alignment. *J. Thromb. Haemost.* **2010**, *8*, 2826–2828.
8. Campbell, R.A.; Aleman, M.M.; Gray, L.D.; Falvo, M.R.; Wolberg, A.S. Flow profoundly influences fibrin network structure: Implications for fibrin formation and clot stability in haemostasis. *Thromb. Haemost.* **2010**, *104*, 1281–1284.
9. Weiss, H.J.; Turitto, V.T.; Baumgartner, H.R. Role Of Shear Rate and Platelets in Promoting Fibrin Formation on Rabbit Subendothelium. Studies Utilizing Patients With Quantitative and Qualitative Platelet Defects. *J. Clin. Investig.* **1986**, *78*, 1072–1082.

10. Lemos, M.; Pozo, R.M.K.; Montes, G.S.; Saldiva, P.H.N. Organization of collagen and elastic fibers studied in stretch preparations of whole mounts of human visceral pleura. *Ann. Anat. Anat. Anz.* **1997**, *179*, 447–452.

11. Römgens, A.M.; van Donkelaar, C.C.; Ito, K. Contribution of collagen fibers to the compressive stiffness of cartilaginous tissues. *Biomech. Model. Mechanobiol.* **2013**, *12*, 1221–1231.

12. Munevar, S.; Wang, Y.L.; Dembo, M. Traction force microscopy of migrating normal and H-ras transformed 3T3 fibroblasts. *Biophys. J.* **2001**, *80*, 1744–1757.

13. Shao, Y.; Tan, X.Y.; Novitski, R.; Muqaddam, M.; List, P.; Williamson, L.; Fu, J.P.; Liu, A.P. Uniaxial cell stretching device for live-cell imaging of mechanosensitive cellular functions. *Rev. Sci. Instrum.* **2013**, *84*, 114304, doi:10.1063/1.4832977.

14. Richardson, W.J.; Metz, R.P.; Moreno, M.R.; Wilson, E.; Moore, J.E. A Device to Study the Effects of Stretch Gradients on Cell Behavior. *J. Biomech. Eng.* **2011**, *133*, 101008, doi:10.1115/1.4005251.

15. Liu, W.; Carlisle, C.R.; Sparks, E.A.; Guthold, M. The mechanical properties of single fibrin fibers. *J. Thromb. Haemost.* **2010**, *8*, 1030–1036.

16. Liu, W.; Jawerth, L.M.; Sparks, E.A.; Falvo, M.R.; Hantgan, R.R.; Superfine, R.; Lord, S.T.; Guthold, M. Fibrin fibers have extraordinary extensibility and elasticity. *Science* **2006**, *313*, 634, doi:10.1126/science.1127317.

17. Carlisle, C.R.; Coulais, C.; Guthold, M. The mechanical stress–strain properties of single electrospun collagen type I nanofibers. *Acta Biomater.* **2010**, *6*, 2997–3003.

18. Baker, S.; Sigley, J.; Helms, C.C.; Stitzel, J.; Berry, J.; Bonin, K.; Guthold, M. The mechanical properties of dry, electrospun fibrinogen fibers. *Mater. Sci. Eng. C* **2012**, *32*, 215–221.

19. Varjú, I.; Sótonyi, P.; Machovich, R.; Szabó, L.; Tenekedjiev, K.; Silva, M.M.C.G.; Longstaff, C.; Kolev, K. Hindered dissolution of fibrin formed under mechanical stress. *J. Thromb. Haemost.* **2011**, *9*, 979–986.

20. Hong, S.Y.; Yoon, J.; Jin, S.W.; Lim, Y.; Lee, S.J.; Zi, G.; Ha, J.S. High-Density, Stretchable, All-Solid-State Microsupercapacitor Arrays. *ACS Nano* **2014**, *8*, 8844–8855.

21. Lim, B.Y.; Yoon, J.; Yun, J.; Kim, D.; Hong, S.Y.; Lee, S.-J.; Zi, G.; Ha, J.S. Biaxially stretchable, integrated array of high performance microsupercapacitors. *ACS Nano* **2014**, *8*, 11639–11650.

22. Lazarus, N.; Meyer, C.D.; Bedair, S.S.; Nochetto, H.; Kierzewski, I.M. Multilayer liquid metal stretchable inductors. *Smart Mater. Struct.* **2014**, *23*, 085036, doi:10.1088/0964-1726/23/8/085036.

23. Robinson, A.; Aziz, A.; Liu, Q.; Suo, Z.; Lacour, S.P. Hybrid stretchable circuits on silicone substrate. *J. Appl. Phys.* **2014**, *115*, 143511, doi:10.1063/1.4871279.

24. Xu, P.; Mark, J.E. Elasticity Measurements on Bimodal Networks in Elongation and Compression: Networks Crosslinked in Solution and Studied Unswollen, and Networks Crosslinked in the Undiluted State and Studied Swollen. *Polymer* **1992**, *33*, 1843–1848.

25. Serrell, D.B.; Oreskovic, T.L.; Slifka, A.J.; Mahajan, R.L.; Finch, D.S. A uniaxial bioMEMS device for quantitative force-displacement measurements. *Biomed. Microdevices* **2007**, *9*, 267–275.

26. Yang, S.Y.; Saif, T. Reversible and repeatable linear local cell force response under large stretches. *Exp. Cell Res.* **2005**, *305*, 42–50.

27. Sim, G.D.; Won, S.; Jin, C.Y.; Park, I.; Lee, S.B.; Vlassak, J.J. Improving the stretchability of as-deposited Ag coatings on poly-ethylene-terephthalate substrates through use of an acrylic primer. *J. Appl. Phys.* **2011**, *109*, 073511, doi:10.1063/1.3567917.

28. Jahanshahi, A.; Salvo, P.; Vanfleteren, J. Stretchable biocompatible electronics by embedding electrical circuitry in biocompatible elastomers. In Proceedings of the 2012 IEEE Annual Conference on Engineering in Medicine and Biology Society, San Diego, CA, USA, 28 August 2012; pp. 6007–6010.

29. Habrard, F.; Patscheider, J.; Kovacs, G. Stretchable Metallic Electrodes for Electroactive Polymer Actuators. *Adv. Eng. Mater.* **2014**, *16*, 1133–1139.

30. Leong, K.W.; Brott, B.C.; Langer, R. Bioerodible Polyanhydeides as Drug-Carrier Matrices. I. Characterization, Degradation, and Release Characteristics. *J. Biomed. Mater. Res.* **1985**, *19*, 941–955.

31. Li, J.; Jiang, G.Q.; Ding, F.X. The effect of pH on the polymer degradation and drug release from PLGA-mPEG microparticles. *J. Appl. Polym. Sci.* **2008**, *109*, 475–482.

32. Grayson, A.C.R.; Shawgo, R.S.; Johnson, A.M.; Flynn, N.T.; Li, Y.W.; Cima, M.J.; Langer, R. A BioMEMS review: MEMS technology for physiologically integrated devices. *Proc. IEEE* **2004**, *92*, 6–21.

33. Heo, Y.J.; Kan, T.; Iwase, E.; Matsumoto, K.; Shimoyama, I. Stretchable cell culture platforms using micropneumatic actuators. *Micro Nano Lett.* **2013**, *8*, 865–868.

34. Kamotani, Y.; Bersano-Begey, T.; Kato, N.; Tung, Y.C.; Huh, D.; Song, J.W.; Takayama, S. Individually programmable cell stretching microwell arrays actuated by a Braille display. *Biomaterials* **2008**, *29*, 2646–2655.

35. Yang, S.; Saif, M.T.A. Microfabricated Force Sensors and Their Applications in the Study of Cell Mechanical Response. *Exp. Mech.* **2009**, *49*, 135–151.

36. Wipff, P.J.; Majd, H.; Acharya, C.; Buscemi, L.; Meister, J.J.; Hinz, B. The covalent attachment of adhesion molecules to silicone membranes for cell stretching applications. *Biomaterials* **2009**, *30*, 1781–1789.

Prosthetic Meshes for Repair of Hernia and Pelvic Organ Prolapse: Comparison of Biomechanical Properties

Manfred M. Maurer, Barbara Röhrnbauer, Andrew Feola, Jan Deprest and Edoardo Mazza

Abstract: This study aims to compare the mechanical behavior of synthetic meshes used for pelvic organ prolapse (POP) and hernia repair. The analysis is based on a comprehensive experimental protocol, which included uniaxial and biaxial tension, cyclic loading and testing of meshes in dry conditions and embedded into an elastomer matrix. Implants are grouped as POP or hernia meshes, as indicated by the manufacturer, and their stiffness in different loading configurations, area density and porosity are compared. Hernia meshes might be expected to be stiffer, since they are implanted into a stiffer tissue (abdominal wall) than POP meshes (vaginal wall). Contrary to this, hernia meshes have a generally lower secant stiffness than POP meshes. For example, DynaMesh PRS, a POP mesh, is up to two orders of magnitude stiffer in all tested configurations than DynaMesh ENDOLAP, a hernia mesh. Additionally, lighter, large pore implants might be expected to be more compliant, which was shown to be generally not true. In particular, Restorelle, the lightest mesh with the largest pores, is less compliant in the tested configurations than Surgipro, the heaviest, small-pore implant. Our study raises the question of defining a meaningful design target for meshes in terms of mechanical biocompatibility.

Reprinted from *Materials*. Cite as: Maurer, M.M.; Röhrnbauer, B.; Feola, A.; Deprest, J.; Mazza, E. Prosthetic Meshes for Repair of Hernia and Pelvic Organ Prolapse: Comparison of Biomechanical Properties. *Materials* **2015**, *8*, 2794-2808.

1. Introduction

Mechanical biocompatibility of prosthetic meshes for hernia and pelvic organ prolapse (POP) is related to the ability of implants to display a mechanical behavior compatible with its function and favoring its integration into the surrounding native tissue [1–6]. This approach for implant assessment has received increased attention in recent years. While initial investigations focused on the ability of a mesh to provide sufficient strength and resistance to maximum loads [7–12], it recently became clear that the deformation behavior in a physiological range, also called "comfort zone", is of major importance [13,14]. A mismatch of mechanical properties of the implants compared to native tissue has been associated with clinical complications [15–19], although none of these works explicitly link mechanical properties with clinical outcome. It has recently been suggested that meshes designed to mimic the biomechanical properties of the area of application are advantageous [2,14,20]. These investigations are further motivated by an FDA safety communication [21] pointing at risks associated with existing prosthetic meshes and corresponding surgery procedures for repair of POP.

A wealth of studies has been conducted analyzing either hernia or POP meshes (see [3,22] for an extensive literature overview). However, little work was performed to compare the mechanical response of these two groups, which may shed light onto the prevalent clinical complications. The mechanical environment and loading conditions these implants are exposed to differ significantly between the abdominal wall and pelvic floor. Physiological loads in terms of membrane tension were calculated based on Laplace's law to be around 0.035 N/mm in the pelvic region and 0.136 N/mm at the abdominal wall at rest, but can be orders of magnitude higher at increased intra-abdominal pressures [5,7,23]. This gives an indication of the range of load at which mesh implants should work best in supporting and mimicking native tissue, thus ensuring mechanical biocompatibility.

Based on the experimental study presented in [22], the data analysis in the present investigation is extended to compare the mechanical properties of hernia and POP mesh implants with respect to physiological loading conditions.

2. Experimental Section

Nine mesh implants were investigated. They were grouped into hernia ($n = 5$) and POP ($n = 4$) implants based on the manufacturer's information, available on their respective websites and analyzed accordingly. All products are described in Table 1.

The mechanical testing procedure has been previously described in detail [22]. In short, each mesh type was tested in eight different configurations: 2 (uniaxial tension or biaxial tension) × 2 (dry or embedded) × 2 (0° or 90° direction). These test configurations represent the *in vivo* loading and environmental conditions of the mesh implants. Long, narrow strips of meshes used in "line-type" suspensions are mainly loaded in uniaxial tension, whereas wider sheets, such as for hernia repair, are typically subjected to multiaxial tension states. Our earlier study examined the anisotropic behavior of these meshes along two perpendicular directions following the main knitting patterns. However, here, the focus is on the stiffer of the two directions on a per mesh basis. A dry mesh is tested as delivered, whereas embedded infers a specimen being embedded into a soft elastomer matrix (Young's modulus 0.0276 N/mm^2 [25]), mimicking *in vivo*, ingrown conditions.

Experiments with uniaxial tension and biaxial tension (realized as uniaxial strain test, also called "strip biaxial"; [22,26]) were performed on the same tensile test machine. In the uniaxial strain test, lateral contraction of the specimen is constrained, leading to stresses in the direction perpendicular to the loading axis, thus subjecting the sample to a biaxial state of tension. Test piece dimensions were selected to generate a free area of 30 mm × 15 mm (uniaxial) and 50 mm × 15 mm (biaxial). Each specimen was loaded to a maximum of 30% nominal strain (loading rate ~10^{-3}s^{-1}) and unloaded back to a pre-force threshold of 0.01 N for 10 cycles.

Deformation analysis was performed in an optical, non-contact procedure in the center of the specimen, allowing for extraction of local strains (ε_{loc}) as the result of an image analysis algorithm, thus avoiding edge and clamp effects at the specimen boundaries. Force measurements at the clamps were converted to nominal membrane tension (M_t (N/mm)) by dividing by the undeformed width of the sample. For a detailed description of the loading protocol and data extraction, refer to [22].

Table 1. List of mesh types used for the present investigation, with their weight classified as ultralight, light or standard according to [24]. Principal directions of testing are marked in red. Scale bar (lower right): 5 mm. Their clinical application is listed as pelvic organ prolapse (POP) or hernia repair, as specified by the manufacturer.

Image	Mesh	Application	Material, Weight
	Bard™ Mesh Marlex (BM)	Hernia	Polypropylene, standard
	DynaMesh® ENDOLAP (DM)	Hernia	PVDF (polyvinylidene fluoride), standard
	Ethicon Physiomesh® (PM)	Hernia	Polypropylene, ultralight
	Surgipro™ Polypropylene Monofilament Mesh (SPMM)	Hernia	Polypropylene, standard
	Ethicon Ultrapro™ (UP)	Hernia	Polypropylene, light
	DynaMesh® PRS (DMPRS)	POP	PVDF, standard
	Gynecare PROLIFT™ (PE)	POP	Polypropylene, ultralight
	Coloplast Restorelle™ (Rest)	POP	Polypropylene, ultralight
	Parietex Ugytex® (UT)	POP	Polypropylene, light

The resulting $M_\Gamma\!-\!\varepsilon_{loc}$ curves of each of the 8 specimens of each type, as well as the area density and porosity measurements [27,28] form the basis for the analysis and comparison of mesh groups.

Dry mesh samples of known dimensions were weighted before mechanical testing using a high-resolution balance, and their area density was calculated as weight per area (kg/m^2). Porosity is determined as the ratio of open area to the total area, including filaments of one undeformed unit cell of the knitting pattern [27,28].

From the M_t–ε_{loc}, curves the secant stiffness K (N/mm) in the stiffer direction at the reference membrane tension $M_t^{ref} = 0.035$ N/mm (for hernia, as well as POP meshes) was extracted, for both the 1st and 10th cycle (see Figure 1). It is defined as:

$$K = \frac{M_t^{ref}}{\Delta \varepsilon} \tag{1}$$

where $\Delta \varepsilon$ is the difference of local strain at the reference membrane tension M_t^{ref} and at the beginning of the current cycle.

The specific value of M_t was chosen as a load representative of the membrane tension in the pelvic region under physiological intra-abdominal pressure (IAP) at rest [23]. Each mesh is thus characterized by 10 parameters, *i.e.*, a secant stiffness value for each of the tested configurations (uniaxial and biaxial tension, dry and embedded) in 1st and 10th cycle, as well as area density and porosity.

The implants are grouped into POP and hernia meshes as indicated on their official product insert. Each parameter is shown in a bar graph, as well as standard box plots in order to visualize the differences between the two groups. To determine the statistical significance, the Wilcoxon rank sum test (equivalent to the Mann–Whitney U-test) is applied for each parameter.

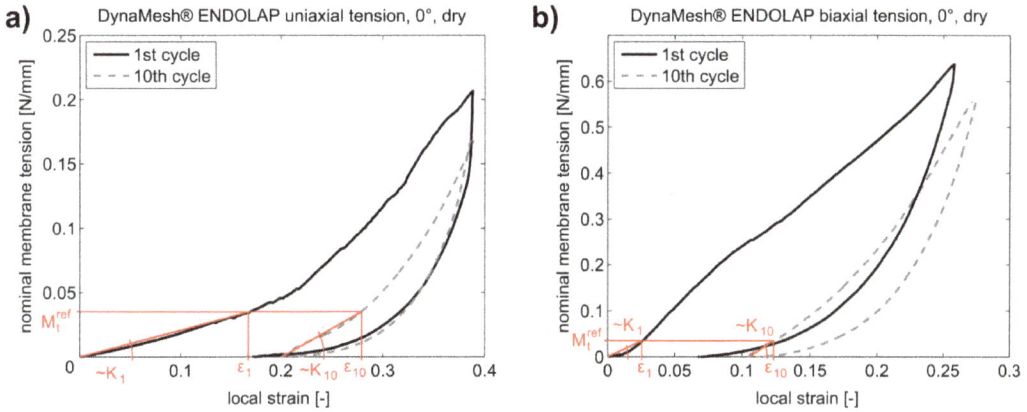

Figure 1. Exemplary tension-strain curves for DynaMesh ENDOLAP (DM) for the illustration of secant stiffness determination. (**a**) Uniaxial tension in the 0° direction, dry mesh, loading and unloading in 1st and 10th cycle; (**b**) Biaxial tension in the 0° direction, dry mesh, loading and unloading in 1st and 10th cycle. Secant stiffness in the 1st and 10th cycle is shown for both.

3. Results and Discussion

3.1. Results

Figure 2 shows the uniaxial secant stiffness K_{uni} (N/mm) of each specimen grouped according to the manufacturer indication into POP (red) and hernia (blue) meshes, with the mean of each group shown in darker red and blue, respectively. The specific testing conditions (dry/embedded and first/10th cycle) are indicated in each subgraph. Figure 3 represents the corresponding biaxial secant stiffness K_{bi} (N/mm). The respective stiffness values for each configuration are reported in Tables 2 (POP meshes) and 3 (hernia meshes).

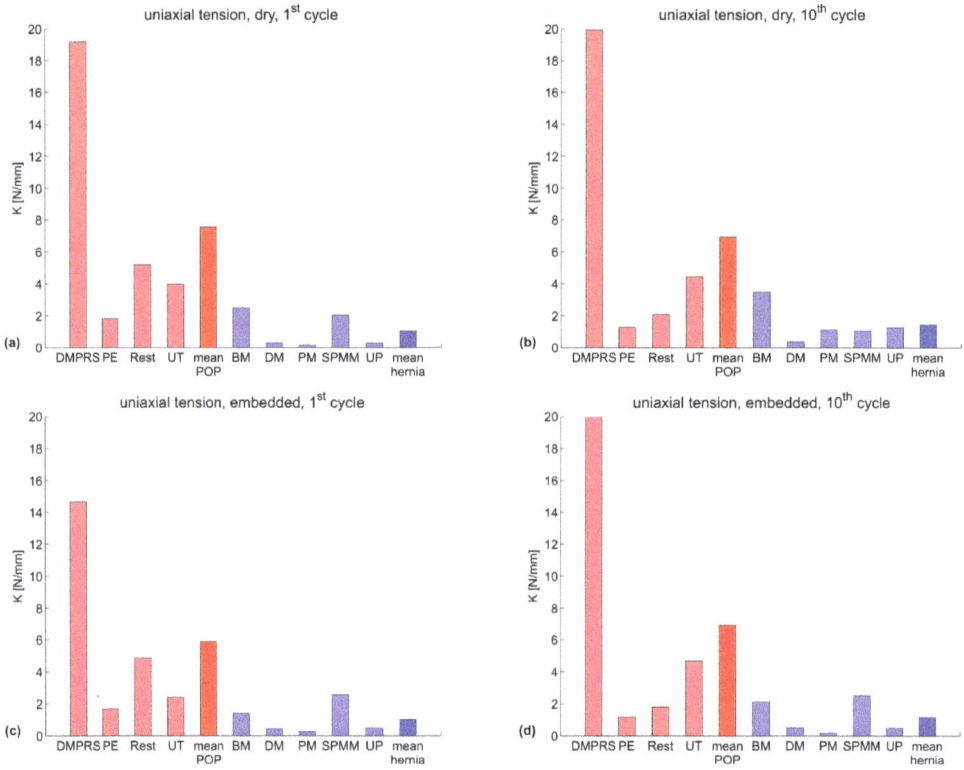

Figure 2. Uniaxial secant stiffness K_{uni} (N/mm) for all meshes in four configurations: (**a**) dry mesh, first cycle; (**b**) dry mesh, 10th cycle; (**c**) embedded mesh, first cycle; (**d**) embedded mesh, 10th cycle. POP meshes are shown in red; hernia meshes in blue. The mean of each group is plotted darker.

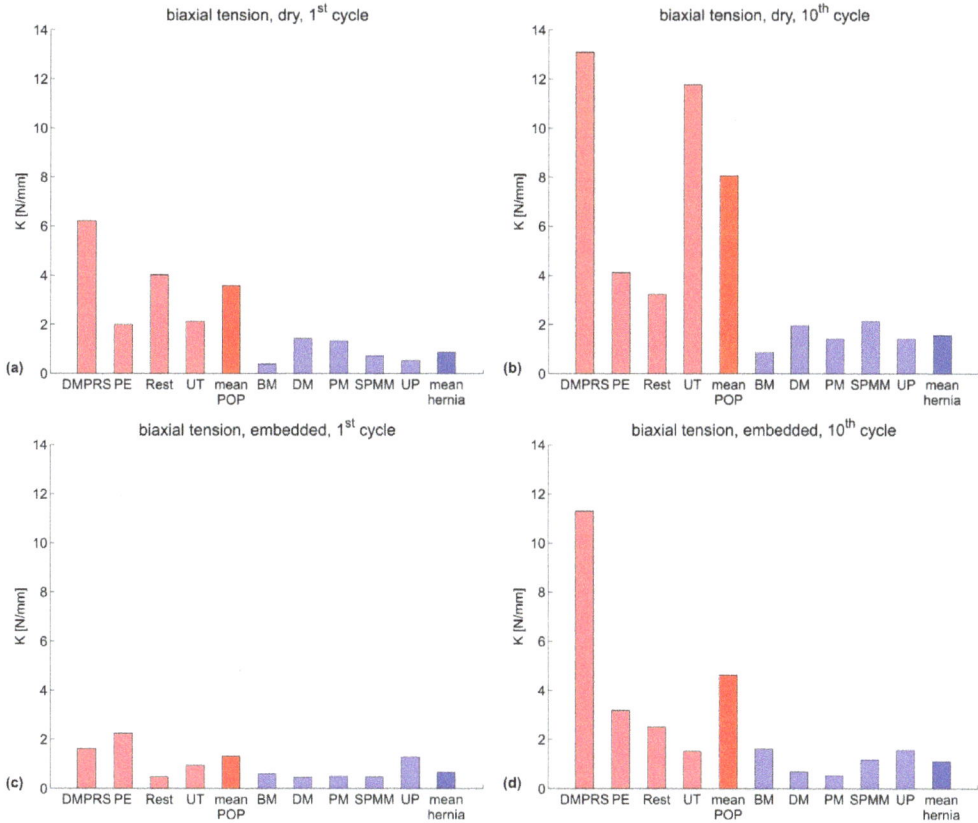

Figure 3. Biaxial secant stiffness K_{bi} (N/mm) for all meshes in four configurations: (**a**) dry mesh, first cycle; (**b**) dry mesh, 10th cycle; (**c**) embedded mesh, first cycle; (**d**) embedded mesh, 10th cycle. POP meshes are shown in red; hernia meshes in blue. The mean of each group is plotted darker.

Table 2. Numerical values of stiffness for all POP meshes and all tested configurations.

Testing Configuration	K (N/mm) POP Meshes				
	DMPRS	PE	Restorelle	UT	mean
uniaxial, dry, 1st cycle	19.2	1.8	5.2	4.0	7.5
uniaxial, dry, 10th cycle	19.9	1.3	2.1	4.5	6.9
uniaxial, embedded, 1st cycle	14.6	1.7	4.8	2.4	5.9
uniaxial, embedded, 10th cycle	20.0	1.2	1.8	4.7	6.9
biaxial, dry, 1st cycle	6.2	2.0	4.0	2.1	3.6
biaxial, dry, 10th cycle	13.1	4.1	3.2	11.8	8.1
biaxial, embedded, 1st cycle	1.6	2.2	0.5	0.9	1.3
biaxial, embedded, 10th cycle	11.3	3.2	2.5	1.5	4.6

Table 3. Numerical values of stiffness for all hernia meshes and all tested configurations.

Testing configuration	*K* (N/mm) Hernia Meshes					
	BM	**DM**	**PM**	**SPMM**	**UP**	**mean**
uniaxial, dry, 1st cycle	2.5	0.3	0.2	2.0	0.3	1.1
uniaxial, dry, 10th cycle	3.5	0.4	1.1	1.1	1.3	1.5
uniaxial, embedded, 1st cycle	1.4	0.5	0.3	2.6	0.5	1.0
uniaxial, embedded, 10th cycle	2.1	0.5	0.2	2.5	0.5	1.2
biaxial, dry, 1st cycle	0.4	1.4	1.3	0.7	0.5	0.9
biaxial, dry, 10th cycle	0.9	2.0	1.4	2.1	1.4	1.6
biaxial, embedded, 1st cycle	0.6	0.5	0.5	0.5	1.3	0.7
biaxial, embedded, 10th cycle	1.6	0.7	0.5	1.2	1.6	1.1

Figures 4 and 5 depict the summarizing box plots for the two groups, for uniaxial and biaxial stiffness in each configuration.

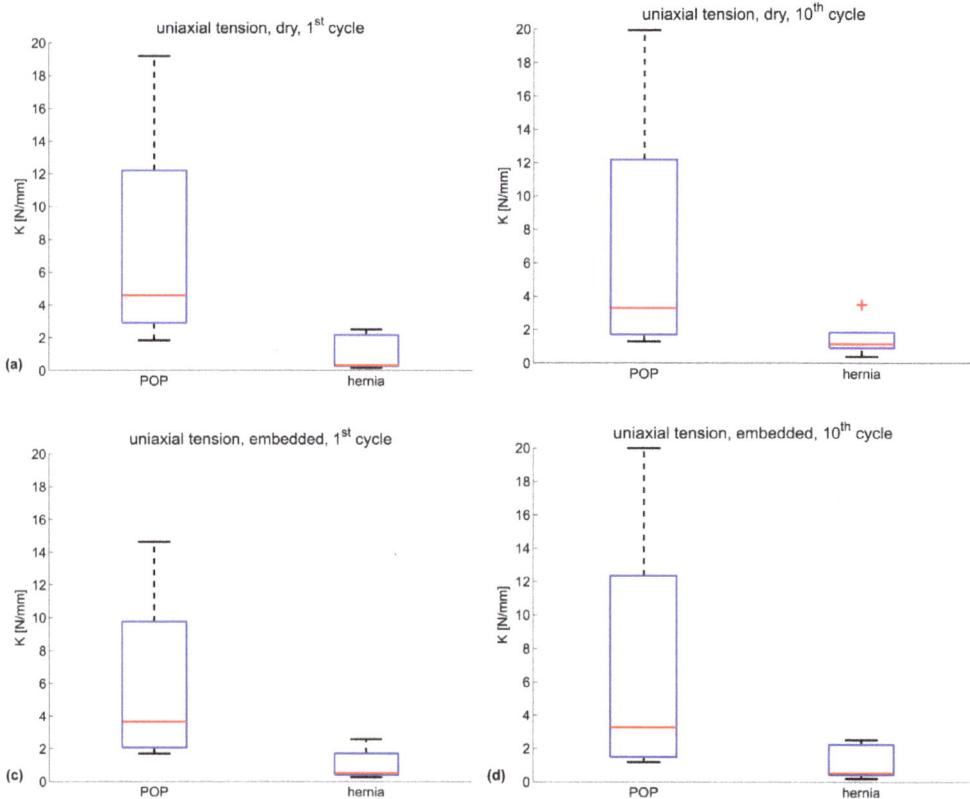

Figure 4. Box plots for uniaxial secant stiffness K_{uni} (N/mm) for all meshes in four configurations: (**a**) dry mesh, first cycle; (**b**) dry mesh, 10th cycle; (**c**) embedded mesh, first cycle; (**d**) embedded mesh, 10th cycle. The red line marks the median of the group; the box represents the 25th and 75th percentile; the extended whiskers the most extreme data points. The + sign indicates an outlier.

The variability for the POP group is very large for all parameters, thus affecting the statistical significance of the differences observed. The Wilcoxon rank sum test indicates a statistically-significant difference between the POP and hernia groups for the biaxial stiffness in the dry condition, both at the first and 10th cycle ($p = 0.016$ for both); see Figures 4a,b and 5a,b. The POP meshes were four- or five-fold stiffer than the hernia meshes in the first cycle and 10th cycle, respectively.

When comparing the mean and median stiffness for all configurations, POP implants are overall less compliant than hernia implants. Since the abdominal wall is known to be stiffer than vaginal tissue [9,29] and if mechanical biocompatibility were mainly dependent on similar properties to the implant area, one would expect a more compliant design for implants for POP compared to hernia. Embedding a mesh into a polymer matrix, thus reflecting the interaction with native tissue, affects the mechanical response of the implants. The differences between the groups are still evident also for this case (see Figures 2–4 and Figure 5c,d).

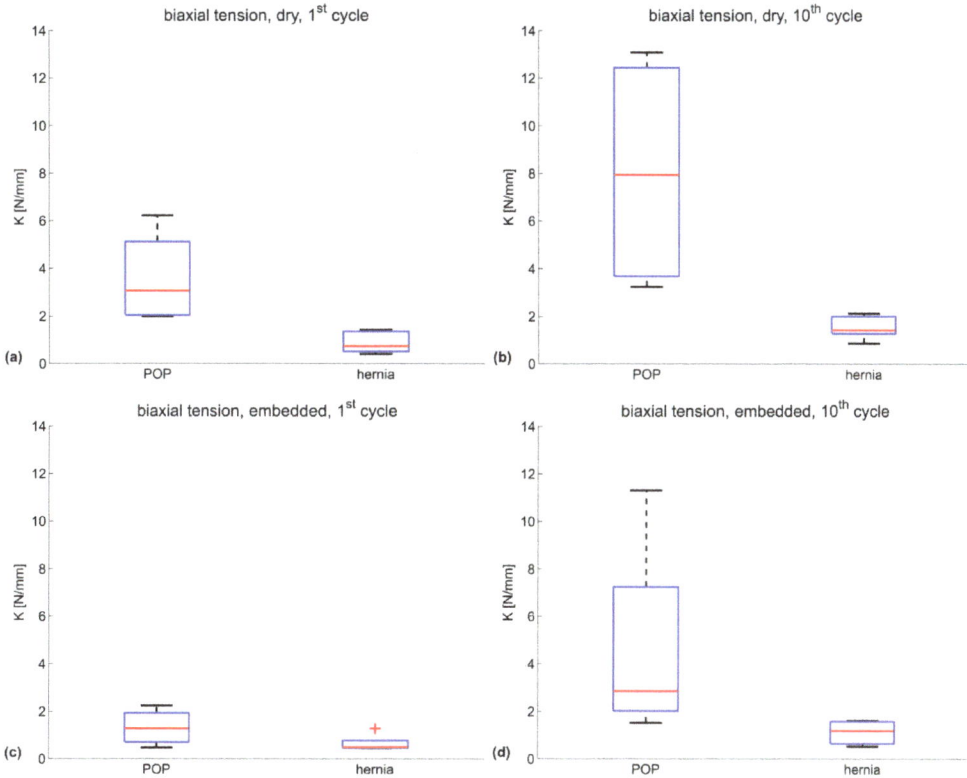

Figure 5. Box plots for biaxial secant stiffness K_{bi} (N/mm) or all meshes in four configurations: (**a**) dry mesh, first cycle; (**b**) dry mesh, 10th cycle; (**c**) embedded mesh, first cycle; (**d**) embedded mesh, 10th cycle. The red line marks the median of the group; the box represents the 25th and 75th percentile; the extended whiskers the most extreme data points. The + sign indicates an outlier.

High density and small pores are often linked to high stiffness in prosthetic meshes [30,31]. When comparing the POP and hernia groups, no statistically-significant difference can be found in these parameters (see Figures 6 and 7). However, tendencies can be seen with the hernia meshes being heavier (and similarly porous), while still being in general more compliant.

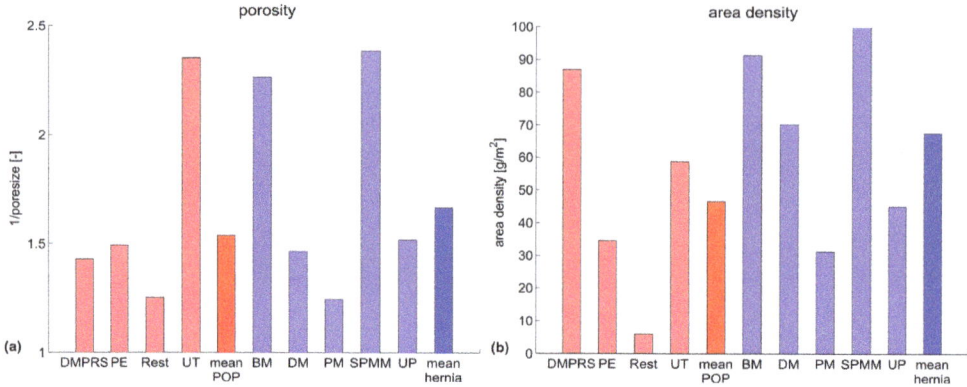

Figure 6. (**a**) Porosity and (**b**) area density of all meshes. POP meshes are shown in red, hernia meshes in blue, with their respective mean values plotted darker. Porosity is shown as the inverse of pore size, *i.e.*, higher values represent smaller pores.

3.2. Discussion

Better clinical outcome might be expected from meshes designed to mimic the physiologically-relevant deformation behavior of the underlying native tissue, thus ensuring mechanical biocompatibility. This entails a meaningful stiffness reference target for mesh design. However, the physiological loading configuration, as well as the range of load levels in terms of membrane tension in the abdominal and vaginal wall still remain largely uncertain. While membrane tensions in the abdominal wall are generally higher [7] than in the pelvic region (simply due to geometric reasons, as shown in [23]), increasing the level of membrane tension at which the secant stiffness is evaluated for the hernia meshes to a level of 0.136 N/mm (reported in [23] as a tension at rest in the abdominal wall) only marginally increases their stiffness and does not change the trends reported in Figures 2–7.

The range of stiffness values for native tissue reported in the literature shows large variation and is mostly based on uniaxial tensile tests, while the predominant loading state *in vivo* is biaxial. Song *et al.* [32] report a Young's modulus of 0.042 N/mm^2 and 0.0225 N/mm^2 in the transverse and sagittal plane, respectively, for human abdominal wall during *in vivo* insufflation, which would translate to membrane stiffness values of 1.26 N/mm and 0.675 N/mm, respectively, multiplying by the reported thickness of around 30 mm [32]. Analyzing the uniaxial stress-strain graphs shown in [33], abdominal skin has a secant stiffness of 1.7 N/mm at a membrane tension reference of 0.136 N/mm, whereas vaginal wall stiffness is 1.45 N/mm at a reference of 0.035 N/mm membrane tension. Rabbits are one model system for mesh performance evaluation. Analysis of the stress-strain curves in [34] yields uniaxial stiffness values for the abdominal wall complex of rabbits of 0.87–0.98 N/mm, whereas [35] report 0.28 N/mm at the same reference membrane tension of 0.136 N/mm. Biaxial stiffness under inflation, however, increases by an order of magnitude to 2.41 N/mm. Similarly, vaginal wall tissue stiffness values at a reference membrane tension of 0.035 N/mm reach from 0.155 N/mm (prolapsed tissue; [29]), 0.675 N/mm (healthy tissue; [33]), 1.4 N/mm (prolapsed tissue; [11]), up to 6.47 N/mm (healthy tissue; [11]).

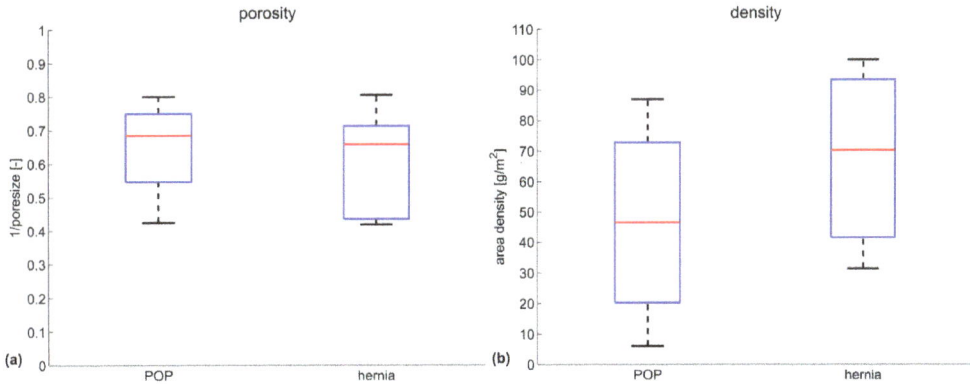

Figure 7. Box plots for **(a)** porosity and **(b)** area density of all meshes. The red line marks the median of the group; the box represents the 25th and 75th percentile; the extended whiskers the most extreme data points. Porosity is shown as the inverse of pore size, *i.e.*, higher values represent smaller pores.

This variability is due to differences in experimental methodology, *in vivo vs. ex vivo* testing, cadaver testing, animal tissue *vs.* human tissue, pathological *vs.* healthy tissue, as well as the inherent variability of biological soft tissues. This scatter poses a significant problem in determining meaningful mechanical design targets for prosthetic mesh implants and warrants further investigation. Focus should be on the definition of consistent testing procedures based on physiological, *in vivo* loading and stress magnitude conditions.

It has to be noted that the approach of mimicking the ingrown state of the mesh using an elastomer matrix is only partially representative of the *in vivo* condition. This is mainly due to the differences between the embedding procedure and the process of tissue ingrowth. A mesh sample is simply laid into liquid elastomer and the elastomer left to cure. This results in an elastomer-mesh complex that is formed in an unloaded initial configuration, while *in vivo* ingrowth might be expected to happen in a loaded state. Due to this discrepancy, the definition of a secant stiffness between tension values chosen here (pre-force threshold as defined in the experimental protocol and reference membrane tension) might not be representative of the actual *in vivo* load range. This further highlights the need to investigate the expected *in vivo* loading conditions of mesh implants, so to define testing protocols that reproduce physiological states.

When comparing individual meshes in terms of their physical and mechanical parameters, an instructive example can be seen in two meshes manufactured by FEG Textiltechnik: DynaMesh ENDOLAP (DM), used in hernia repair, and DynaMesh PRS (DMPRS), used for pelvic prolapse repair. While their porosity is similar and DM is indeed the heavier of the two, as expected for a hernia mesh, DMPRS is much stiffer in all tested configurations; and up to two orders of magnitude in the case of uniaxial tension.

Similarly, Restorelle (a POP mesh) is the lightest implant with the largest pores; however, its stiffness is clearly above average for most tested configurations. Restorelle is less compliant in the configurations tested here than SPMM, the heaviest, small-pore implant. Note that in previous studies [13,36,37], Restorelle was shown to be more compliant than some of the meshes tested here. This can be attributed to the differences in the experimental protocols. In particular, the present experiments evaluate the response of meshes at a physiological tension level. Thus the low deformation regime determines the measured stiffness in the present work, whereas the ball burst test in [36] depends more on the higher deformation regime.

Lightweight and large porous meshes might be expected to be more compliant than heavy, small porous implants [36]. However, the present analysis shows that while POP meshes are on average indeed lighter and often have larger pores, they are generally stiffer in the physiological loading regime. In these loading configurations, porosity and density alone cannot be predictors for mesh stiffness. Their specific knitting pattern and microstructure can lead to mechanism-like behavior in a physiological loading range, effects that determine their compliance. This calls for a careful evaluation of the mechanical properties of each mesh on several length scales, in conditions representative of those expected *in vivo*.

The present experimental results do not take into account loads in directions other than the main knitting patterns. In fact, the meshes are usually implanted such that their knitting pattern aligns with principal loading directions, such as in line type suspensions (e.g., urethral slings) or sacrocolpopexy procedures. Some meshes, such as DM, Ultrapro (UP) and PE, even have colored filaments interweaved, guiding the physician during implantation. Deviation from this rule might lead to a mechanical response that strongly differs from the data reported here. Similarly, while each mesh was tested in two perpendicular directions, only the stiffer of the two is considered for the present analysis, being indicative of an upper bound of stiffness. Knitted meshes do indeed tend to behave in an anisotropic way, as investigated in [22], with anisotropy indices ranging from 1.0 for SPMM (similar stiffness in both evaluated directions) to 8.0 for DMPRS.

In addition, only one sample per configuration has been tested due to the limited availability of raw mesh material, which also limited the sample size. However, the level of variability for mesh implants reported in the literature [18,22,37,38] is low, justifying the analysis conducted.

4. Conclusions

The mechanical biocompatibility of prosthetic mesh implants for hernia and POP repair very likely is an important factor in ensuring their functionality and integration into the host tissue. We see matching mechanical properties in a physiological loading range as desirable and an important step towards reducing clinical complications.

This study has shown that some meshes designated as suited for POP repair tend to be stiffer than those used for hernia repair, even though the abdominal wall has been shown to be less compliant than the vaginal wall. Additionally, the expectation of lightweight, large pore meshes being more compliant than their counterparts was contradicted by the presented data and specific testing configurations, indicating that a biomechanical analysis of each product is necessary to determine its mechanical suitability. Knowledge of the physiological, *in vivo* mechanical environment in terms of loading configuration and magnitude is required in order to define a suitable design target for optimization of implants. Data reported in the literature show large variations in testing configurations and corresponding stiffness values for pelvic organs and abdominal wall tissue. The consensus for a standardized, physiological mechanical testing procedure is needed for native tissues and implants, opening the path for a conscious mechanical design of prosthetic meshes.

Acknowledgments

Manfred M. Maurer and Edoardo Mazza acknowledge internal funding by ETH Zurich for the support of this research.

Author Contributions

Manfred M. Maurer, Barbara Röhrnbauer, Jan Deprest and Edoardo Mazza conceived of and designed the study. Manfred M. Maurer, Barbara Röhrnbauer and Andrew Feola performed the experiments. Manfred M. Maurer and Edoardo Mazza analyzed the data. Barbara Röhrnbauer and Jan Deprest contributed materials. All authors contributed to writing the paper.

Conflicts of Interest

The authors declare no conflict of interest.

References

1. Röhrnbauer, B. *Mechanical Characterization and Modeling of Prosthetic Meshes*; ETH Zurich: Zurich, Switzerland, 2013.
2. Röhrnbauer, B.; Mazza, E. Uniaxial and biaxial mechanical characterization of a prosthetic mesh at different length scales. *J. Mech. Behav. Biomed. Mater.* **2014**, *29*, 7–19.
3. Mazza, E.; Ehret, A.E. Mechanical biocompatibility of highly deformable biomedical materials. *J. Mech. Behav. Biomed. Mater.* **2015**, *48*, 100–124.
4. Liang, R.; Abramowitch, S.; Knight, K.; Palcsey, S.; Nolfi, A.; Feola, A.; Stein, S.; Moalli, P.A. Vaginal degeneration following implantation of synthetic mesh with increased stiffness. *BJOG Int. J. Obstet. Gynaecol.* **2013**, *120*, 233–243.
5. Mangera, A.; Bullock, A.J.; Chapple, C.R.; Macneil, S. Are biomechanical properties predictive of the success of prostheses used in stress urinary incontinence and pelvic organ prolapse? A systematic review. *Neurourol. Urodyn.* **2012**, *31*, 13–21.
6. Konerding, M.A.; Chantereau, P.; Delventhal, V.; Holste, J.-L.; Ackermann, M. Biomechanical and histological evaluation of abdominal wall compliance with intraperitoneal onlay mesh implants in rabbits: A comparison of six different state-of-the-art meshes. *Med. Eng. Phys.* **2012**, *34*, 806–816.

7. Klinge, U.; Conze, J.; Limberg, W.; Brucker, C.; Ottinger, A.P.; Schumpelick, V. Pathophysiology of the abdominal wall. *Der Chirurg* **1996**, *67*, 229–233.

8. Choe, J.M.; Kothandapani, R.; James, L.; Bowling, D. Autologous, cadaveric, and synthetic materials used in sling surgery: Comparative biomechanical analysis. *Urology* **2001**, *58*, 482–486.

9. Junge, K.; Klinge, U.; Prescher, A.; Giboni, P.; Niewiera, M.; Schumpelick, V. Elasticity of the anterior abdominal wall and impact for reparation of incisional hernias using mesh implants. *Hernia* **2001**, *5*, 113–118.

10. Cosson, M.; Debodinance, P.; Boukerrou, M.; Chauvet, M.P.; Lobry, P.; Crépin, G.; Ego, A. Mechanical properties of synthetic implants used in the repair of prolapse and urinary incontinence in women: Which is the ideal material? *Int. Urogynecol. J.* **2003**, *14*, 169–178.

11. Rubod, C.; Boukerrou, M.; Brieu, M.; Jean-Charles, C.; Dubois, P.; Cosson, M. Biomechanical properties of vaginal tissue: Preliminary results. *Int. Urogynecol. J.* **2008**, *19*, 811–816.

12. Claerhout, F.; Verbist, G.; Verbeken, E.; Konstantinovic, M.; de Ridder, D.; Deprest, J. Fate of collagen-based implants used in pelvic floor surgery: A 2-year follow-up study in a rabbit model. *Am. J. Obst. Gyn.* **2008**, *198*, 91–96.

13. Jones, K.A.; Feola, A.; Meyn, L.; Abramowitch, S.D.; Moalli, P.A. Tensile properties of commonly used prolapse meshes. *Int. Urogynecol. J.* **2009**, *20*, 847–853.

14. Ozog, Y.; Konstantinovic, M.L.; Werbrouck, E.; de Ridder, D.; Edoardo, M.; Deprest, J. Shrinkage and biomechanical evaluation of lightweight synthetics in a rabbit model for primary fascial repair. *Int. Urogynecol. J.* **2011**, *22*, 1099–1108.

15. Fenner, D.E. New surgical mesh. *Clin. Obstet. Gynecol.* **2000**, *43*, 650–658.

16. Dietz, H.P.; Vancaillie, P.; Svehla, M.; Walsh, W.; Steensma, A.B.; Vancaillie, T.G. Mechanical properties of urogynecologic implant materials. *Int. Urogynecol. J.* **2003**, *14*, 239–243.

17. Cobb, W.S.; Kercher, K.W.; Heniford, B.T. The argument for lightweight polypropylene mesh in hernia repair. *Surg. Innov.* **2005**, *12*, 63–69.

18. Feola, A.; Abramowitch, S.; Jallah, Z.; Stein, S.; Barone, W.; Palcsey, S.; Moalli, P. Deterioration in biomechanical properties of the vagina following implantation of a high-stiffness prolapse mesh. *BJOG Int. J. Obstet. Gynaecol.* **2013**, *120*, 224–232.

19. Lowman, J.K.; Jones, L.A.; Woodman, P.J.; Hale, D.S. Does the prolift system cause dyspareunia? *Am. J. Obstet. Gynecol.* **2008**, *199*, 701–706.

20. Ozog, Y.; Konstantinovic, M.; Werbrouck, E.; de Ridder, D.; Mazza, E.; Deprest, J. Persistence of polypropylene mesh anisotropy after implantation: An experimental study. *BJOG Int. J. Obstet. Gynaecol.* **2011**, *118*, 1180–1185.

21. FDA Centre for Devices and Radiological Health. Fda Safety Communication: Update on Serious Complications Associated with Transvaginal Placement of Surgical Mesh for Pelvic Organ Prolapse. Avaiable online: http://www.fda.gov/medicaldevices/safety/alertsandnotices/ucm262435.htm (accessed on 23 March 2015).

22. Maurer, M.M.; Roehrnbauer, B.; Feola, A.; Deprest, J.; Mazza, E. Mechanical biocompatibility of prosthetic meshes: A comprehensive protocol for mechanical characterization. *J. Mech. Behav. Biomed. Mater.* **2014**, *40*, 42–58.

23. Ozog, Y.; Deprest, J.; Haest, K.; Claus, F.; de Ridder, D.; Mazza, E. Calculation of membrane tension in selected sections of the pelvic floor. *Int. Urogynecol. J.* **2014**, *25*, 499–506.

24. Coda, A.; Lamberti, R.; Martorana, S. Classification of prosthetics used in hernia repair based on weight and biomaterial. *Hernia J. Hernias Abdom. Wall Surg.* **2012**, *16*, 9–20.

25. Farine, M. *Instrumented Indentation of Soft Materials And Biological Tissues*; ETH Zurich: Zurich, Switzerland, 2013.

26. Hollenstein, M.; Ehret, A.E.; Itskov, M.; Mazza, E. A novel experimental procedure based on pure shear testing of dermatome-cut samples applied to porcine skin. *Biomech. Model. Mechanobiol.* **2011**, *10*, 651–661.

27. Chu, C.C.; Welch, L. Characterization of morphologic and mechanical properties of surgical mesh fabrics. *J. Biomed. Mater. Res.* **1985**, *19*, 903–916.

28. Pourdeyhimi, B. Porosity of surgical mesh fabrics: New technology. *J. Biomed. Mater. Res.* **1989**, *23*, 145–152.

29. Peña, E.; Calvo, B.; Martínez, M.A.; Martins, P.; Mascarenhas, T.; Jorge, R.M.N.; Ferreira, A.; Doblaré, M. Experimental study and constitutive modeling of the viscoelastic mechanical properties of the human prolapsed vaginal tissue. *Biomech. Model. Mechanobiol.* **2010**, *9*, 35–44.

30. Klinge, U.; Klosterhalfen, B.; Birkenhauer, V.; Junge, K.; Conze, J.; Schumpelick, V. Impact of polymer pore size on the interface scar formation in a rat model. *J. Surg. Res.* **2002**, *103*, 208–214.

31. Klosterhalfen, B.; Junge, K.; Klinge, U. The lightweight and large porous mesh concept for hernia repair. *Expert Rev. Med. Devices* **2005**, *2*, 103–117.

32. Song, C.; Alijani, A.; Frank, T.; Hanna, G.B.; Cuschieri, A. Mechanical properties of the human abdominal wall measured *in vivo* during insufflation for laparoscopic surgery. *Surg. Endosc.* **2006**, *20*, 987–990.

33. Gabriel, B.; Rubod, C.; Brieu, M.; Dedet, B.; de Landsheere, L.; Delmas, V.; Cosson, M. Vagina, abdominal skin, and aponeurosis: Do they have similar biomechanical properties? *Int. Urogynecol. J.* **2011**, *22*, 23–27.

34. Hernandez, B.; Pena, E.; Pascual, G.; Rodriguez, M.; Calvo, B.; Doblare, M.; Bellon, J.M. Mechanical and histological characterization of the abdominal muscle. A previous step to modelling hernia surgery. *J. Mech. Behav. Biomed. Mater.* **2011**, *4*, 392–404.

35. Rohrnbauer, B.; Ozog, Y.; Egger, J.; Werbrouck, E.; Deprest, J.; Mazza, E. Combined biaxial and uniaxial mechanical characterization of prosthetic meshes in a rabbit model. *J. Biomech.* **2013**, *46*, 1626–1632.

36. Feola, A.; Barone, W.; Moalli, P.; Abramowitch, S. Characterizing the *ex vivo* textile and structural properties of synthetic prolapse mesh products. *Int. Urogynecol. J.* **2013**, *24*, 559–564.

37. Shepherd, J.P.; Feola, A.J.; Abramowitch, S.D.; Moalli, P.A. Uniaxial biomechanical properties of seven different vaginally implanted meshes for pelvic organ prolapse. *Int. Urogynecol. J.* **2012**, *23*, 613–620.

38. Edwards, S.L.; Werkmeister, J.A.; Rosamilia, A.; Ramshaw, J.A.M.; White, J.F.; Gargett, C.E. Characterisation of clinical and newly fabricated meshes for pelvic organ prolapse repair. *J. Mech. Behav. Biomed. Mater.* **2013**, *23*, 53–61.

Reinforcement Strategies for Load-Bearing Calcium Phosphate Biocements

Martha Geffers, Jürgen Groll and Uwe Gbureck

Abstract: Calcium phosphate biocements based on calcium phosphate chemistry are well-established biomaterials for the repair of non-load bearing bone defects due to the brittle nature and low flexural strength of such cements. This article features reinforcement strategies of biocements based on various intrinsic or extrinsic material modifications to improve their strength and toughness. Altering particle size distribution in conjunction with using liquefiers reduces the amount of cement liquid necessary for cement paste preparation. This in turn decreases cement porosity and increases the mechanical performance, but does not change the brittle nature of the cements. The use of fibers may lead to a reinforcement of the matrix with a toughness increase of up to two orders of magnitude, but restricts at the same time cement injection for minimal invasive application techniques. A novel promising approach is the concept of dual-setting cements, in which a second hydrogel phase is simultaneously formed during setting, leading to more ductile cement–hydrogel composites with largely unaffected application properties.

Reprinted from *Materials*. Cite as: Geffers, M.; Groll, J.; Gbureck, U. Reinforcement Strategies for Load-Bearing Calcium Phosphate Biocements. *Materials* **2015**, *8*, 2700-2717.

1. Introduction

Self-setting cements based on calcium phosphate chemistry combine the advantages of the high biocompatibility of calcium phosphates with the free mouldability of cements and the mechanical stability of ceramic implants [1,2]. Such calcium phosphate cements (CPC) are usually based on freshly prepared mixtures of crystalline or amorphous calcium orthophosphate, calcium hydroxide or calcium carbonate powders with an aqueous solution, which undergo setting in a continuous dissolution–precipitation reaction. Although various mixtures of calcium and phosphate sources can serve as raw materials, there are in principle only two cement types as products of the setting reaction: At neutral or basic pH the calcium phosphate cement sets to nanocrystalline hydroxyapatite (HA, with a variable stoichiometric composition between $Ca_9(PO_4)_5HPO_4OH$–$Ca_{10}(PO_4)_6(OH)_2$), while at low pH < 4.2, orthophosphate ions are protonated and the secondary phosphates brushite ($CaHPO_4·2H_2O$, DCPD) and monetite ($CaHPO_4$, DCPA) are the least soluble calcium phosphates [3,4] and hence precipitated during setting of acidic cement pastes until an end pH of close to 5 [1,2,5–7]. Detailed reviews about CPCs reflecting their synthesis, setting reaction, rheological properties or biological performance can be found in literature [2,8,9]. CPC are resorbed *in vivo* and replaced by new bone tissue [10,11], whereas the speed of degradation depends on the final composition of the cement matrix. Hydroxyapatite forming cements degrade only slowly within years since the surrounding extracellular fluid ($[Ca^{2+}] \sim 2.5$ mmoL/L, $[HPO_4^{2-}] \sim 1$ mmoL/L [12]) is supersaturated regarding HA (solubility of hydroxyapatite ~ 0.2–0.3 mg/L) [2]. HA forming cements degrade solely by osteoclastic bone remodeling, which is limited to surface degradation since cells cannot penetrate the microporous cement structure. Osteoclastic cells resorb the cement by providing a local acidic environment increasing the solubility of the mineral [13–16]. In contrast, cements forming brushite or monetite have a higher solubility (calculated solubility in water for monetite: 41–48 mg/L, brushite: 85–88 mg/L [17]) and many studies have demonstrated the bone remodelling capacity of such cements in various animal models within a time period of 8–52 weeks [18–21]. A passive resorption of such cements by simple chemical dissolution is a topic of contention in the literature, whereas some authors postulate that the extracellular liquid is in equilibrium with brushite [22], while others have calculated a thermodynamic instability of brushite in simulated body fluid [23]. The latter is supported by the fact that brushite forming cements are indeed dissolved *in vivo* even in the absence of osteoclastic cells (e.g., after intramuscular implantation) [24]. Worth noting is that for brushite forming cements a phase transformation into lower soluble minerals like octacalcium phosphate, hydroxyapatite or whitlockite can occur *in vivo* by a dissolution–reprecipitation reaction, which slows down biodegradation [25,26].

Calcium phosphate bone cements have been shown to provide compressive strength of up to 80 MPa measured under application near conditions without a precompaction of the cement paste leading to lower porosity/higher strength, since this is not applicable under *in vivo* conditions [27]. Set CPC can be considered as porous ceramic materials with an inherent brittleness and comparatively low flexural strength compared to natural hard tissues such as bone or teeth. A comprehensive characterization of the elastic and failure properties for both hydroxyapatite and brushite forming CPC by Charrière *et al.* [28] indicated brushite cements to be suitable as bone fillers, while hydroxyapatite cements were attributed to having the potential to be a structural biomaterial. The low fracture toughness restricts the use of CPC to non-load-bearing defects [29]. Typical applications are the treatment of maxillofacial defects or deformities [1] or the repair of craniofacial defects [30]. An extension of the application of calcium phosphate cements to load-bearing defects, e.g., in vertebroplasty or kyphoplasty [31–33], would require less brittle cements with an increased fracture toughness. This is of high interest since the application of commonly used polymeric cements have strong drawbacks near the spinal cord due to their strong exothermic setting reaction and cytotoxic monomer release [34–36]. Common approaches to reduce brittleness of CPC and to improve their mechanical performance for load-bearing applications cover the modification of the cement liquid with polymeric additives such as collagen [37–40], the addition of fibres to the cement matrix [41,42] or the use of dual-setting cements in which a dissolved monomer is simultaneously cross-linked during cement setting [43–45] (Figure 1). This article aims to feature the most significant reinforcement strategies for calcium phosphate cements based on either intrinsic (porosity) or extrinsic (fiber addition, dual setting cement) material modifications.

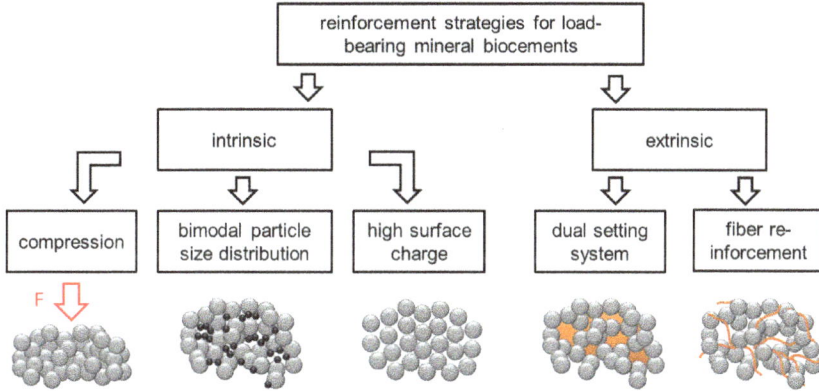

Figure 1. Strategies to reinforce mineral biocement for load-bearing applications.

2. Porosity Reduction for Strength Improvement of CPC

Calcium phosphate biocements set by a dissolution–precipitation reaction, during which the cement raw material continuously dissolves to form a supersaturated solution with regard to the setting product. The latter is precipitated from the aqueous cement phase and forms an entangled cementitious crystal matrix. The mechanical strength of a cement matrix is a direct result of this crystal entanglement and several factors determine the final strength of the matrix, such as degree of conversion, setting product or porosity. The latter is likely the most important factor and it is known from literature that porosity reduction in cements from 50% to 31% by compression can increase compressive strength by nearly an order of magnitude [46]. Porosity in biocements predominately originates from the presence of unreacted cement liquid after setting located in the voids between the entangled crystal matrix. Since any excess of water used for paste mixing, which is not consumed during the setting reaction creates porosity, the main influencing parameter on the total cement porosity is the powder to liquid ratio (PLR) used for cement processing. Pore sizes in CPC typically have a diameter range spanning from a few nanometers to several micrometers [47,48] and are occupying about 22–55 vol% of cements without further paste manipulation (e.g., compaction, porogen addition) [49,50]. Generally, pores in hydroxyapatite cements are smaller than in brushite cements (due to smaller crystal size of HA), whereas the total porosity is mostly smaller for brushite cement. The latter is a result of an increased water consumption during brushite cement setting.

Porosity considerably lowers the strength and stiffness (Young's modulus) of the cements matrix with an inverse exponential relationship between cement porosity and compressive strength:

$$CS = CS_0 \exp^{-2KP} \tag{1}$$

where CS is the compressive strength at a given porosity; CS_0 is the maximum theoretical strength of the material; K is a constant; and P is porosity [46]. Porosity is usually measured by helium pycnometry [51], mercury intrusion porosimetry (MIP) [52] or it is calculated based on the phase composition of the set cements and their densities [53]. Due to the disadvantages of these methods (destructive, long analysis times, toxicity of mercury, misleading results due to amorphous phases), Unosson *et al.* [54] have investigated a method which is based on the assumption that the evaporated water from a dried cement sample equals to the volume of pores within the cement. Since the accuracy of this method depends on a quantitative drying of samples without affecting the phase composition, the authors evaluated several drying conditions (vacuum, elevated temperature) for cement samples and compared the results with porosity determined by the above mentioned methods. Since the measured porosity was found to vary between the different methods, the authors recommended using more than one method to determine cement porosity, whereas the water evaporation method (24 h in vacuum) proved to be fast, easy and precise in estimating the porosity of CPCs.

Porosity reduction by decreasing the amount of cement liquid used for mixing is a key parameter to increase the intrinsic strength of any biocement matrix. This, however, is limited, since every cement powder requires a formulation specific minimum amount of water ("plastic limit") for surface wetting of all cement particles and for filling the space between the particles [55]. A correlation between the powder to liquid ratio used for forming a cement paste and the resulting porosity/compressive strength is displayed in Figure 2 for both HA and brushite forming cements. An effective method to reduce cement porosity is based on both creating a bimodal size distribution of cement raw materials and the creation of a high surface charge (zeta-potential) of the particles. A bimodal size distribution is thought to fill space in cement pastes normally occupied by water. The possibility to reduce porosity has been demonstrated for both hydroxyapatite [27] and brushite [49,50] forming biocements. In addition, a high surface charge (zeta-potential) will help to disperse agglomerates of fine sized particles by reducing attractive interparticulate forces. The zeta-potential can be influenced by using multiple charged ions as additives to the cement liquid, e.g., tatrates or citrates [56], which adsorb at the particle surface and increase the zeta-potential to values of $\sim$-40 to -50 mV. Applying these two principles to a matrix of α-tricalcium phosphate (monomodal size distribution with d_{50} ~9.8 μm) by using 13–33 wt% fine sized $CaHPO_4$ filler (d_{50} ~ 1.16 μm) and 0.5 M trisodium citrate solution increase the plastic limit of the cements from 3.5 to 5.0 g/mL. At the same time, porosity was decreased from 37% to 25% and a strength improvement from 50 to 79 MPa could be found [27]. Another study by Engstrand et al. [49] investigated the effect of β-TCP filler particles on the mechanical properties of a brushite forming cement (β-TCP-MCPM system). The results showed that the addition of low amounts of a filler (up to 10%) in combination with 0.8 M citric acid solution can effectively increase the powder to liquid ratio and hence decrease porosity from ~30% to ~23%. This strongly affects compressive strength of the cements with an increase from ~23 MPa (no filler and citric acid) to ~42 MPa. Space in cement pastes may also be filled by using hard agglomerates similar to civil engineering Portland cements as shown by Gu et al. [57]. In this study, the dispersion of 20% high-strength β-tricalcium phosphate granules with a size of 200–450 μm in the cement showed an increase of the compressive strength by 70%, while maintaining the rheological properties (injectability through 2.2 mm needle by applying a 5 kg weight on the syringe plunger) of the cement paste.

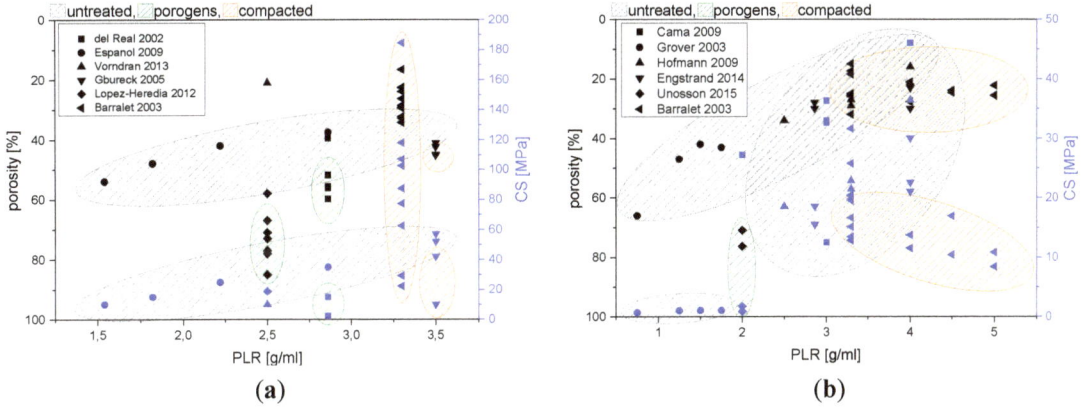

Figure 2. Correlation between powder to liquid ratio and porosity/compressive strength for (**a**) hydroxyapatite and (**b**) brushite cement from different studies. Cements were either set without compacting manipulation (untreated), processed by pre-compaction or porogens were added to create artificial macroporosity. Data were obtained from: (**a**) del Real 2002 [58], Espanol 2009 [47], Vorndran 2013 [59], Gbureck 2005 [27], Lopez-Heredia 2012 [60], Barralet 2003 [61], (**b**) Cama 2009 [62], Grover 2003 [51], Hofmann 2009 [50], Engstrand 2014 [49], Unosson 2015 [63], Barralet 2003 [64].

Caution must be exercised when comparing the obtained strength values from different studies, since many parameters during cement sample preparation and testing can affect the results. Unlike polymeric polymethylmethacrylate (PMMA) based bone cements [65], testing of calcium phosphate bone cement is not regulated, and our own experiences show that strength of set cement can vary by several times depending on the sample preparation and testing conditions. Generally, strength of dried samples is superior to that of (application near) wet specimen, mainly because water acts as a lubricant between the entangled crystals of the precipitated matrix. In addition, sample preparation may cause changes of cement porosity, e.g., by precompacting the paste in a mold. This ejects liquid from the paste (through the narrow gap between mold and plunger) leading to a lower porosity and hence a higher strength compared to uncompacted samples [61,66,67].

3. Fiber Reinforcement of CPC

Similar to reinforcement approaches of sintered hydroxyapatite ceramics [68], the addition of fibers to CPC is one of the most successful reinforcement technique [41,69]. The mechanical behavior of such fiber reinforced calcium phosphate cements (FRCPC) is a result of the complex interaction between all of the composite constituents. Contributions to the macroscopic behavior come from strength and stiffness of both fiber and cementitious matrix, matrix toughness, mechanical interaction between fibers and matrix as well as supplementary effects of polymeric additives or aggregates [69].

Fiber reinforcement studies have been performed with many different types of fibres (degradable *vs.* non-degradable, see Table 1 showing a strong increase of the mechanical strength depending on several parameters such as (1) matrix composition and strength, (2) fibre volume fraction, orientation, aspect ratio and tensile modulus as well as (3) the interface properties between matrix and fibres [69]. In addition to an increase of the bending strength from approx. 10–15 MPa for pure CPC to a maximum strength of 45 MPa (polyglactin fibers)—60 MPa (carbon fibers), especially the work of fracture for fiber reinforced cement composites usually increases by at least one order of magnitude (Table 1).

As illustrated in Figure 3, there is not only a complex interaction of factors, but in clinical application the properties of the fiber–cement composites are also time dependent since both the cement matrix and the fibers may degrade during tissue regeneration.

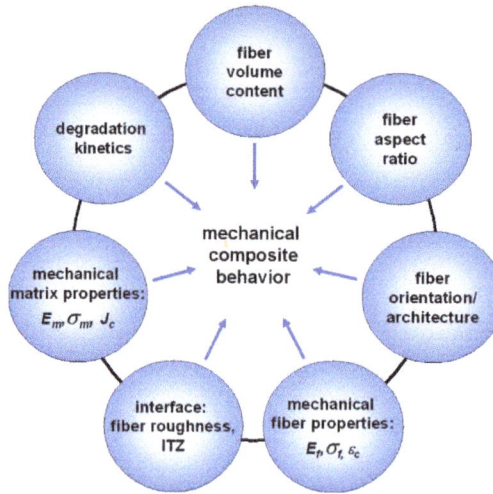

Figure 3. Interaction of material parameters which influence the time dependent mechanical behavior of the FRCPC composite. Reprinted with permission from [69].

Generally, the load-bearing capacity of fibers increases with their Young's modulus, whereas the maximum tensile stress within the fiber is determined by the fiber's modulus and the matrix strain [70]:

$$\sigma_f^{max} = E_f \varepsilon_m \tag{2}$$

When the composite is loaded, differences between Young's moduli of fiber and matrix lead to additional strain near the interface, mainly in the softer material [71]. The diameter of the fibers directly influences the total interface area between fibers and matrix for a given fiber volume fraction and affects both homogeneity and processability of the fiber–cement mixtures. Most biomedical composites are reinforced by discontinuous fibers. Their length and diameter are of great relevance, since substantial load has to be transferred from the matrix to the fiber via the interface for a reinforcing effect. Load is predominantly transferred by shear stresses at the lateral surface of the fibers rather than via the end faces of the fibers. Reinforcement effects are only observed, if the fiber length exceeds a critical value l_c, which can be calculated based on the assumption that the fiber is loaded up to the fracture strength:

$$l_c = \frac{d\sigma_{f,B}}{2\tau_i} \qquad (3)$$

where σ_{fB} and τ_i denote fracture strength of the fiber and shear stress at the interface and d is the diameter of the fiber. Optimum fiber volume content has been addressed by many researchers. Civil engineering concretes typically are reinforced with <5 vol% of steel, glass, natural or synthetic polymer fibers [72]. In many studies on medical FRCPC, the fiber content is one order of magnitude higher than in fiber reinforced cements for civil engineering. This is attributed to a frequently observed trend in FRCPC research [73–75] that strength and ductility of the composites increased with fiber content. Moderate load transfer due to non-optimized interface strength and low modulus of the fibers require such high fiber volume fraction. Furthermore, fiber costs are not such a limiting factor, at least in the research stage.

Table 1. Examples for the reinforcement of calcium phosphate cements with either degradable or non-degradable fibres. (3 p.b.: Three point bending, 4 p. b.: Four point bending. [#] UD: Unidirectional fibers. TTCP: Tetracalcium phosphate. HA$_w$: Hydroxyapatite whiskers) [69].

Composition Fiber/Additive/Matrix	Fiber Volume Fraction	Strength [MPa]	Work of Fracture [kJ/m²]	Test Method	Ref.
DEGRADABLE FIBRES					
HA matrix (TTCP + DCPA (+ Na$_2$HPO$_4$ − solution))	-	10–15	0.032–0.05	3 p. b.	[76,77]
Polyglactin 910/-/HA (TTCP + DCPA)	25 vol%	17.5–25	2.6–3.6	3 p. b.	[76]
Polyglactin 910/-/HA (TTCP + DCPA)	Mesh multilayer	8.5–24.5	0.75–3.1	3 p. b.	[78]
Polyglactin 910/chitosan lactate/HA (TTCP + DCPA)	45 vol%	41	11	3 p. b.	[74]
Polyglactin 910/chitosan lactate/HA (TTCP + DCPA)	Mesh multilayer	43	9.8	3 p. b.	[79]
Polyglactin 910/(poly(caprolactone))/ brushite (β-TCP + H$_3$PO$_4$)	24 vol% random short 6–25 long fibers UD [#]	7.5–20	n.a.	4 p. b.	[80]
NON-DEGRADABLE FIBRES					
Carbon/-/HA (TTCP + DCPA)	2–10 vol%	32–60	3.5–6.5	3 p. b.	[72]
CNT/-/HA (α-TCP + HA)	0.2–1.0 wt%	8.2–10.5	n.a.	3 p. b.	[81]
Aramid/-/macroporous HA (TTCP + DCPA + Na$_2$HPO$_4$)	6 vol%	7.5–13.5	0.8–6.5	3 p. b.	[75]
HA$_w$/-/HA (TTCP + DCPA)	10–40 vol%	5.4–7.4	57–102	4 p. b.	[82]

Biodegradable polylactic-co-glycolic acid (PLGA) is one of the most frequently used reinforcement fiber materials for CPC. For a high fiber volume, considerable increase in bending strength has been reported, e.g., from 2.7 MPa (unreinforced CPC) to 17.7 MPa for CPC with 45 vol% polyglactin fibers [74]. This strengthening effect can be further enhanced to 40.5 MPa by incorporation of chitosan lactate into the matrix. The synergistic strengthening of the CPC by chitosan and fibers together is stronger than from either suture fibers or chitosan alone [74], which was explained by both a much stronger cement matrix after chitosan incorporation supporting the suture fibers to better resist cracking as well as an improved suture-matrix bonding [72,74]. Generally, strength increases with length to diameter aspect ratio of fibers, whereas occurrence of fiber aggregation leading to inhomogenities in fiber distribution represents the practical upper limit for the aspect ratio. Xu and co-workers [72] systematically varied the length of carbon fibers in HA cement and found a continuous increase of strength between 3 and 75 mm fiber length (aspect ratio of 1000 and 9000), which was followed by a strength decrease for 200 mm long fibers (aspect ratio 25,000). While the use of such long fibers strongly alters the workability of cement pastes and impedes a minimal invasive application by injection, cement pastes filled with short fibers have been demonstrated to maintain their injection properties up to a fiber length of 1 mm and a fiber volume of 7.5% [83].

Cements may also be modified by using fiber meshes instead of single fibers, especially in cases where biomechanical stresses will primarily be oriented linearly or biaxially to the cement implant. Meshes provide a strength enhancement (in linear or biaxial direction) beyond that of randomly directed fibers and have the advantage that even thin bony structures (e.g., malar, orbital bones) or extensive cranial deficiencies can be reconstructed [30,78,84]. Von Gonten [30] could demonstrate that such a polyglactin mesh–CPC composites have a similar work of fracture to PMMA cements up to seven days' immersion in a buffered electrolyte, which was considered to have potential for structural repair of bone defects.

Most of the studies about FRCPC deal with both non-degradable fibers and with a poorly soluble hydroxyapatite cement matrix (see Table 1 and references [41,69]). This will initially result in long term stable cement composites with only minor changes of mechanical properties. However, even the slow matrix degradation by osteoclastic cells will dissect fibers in a longer time frame, which will be encapsulated in newly formed tissue with the possibility of foreign body reactions. Especially, approaches using technical fiber types (e.g., carbon fibers) or even carbon nanotubes are questionable regarding this point due to their low biocompatibility. The use of degradable fibers in FRCPC may solve this problem and the *in vivo* behavior of such FRCPC has been proven in various studies and is part of a recent review article by Krüger *et al.* [69]. However, at the same time, the use of degradable fibers will result in a time dependent loss of the reinforcement effect due to dissolution of the fibers in an aqueous environment. This effect of fiber degradation on the composite strength was simulated for polyglactin/PLGA fiber material by immersion of reinforced hydroxyapatite cement in a simulated physiological solution [73,74]. These studies confirmed a strength decrease of the reinforced composite after 4–6 weeks' immersion [73], which could be compensated by a simultaneous chitosan infiltration of the cement matrix [74]. As a solution to the above mentioned problems of either a loss of mechanical properties during fiber degradation or the release of non-degraded fibers, the different degradation kinetics of fibers and cement matrix need to be adjusted. An approach is the use of more degradable cements based on the formation of dicalcium phosphate dehydrate (brushite) in conjunction with PLGA fibers [82]. Other promising works are dealing with degradable magnesium phosphate cements, which are reinforced with magnesium metal wires [85]. Especially, the latter provides strong reinforcement effects with a maximum bending strength of the composites of 139 MPa.

4. Dual Setting Cements

While the addition of non-reactive polymers (e.g., collagen, chitosan, hyaluronic acid, cellulose derivates) [37–39,86,87] is commonly used for improving cement cohesion or biological performance, it is only of small benefit for the mechanical cement performance. A reduction of the brittleness of CPC and an increase of strength can be achieved by using polymeric compounds which can be cross-linked by binding calcium ions due to a high density of either carboxylic acid or organic phosphate moieties in the polymer chain, e.g., polyacrylic acid [88–92], polymethyl vinyl ether maleic acid [89,93], poly[bis(carboxylatophenoxy)phosphazene] [94] or poly(vinyl phosphonate) [95]. Such polymer modified cements set both by the aforementioned dissolution–precipitation mechanism as well as by deprotonating the organic acid following the formation of intra- or inter-chained bonding Ca^{2+}–Acid chelates [94] with a highly reactive cement component (mostly tetracalcium phosphate) from the cement powder. Processing of such polymer–cement composites is either possible by reacting an aqueous solution at ambient conditions with the cement powder or by reacting dry cement/polymer mixtures at elevated temperature/pressure in a solid state reaction.

An alternative approach is the use of reactive monomer systems, which are dissolved in the cement liquid and simultaneously react during cement setting by a gelation–polymerisation process. This forms within several minutes a hydrogel matrix with embedded cement particles, which are subsequently converted into the setting product by a continuous dissolution–precipitation reaction. The result is finally an interconnecting hydrogel matrix within the porous cement structure as shown in Figure 4. The advantages of this strategy are the possibility of a high polymer loading of the cement (and hence a large strength and toughness increase) as well as practically unchanged rheological properties of the fresh cement paste. Both are related to the fact that the dissolved monomers are commonly small, water miscible liquids with low viscosity such that even high monomer concentrations are not strongly altering the initial cement viscosity.

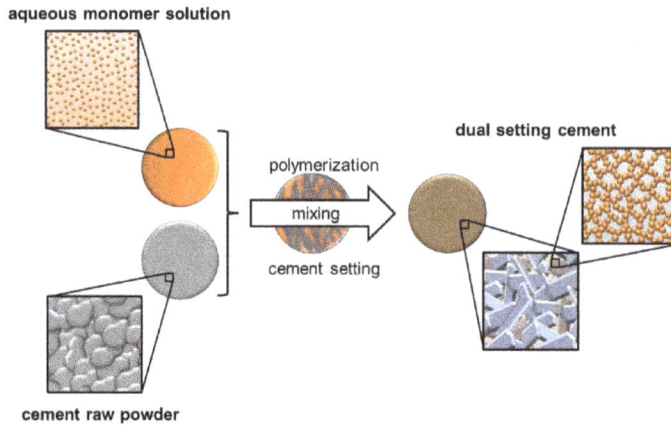

aqueous monomer solution

dual setting cement

polymerization

mixing

cement setting

cement raw powder

Figure 4. Hardening mechanism of dual-setting cements with the formation of interconnected matrices of hydrogel and precipitated cement crystals.

An early study regarding this concept was using mixtures of triethyleneglycol-dimethacrylate (TEGDMA), Bisphenol-A-dimethacrylate (Bis-GMA), hydroxyethylmethacrylate (HEMA) and 10% water as cement liquid. After adding this liquid to an equimolar mixture of TTCP and DCPA, polymerization was initiated by benzoylperoxide (coating on cement particles) and di-(N,N)-2-hydroxyethyl-p-toluidine (added to the cement liquid) [96]. Although this study revealed high diametral tensile strengths of up to 26 MPa for such composites, no hydroxyapatite formation of the cement was observed even after 30 d storage in water. This was attributed to the low water content of the cement liquid as well as to an adsorption of the hydrogel on TTCP/DCPA cement particles. This problem was overcome by Dos Santos *et al.* [43,44,97], who modified the cement liquid of an α-tricalcium phosphate cement by the addition of 5%–20% acrylamide and 1% ammonium polyacrylate. While the latter was used to increase initial cement viscosity and to reduce cement wash out in an aqueous environment, the acrylamide was chemically polymerised during cement setting by the use of 0.25% of N,N,N',N'-tetramethylethylenediamide (TEMED) and 0.01% ammonium persulfate. This modification doubled the compressive strength of the set cement from 25–50 MPa while the tensile strength was increased from 9 MPa to <21 MPa. At the same time, the high water content of the cement liquid enabled setting of α-TCP cement particles to calcium deficient hydroxyapatite within seven days. A follow up study by the same authors extended the approach to a fiber reinforced–double setting cement matrix [43] by using 1–4 wt% of 4–10 mm long carbon, nylon and polypropylene fibers. The addition of the fibers was found to reduce the compressive strength of the cement, which was attributed to an increase of porosity. However, this was compensated by strong increase of the cement toughness and tensile strength, which increased from 17–28 MPa.

A major concern about this matrix is the toxicity of non-reacted acrylamide monomer. To overcome this problem, Christel *el al.* [98] investigated the modification of alpha-tricalcium phosphate cement (α-TCP) with 30%–70% of less-toxic 2-hydroxyethylmethacrylate (HEMA), which also resulted in mechanically stable polymer-ceramic composites with interpenetrating organic and inorganic networks. Four-point bending strength was found to increase from 9 MPa to more than 14 MPa when using 50% HEMA, and the bending modulus decreased from 18 GPa to approx. 4 GPa. In addition, cement composites with $\geq$50% HEMA showed strongly reduced brittle fracture behaviour with an increase of the work of fracture by more than an order of magnitude. While bending of pure ceramic samples was possible only to a maximum of 0.07 mm, samples with 50% or more HEMA monomer had a higher flexibility and bending was possible for 0.4–1.5 mm until fracture. At the same time, the authors could prove that important cement characteristics such as compressive strength or injectability were not significantly altered by using HEMA modification. Another study by Wang *et al.* [45] used methacrylate modified dextran as monomer in a cement matrix of tetracalcium phosphate/dicalcium phosphate anhydrous in a weight ratio of 10:1—1:3 (CPC: Meth.-dextran). The results showed an increase of the compressive strength from 24–83 MPa for a polymer content of 16.7%, as well as improvement of the fracture energy by nearly two orders of magnitude from 0.084–8.35 kJ$\cdot$m^{-2}.

Apart from using organic monomers to form a second network in cements, it is also possible to apply the concept of dual setting cements to pure inorganic materials. Silica addition to CPC is a common approach to modify bioactivity, cement paste cohesion and mechanical cement properties [99,100]. However, most studies either used non-reactive silica fillers in cements [101–103] or they added non-reactive calcium phosphate particles to an *in situ* forming silica matrix prepared by sol-gel processing [104–106]. In contrast, Geffers *et al.* [48] modified a brushite forming cement paste with a second inorganic silica based precursor, which was obtained by pre-hydrolysing tetraethyl orthosilicate (TEOS) under acidic conditions. The addition of the cement powder (mixture of β-tricalcium phosphate and monocalcium phosphate) provoked an increase of the pH of the silica precursor such that cement setting by a dissolution–precipitation process, and the condensation reaction of the hydrolysed TEOS occurred simultaneously. This resulted in an interpenetrating phase composite material in which the macro pores of the cement (pore sizes in µm range) were infiltrated by the micro porous silica gel (pore sizes in nm range), leading to a higher density and a compressive strength approximately 5–10 times higher than the CPC reference.

5. Conclusions and Outlook

This article features reinforcement strategies of biocements to improve their strength and toughness for an application at load-bearing defect sides. While porosity reduction is based on the optimization of an intrinsic cement property leading to higher strength, the addition of fibers or the creation of dual setting cement matrixes are extrinsic approaches not only improving strength but also toughness of the matrix. Surprisingly, most studies devoted to the mechanical properties of calcium phosphate biocements only deal with one of the presented strategies. Here, the simultaneous application of the different methods will definitely bring further improvements such that those optimized cements can likely be applied for load bearing defects. Desired mechanical properties would be likely similar to those of polymeric PMMA cements (bending strength $\geq$50 MPa, bending modulus $\geq$1800 MPa) compressive strength $\geq$70 MPa according to ISO 5833:2002 [107]), whereas few studies have already reached or even exceeded one of these parameters [27,85]. However, practically all strength values for CPC in literature were obtained by test methods under static conditions and there are only few reports dealing with the fatigue properties of calcium phosphate cements in load-bearing defect models [108,109]. Hence, testing of cement strength under cyclic loading is one of the most important parameters which needs to be addressed in future research. In addition, since most of the studies on mechanically reinforced biocements were performed with only slowly degradable hydroxyapatite cement matrices and poorly or even non-degradable additives (fibers, polymers), the major challenge for the future is a transfer of the presented concepts to fully degradable materials. This is demanding since degradable cements based on the formation of brushite have harsh setting conditions (low pH, heat release, fast crystallization) and consume a considerable amount of water during setting. Especially, the latter may interfere with the formation of a second hydrogel phase, since the formation of hydrogel and hydrated cement setting product will compete for the available water in the cement liquid.

Acknowledgments

The authors would like to acknowledge financial support from the Deutsche Forschungsgemeinschaft (DFG GB1/20-1 and DFG GR3232/3-1).

Author Contributions

Martha Geffers is a PhD student and has performed the literature search for this review article and has written major parts of the manuscript. Uwe Gbureck and Jürgen Groll are her PhD supervisors and have supported with writing and correcting the manuscript.

Conflicts of Interest

The authors declare no conflict of interest.

References

1. Bohner, M.; Gbureck, U.; Barralet, J.E. Technological issues for the development of more efficient calcium phosphate bone cements: A critical assessment. *Biomaterials* **2005**, *26*, 6423–6429.
2. Dorozhkin, S.V. Calcium orthophosphate cements for biomedical application. *J. Mater. Sci.* **2008**, *43*, 3028–3057.
3. Chow, L.C. Calcium phosphate cements. *Monogr. Oral Sci.* **2001**, *18*, 148–163.
4. Nancollas, G.H.; Zawacki, S.J. Calcium phosphate mineralization. *Connect. Tissue Res.* **1989**, *21*, 239–244.
5. Tamimi, F.; Sheikh, Z.; Barralet, J. Dicalcium phosphate cements: Brushite and monetite. *Acta Biomater.* **2012**, *8*, 474–487.
6. Sahin, E.; Ciftcioglu, M. Monetite promoting effect of NaCl on brushite cement setting kinetics. *J. Mater. Chem. B* **2013**, *1*, 2943–2950.
7. Bohner, M.; vanLanduyt, P.; Merkle, H.P.; Lemaitre, J. Composition effects on the pH of a hydraulic calcium phosphate cement. *J. Mater. Sci. Mater. Med.* **1997**, *8*, 675–681.
8. Bohner, M. Design of ceramic-based cements and putties for bone graft substitution. *Eur. Cells Mater.* **2010**, *20*, 1–12.

9. Zhang, J.T.; Liu, W.Z.; Schnitzler, V.; Tancret, F.; Bouler, J.M. Calcium phosphate cements for bone substitution: Chemistry, handling and mechanical properties. *Acta Biomater.* **2014**, *10*, 1035–1049.

10. Constantz, B.R.; Barr, B.M.; Ison, I.C.; Fulmer, M.T.; Baker, J.; McKinney, L.A.; Goodman, S.B.; Gunasekaren, S.; Delaney, D.C.; Ross, J.; Poser, R.D. Histological, chemical, and crystallographic analysis of four calcium phosphate cements in different rabbit osseous sites. *J. Biomed. Mater. Res.* **1998**, *43*, 451–461.

11. Apelt, D.; Theiss, F.; El-Warrak, A.O.; Zlinszky, K.; Bettschart-Wolfisberger, R.; Bohner, M.; Matter, S.; Auer, J.A.; von Rechenberg, B. *In vivo* behavior of three different injectable hydraulic calcium phosphate cements. *Biomaterials* **2004**, *25*, 1439–1451.

12. Tas, A.C. The use of physiological solutions or media in calcium phosphate synthesis and processing. *Acta Biomater.* **2014**, *10*, 1771–1792.

13. Grossardt, C.; Ewald, A.; Grover, L.M.; Barralet, J.E.; Gbureck, U. Passive and active *in vitro* resorption of calcium and magnesium phosphate cements by osteoclastic cells. *Tissue Eng. A* **2010**, *16*, 3687–3695.

14. Detsch, R.; Mayr, H.; Ziegler, G. Formation of osteoclast-like cells on HA and TCP ceramics. *Acta Biomater.* **2008**, *4*, 139–148.

15. Wenisch, S.; Stahl, J.P.; Horas, U.; Heiss, C.; Kilian, O.; Trinkaus, K.; Hild, A.; Schnettler, R. *In vivo* mechanisms of hydroxyapatite ceramic degradation by osteoclasts: Fine structural microscopy. *J. Biomed. Mater. Res. A* **2003**, *67*, 713–718.

16. Grover, L.M.; Gbureck, U.; Wright, A.J.; Tremayne, M.; Barralet, J.E. Biologically mediated resorption of brushite cement *in vitro*. *Biomaterials* **2006**, *27*, 2178–2185.

17. Holzapfel, B.M.; Reichert, J.C.; Schantz, J.T.; Gbureck, U.; Rackwitz, L.; Noth, U.; Jakob, F.; Rudert, M.; Groll, J.; Hutmacher, D.W. How smart do biomaterials need to be? A translational science and clinical point of view. *Adv. Drug Deliv. Rev.* **2013**, *65*, 581–603.

18. Theiss, F.; Apelt, D.; Brand, B.A.; Kutter, A.; Zlinszky, K.; Bohner, M.; Matter, S.; Frei, C.; Auer, J.A.; von Rechenberg, B. Biocompatibility and resorption of a brushite calcium phosphate cement. *Biomaterials* **2005**, *26*, 4383–4394.

19. Tamimi, F.; Torres, J.; Lopez-Cabarcos, E.; Bassett, D.C.; Habibovic, P.; Luceron, E.; Barralet, J.E. Minimally invasive maxillofacial vertical bone augmentation using brushite based cements. *Biomaterials* **2009**, *30*, 208–216.

20. Penel, G.; Leroy, N.; van Landuyt, P.; Flautre, B.; Hardouin, P.; Lemaitre, J.; Leroy, G. Raman microspectrometry studies of brushite cement: *In vivo* evolution in a sheep model. *Bone* **1999**, *25*, 81S–84S.

21. Kuemmerle, J.M.; Oberle, A.; Oechslin, C.; Bohner, M.; Frei, C.; Boecken, I.; von Rechenberg, B. Assessment of the suitability of a new brushite calcium phosphate cement for cranioplasty—An experimental study in sheep. *J. Cranio Maxillofac. Surg.* **2005**, *33*, 37–44.

22. Bohner, M.; Lemaitre, J. Can bioactivity be tested *in vitro* with SBF solution? *Biomaterials* **2009**, *30*, 2175–2179.

23. Lu, X.; Leng, Y. Theoretical analysis of calcium phosphate precipitation in simulated body fluid. *Biomaterials* **2005**, *26*, 1097–1108.

24. Klammert, U.; Ignatius, A.; Wolfram, U.; Reuther, T.; Gbureck, U. *In vivo* degradation of low temperature calcium and magnesium phosphate ceramics in a heterotopic model. *Acta Biomater.* **2011**, *7*, 3469–3475.

25. Kanter, B.; Geffers, M.; Ignatius, A.; Gbureck, U. Control of *in vivo* mineral bone cement degradation. *Acta Biomater.* **2014**, *10*, 3279–3287.

26. Bohner, M.; Theiss, F.; Apelt, D.; Hirsiger, W.; Houriet, R.; Rizzoli, G.; Gnos, E.; Frei, C.; Auer, J.A.; von Rechenberg, B. Compositional changes of a dicalcium phosphate dihydrate cement after implantation in sheep. *Biomaterials* **2003**, *24*, 3463–3474.

27. Gbureck, U.; Spatz, K.; Thull, R.; Barralet, J.E. Rheological enhancement of mechanically activated alpha-tricalcium phosphate cements. *J. Biomed. Mater. Res. B Appl. Biomater.* **2005**, *73*, 1–6.

28. Charriere, E.; Terrazzoni, S.; Pittet, C.; Mordasini, P.; Dutoit, M.; Lemaitre, J.; Zysset, P. Mechanical characterization of brushite and hydroxyapatite cements. *Biomaterials* **2001**, *22*, 2937–2945.

29. Dorozhkin, S.V. Calcium orthophosphate-based biocomposites and hybrid biomaterials. *J. Mater. Sci.* **2009**, *44*, 2343–2387.

30. Von Gonten, A.S.; Kelly, J.R.; Antonucci, J.M. Load-bearing behavior of a simulated craniofacial structure fabricated from a hydroxyapatite cement and bioresorbable fiber-mesh. *J. Mater. Sci. Mater. Med.* **2000**, *11*, 95–100.

31. Blattert, T.R.; Jestaedt, L.; Weckbach, A. Suitability of a calcium phosphate cement in osteoporotic vertebral body fracture augmentation a controlled, randomized, clinical trial of balloon kyphoplasty comparing calcium phosphate versus polymethylmethacrylate. *Spine* **2009**, *34*, 108–114.

32. Maestretti, G.; Cremer, C.; Otten, P.; Jakob, R.P. Prospective study of standalone balloon kyphoplasty with calcium phosphate cement augmentation in traumatic fractures. *Eur. Spine J.* **2007**, *16*, 601–610.

33. Tarsuslugil, S.M.; O'Hara, R.M.; Dunne, N.J.; Buchanan, F.J.; Orr, J.F.; Barton, D.C.; Wilcox, R.K. Development of calcium phosphate cement for the augmentation of traumatically fractured porcine specimens using vertebroplasty. *J. Biomech.* **2013**, *46*, 711–715.

34. Heini, P.F. Vertebroplastie: Ein Update. *Orthopäde* **2010**, *39*, 658–664.

35. Grafe, I.A.; Baier, M.; Noldge, G.; Weiss, C.; da Fonseca, K.; Hillmeier, J.; Libicher, M.; Rudofsky, G.; Metzner, C.; Nawroth, P.; *et al.* Calcium-phosphate and polymethylmethacrylate cement in long-term outcome after kyphoplasty of painful osteoporotic vertebral fractures. *Spine* **2008**, *33*, 1284–1290.

36. Kiyasu, K.; Takemasa, R.; Ikeuchi, M.; Tani, T. Differential blood contamination levels and powder–liquid ratios can affect the compressive strength of calcium phosphate cement (CPC): A study using a transpedicular vertebroplasty model. *Eur. Spine J.* **2013**, *22*, 1643–1649.

37. Moreau, J.L.; Weir, M.D.; Xu, H.H.K. Self-setting collagen-calcium phosphate bone cement: Mechanical and cellular properties. *J. Biomed. Mater. Res. A* **2009**, *91*, 605–613.

38. Schneiders, W.; Reinstorf, A.; Biewener, A.; Serra, A.; Grass, R.; Kinscher, M.; Heineck, J.; Rehberg, S.; Zwipp, H.; Rammelt, S. *In Vivo* Effects of modification of hydroxyapatite/collagen composites with and without chondroitin sulphate on bone remodeling in the sheep tibia. *J. Orthop. Res.* **2009**, *27*, 15–21.

39. Tamimi, F.; Kumarasami, B.; Doillon, C.; Gbureck, U.; le Nihouannen, D.; Cabarcos, E.L.; Barralet, J.E. Brushite-collagen composites for bone regeneration. *Acta Biomater.* **2008**, *4*, 1315–1321.

40. O'Hara, R.M.; Orr, J.F.; Buchanan, F.J.; Wilcox, R.K.; Barton, D.C.; Dunne, N.J. Development of a bovine collagen-apatitic calcium phosphate cement for potential fracture treatment through vertebroplasty. *Acta Biomater.* **2012**, *8*, 4043–4052.

41. Canal, C.; Ginebra, M.P. Fibre-reinforced calcium phosphate cements: A review. *J. Mechan. Behav. Biomed. Mater.* **2011**, *4*, 1658–1671.

42. Dos Santos, L.A.; de Oliveira, L.C.; Rigo, E.C.D.; Carrodeguas, R.G.; Boschi, A.O.; de Arruda, A.C.F. Fiber reinforced calcium phosphate cement. *Artif. Org.* **2000**, *24*, 212–216.

43. Dos Santos, L.A.; Carrodeguas, R.G.; Boschi, A.O.; de Arruda, A.C.F. Fiber-enriched double-setting calcium phosphate bone cement. *J. Biomed. Mater. Res. A* **2003**, *65*, 244–250.

44. Dos Santos, L.A.; Carrodeguas, R.G.; Boschi, A.O.; de Arruda, A.C.F. Dual-setting calcium phosphate cement modified with ammonium polyacrylate. *Artif. Org.* **2003**, *27*, 412–418.

45. Wang, J.; Liu, C.S.; Liu, Y.F.; Zhang, S. Double-network interpenetrating bone cement via in situ hybridization protocol. *Adv. Funct. Mater.* **2010**, *20*, 3997–4011.

46. Barralet, J.E.; Gaunt, T.; Wright, A.J.; Gibson, I.R.; Knowles, J.C. Effect of porosity reduction by compaction on compressive strength and microstructure of calcium phosphate cement. *J. Biomed. Mater. Res.* **2002**, *63*, 1–9.

47. Espanol, M.; Perez, R.A.; Montufar, E.B.; Marichal, C.; Sacco, A.; Ginebra, M.P. Intrinsic porosity of calcium phosphate cements and its significance for drug delivery and tissue engineering applications. *Acta Biomater.* **2009**, *5*, 2752–2762.

48. Geffers, M.; Barralet, J.E.; Groll, J.; Gbureck, U. Dual-setting brushite-silica gel cements. *Acta Biomater.* **2015**, *11*, 467–476.

49. Engstrand, J.; Persson, C.; Engqvist, H. The effect of composition on mechanical properties of brushite cements. *J. Mech. Behav. Biomed. Mater.* **2014**, *29*, 81–90.

50. Hofmann, M.P.; Mohammed, A.R.; Perrie, Y.; Gbureck, U.; Barralet, J.E. High-strength resorbable brushite bone cement with controlled drug-releasing capabilities. *Acta Biomater.* **2009**, *5*, 43–49.

51. Grover, L.M.; Knowles, J.C.; Fleming, G.J.P.; Barralet, J.E. *In vitro* ageing of brushite calcium phosphate cement. *Biomaterials* **2003**, *24*, 4133–4141.

52. Ginebra, M.P.; Delgado, J.A.; Harr, I.; Almirall, A.; del Valle, S.; Planell, J.A. Factors affecting the structure and properties of an injectable self-setting calcium phosphate foam. *J. Biomed. Mater. Res. A* **2007**, *80*, 351–361.

53. Burguera, E.F.; Guitian, F.; Chow, L.C. A water setting tetracalcium phosphate-dicalcium phosphate dihydrate cement. *J. Biomed. Mater. Res. A* **2004**, *71*, 275–282.

54. Unosson, J.E.; Persson, C.; Engqvist, H. An evaluation of methods to determine the porosity of calcium phosphate cements. *J. Biomed. Mater. Res. B Appl. Biomater.* **2015**, *103*, 62–71.

55. Bohner, M.; Baroud, G. Injectability of calcium phosphate pastes. *Biomaterials* **2005**, *26*, 1553–1563.

56. Barralet, J.E.; Tremayne, M.; Lilley, K.J.; Gbureck, U. Modification of calcium phosphate cement with alpha-hydroxy acids and their salts. *Chem. Mater.* **2005**, *17*, 1313–1319.

57. Gu, T.; Shi, H.; Ye, J. Reinforcement of calcium phosphate cement by incorporating with high-strength ß-tricalcium phosphate aggregates. *J. Biomed. Mater. Res. B Appl. Biomater.* **2012**, *100*, 350–359.

58. Del Real, R.P.; Wolke, J.G.C.; Vallet-Regi, M.; Jansen, J.A. A new method to produce macropores in calcium phosphate cements. *Biomaterials* **2002**, *23*, 3673–3680.

59. Vorndran, E.; Geffers, M.; Ewald, A.; Lemm, M.; Nies, B.; Gbureck, U. Ready-to-use injectable calcium phosphate bone cement paste as drug carrier. *Acta Biomater.* **2013**, *9*, 9558–9567.

60. Lopez-Heredia, M.A.; Sariibrahimoglu, K.; Yang, W.; Bohner, M.; Yamashita, D.; Kunstar, A.; van Apeldoorn, A.A.; Bronkhorst, E.M.; Lanao, R.P.F.; Leeuwenburgh, S.C.G.; *et al.* Influence of the pore generator on the evolution of the mechanical properties and the porosity and interconnectivity of a calcium phosphate cement. *Acta Biomater.* **2012**, *8*, 404–414.

61. Barralet, J.E.; Hofmann, M.; Grover, L.M.; Gbureck, U. High-strength apatitic cement by modification with alpha-hydroxy acid salts. *Adv. Mater.* **2003**, *15*, 2091–2094.

62. Cama, G.; Barberis, F.; Botter, R.; Cirillo, P.; Capurro, M.; Quarto, R.; Scaglione, S.; Finocchio, E.; Mussi, V.; Valbusa, U. Preparation and properties of macroporous brushite bone cements. *Acta Biomater.* **2009**, *5*, 2161–2168.

63. Unosson, J.E.; Montufar, E.B.; Engqvist, H.; Ginebra, M.P.; Persson, C. Brushite foams—The effect of Tween 80 and Pluronic F-127 on foam porosity and mechanical properties. *J. Biomed. Res. B Appl. Biomater.* **2015**, doi:10.1002/jbm.b.33355.

64. Barralet, J.E.; Grover, L.M.; Gbureck, U. Ionic modification of calcium phosphate cement viscosity. Part II: hypodermic injection and strength improvement of brushite cement. *Biomaterials* **2004**, *25*, 2197–2203.

65. Gravius, S.; Wirtz, D.C.; Marx, R.; Maus, U.; Andereya, S.; Mueller-Rath, R.; Mumme, T. Mechanical *in vitro* testing of fifteen commercial bone cements based on polymethylmethacrylate. *Z. Orthop. Unfall.* **2007**, *145*, 579–585.

66. Chow, L.C.; Hirayama, S.; Takagi, S.; Parry, E. Diametral tensile strength and compressive strength of a calcium phosphate cement: Effect of applied pressure. *J. Biomed. Mater. Res.* **2000**, *53*, 511–517.

67. Ishikawa, K.; Asaoka, K. Estimation of ideal mechanical strength and critical porosity of calcium-phosphate cement. *J. Biomed. Mater. Res.* **1995**, *29*, 1537–1543.

68. Dewith, G.; Corbijn, A.J. Metal fiber reinforced hydroxy-apatite ceramics. *J. Mater. Sci.* **1989**, *24*, 3411–3415.

69. Krueger, R.; Groll, J. Fiber reinforced calcium phosphate cements—On the way to degradable load bearing bone substitutes? *Biomaterials* **2012**, *33*, 5887–5900.

70. Rösler, J.; Harders, H.; Bäker, M. *Mechanisches Verhalten der Werkstoffe*, 2nd ed.; Teubner Verlag: Wiesbaden, Germany, 2008.

71. Callister, W.D.; Rethwisch, D.G. *Materials Science and Engineering: An Introduction*, 8th ed.; John Wiley & Sons, Inc.: Hoboken, NJ, USA, 2010.

72. Brandt, A.M. *Cement-Based Composites—Materials, Mechanical Properties and Performance*, 2nd ed.; Taylor & Francis: Abingdon, UK, 2009.

73. Xu, H.H.K.; Eichmiller, F.C.; Barndt, P.R. Effects of fiber length and volume fraction on the reinforcement of calcium phosphate cement. *J. Mater. Sci. Mater. Med.* **2001**, *12*, 57–65.

74. Xu, H.H.K.; Eichmiller, F.C.; Giuseppetti, A.A. Reinforcement of a self-setting calcium phosphate cement with different fibers. *J. Biomed. Mater. Res.* **2000**, *52*, 107–114.

75. Zhang, Y.; Xu, H.H.K. Effects of synergistic reinforcement and absorbable fiber strength on hydroxyapatite bone cement. *J. Biomed. Mater. Res. A* **2005**, *75*, 832–840.

76. Xu, H.H.K.; Quinn, J.B.; Takagi, S.; Chow, L.C.; Eichmiller, F.C. Strong and macroporous calcium phosphate cement: Effects of porosity and fiber reinforcement on mechanical properties. *J. Biomed. Mater. Res.* **2001**, *57*, 457–466.

77. Xu, H.H.K.; Quinn, J.B. Calcium phosphate cement containing resorbable fibers for short-term reinforcement and macroporosity. *Biomaterials* **2002**, *23*, 193–202.

78. Xu, H.H.K.; Simon, C.G. Self-hardening calcium phosphate cement-mesh composite: Reinforcement, macropores, and cell response. *J. Biomed. Mater. Res.* A **2004**, *69*, 267–278.

79. Xu, H.H.K.; Quinn, J.B.; Takagi, S.; Chow, L.C. Synergistic reinforcement of in situ hardening calcium phosphate composite scaffold for bone tissue engineering. *Biomaterials* **2004**, *25*, 1029–1037.

80. Gorst, N.J.S.; Perrie, Y.; Gbureck, U.; Hutton, A.L.; Hofmann, M.P.; Grover, L.M.; Barralet, J.E. Effects of fibre reinforcement on the mechanical properties of brushite cement. *Acta Biomater.* **2006**, *2*, 95–102.

81. Zhao, P.; Sun, K.; Zhao, T.; Ren, X. Effect of CNTs on property of calcium phosphate cement. *Key Eng. Mater.* **2007**, *336–338*, 1606–1608.

82. Muller, F.A.; Gbureck, U.; Kasuga, T.; Mizutani, Y.; Barralet, J.E.; Lohbauer, U. Whisker-reinforced calcium phosphate cements. *J. Am. Ceram. Soc.* **2007**, *90*, 3694–3697.

83. Maenz, S.; Kunisch, E.; Muehlstaedt, M.; Boehm, A.; Kopsch, V.; Bossert, J.; Kinne, R.W.; Jandt, K.D. Enhanced mechanical properties of a novel, injectable, fiber-reinforced brushite cement. *J. Mech. Behav. Biomed. Mater.* **2014**, *39*, 328–338.

84. Weir, M.D.; Xu, H.H.K.; Simon, C.G. Strong calcium phosphate cement-chitosan-mesh construct containing cell-encapsulating hydrogel beads for bone tissue engineering. *J. Biomed. Mater. Res. A* **2006**, *77*, 487–496.

85. Krueger, R.; Seitz, J.-M.; Ewald, A.; Bach, F.-W.; Groll, J. Strong and tough magnesium wire reinforced phosphate cement composites for load-bearing bone replacement. *J. Mech. Behav. Biomed. Mater.* **2013**, *20*, 36–44.

86. Khairoun, I.; Driessens, F.C.M.; Boltong, M.G.; Planell, J.A.; Wenz, R. Addition of cohesion promoters to calcium phosphate cements. *Biomaterials* **1999**, *20*, 393–398.

87. Alkhraisat, M.H.; Rueda, C.; Marino, F.T.; Torres, J.; Jerez, L.B.; Gbureck, U.; Cabarcos, E.L. The effect of hyaluronic acid on brushite cement cohesion. *Acta Biomater.* **2009**, *5*, 3150–3156.

88. Chen, W.-C.; Ju, C.-P.; Wang, J.-C.; Hung, C.-C.; Lin, J.-H.C. Brittle and ductile adjustable cement derived from calcium phosphate cement/polyacrylic acid composites. *Dent. Mater.* **2008**, *24*, 1616–1622.

89. Watson, K.E.; Tenhuisen, K.S.; Brown, P.W. The formation of hydroxyapatite-calcium polyacrylate composites. *J. Mater. Sci. Mater. Med.* **1999**, *10*, 205–213.

90. Khashaba, R.M.; Moussa, M.; Koch, C.; Jurgensen, A.R.; Missimer, D.M.; Rutherford, R.L.; Chutkan, N.B.; Borke, J.L. Preparation, physical-chemical characterization, and cytocompatibility of polymeric calcium phosphate cements. *Int. J. Biomater.* **2011**, *2011*, 467641.

91. Majekodunmi, A.O.; Deb, S.; Nicholson, J.W. Effect of molecular weight and concentration of poly(acrylic acid) on the formation of a polymeric calcium phosphate cement. *J. Mater. Sci. Mater. Med.* **2003**, *14*, 747–752.

92. Majekodunmi, A.O.; Deb, S. Poly(acrylic acid) modified calcium phosphate cements: The effect of the composition of the cement powder and of the molecular weight and concentration of the polymeric acid. *J. Mater. Sci. Mater. Med.* **2007**, *18*, 1883–1888.

93. Matsuya, Y.; Antonucci, J.M.; Matsuya, S.; Takagi, S.; Chow, L.C. Polymeric calcium phosphate cements derived from poly(methyl vinyl ether-maleic acid). *Dent. Mater.* **1996**, *12*, 2–7.

94. Greish, Y.E.; Brown, P.W.; Bender, J.D.; Allcock, H.R.; Lakshmi, S.; Laurencin, C.T. Hydroxyapatite-polyphosphazane composites prepared at low temperatures. *J. Am. Ceram. Soc.* **2007**, *90*, 2728–2734.

95. Greish, Y.E.; Brown, P.W. Chemically formed HAp-Ca poly(vinyl phosphonate) composites. *Biomaterials* **2001**, *22*, 807–816.

96. Sugawara, A.; Antonucci, J.M.; Takagi, S.; Chow, L.C.; Ohashi, M. Formation of hydroxyapatite in hydrogels from tetracalcium phosphate/dicalcium phosphate mixtures. *J. Nihon Univ. Sch. Dent.* **1989**, *31*, 372–381.

97. Rigo, E.C.S.; dos Santos, L.A.; Vercik, L.C.O.; Carrodeguas, R.G.; Boschi, A.O. alpha-tricalcium phosphate- and tetracalcium phosphate/dicalcium phosphate-based dual setting cements. *Lat. Am. Appl. Res.* **2007**, *37*, 267–274.

98. Christel, T.; Kuhlmann, M.; Vorndran, E.; Groll, J.; Gbureck, U. Dual setting alpha-tricalcium phosphate cements. *J. Mater. Sci. Mater. Med.* **2013**, *24*, 573–581.

99. Zhou, H.; Luchini, T.J.F.; Agarwal, A.K.; Goel, V.K.; Bhaduri, S.B. Development of monetite-nanosilica bone cement: A preliminary study. *J. Biomed. Mater. Res. B Appl. Biomater.* **2014**, *102*, 1620–1626.

100. Ahn, G.; Lee, J.Y.; Seol, D.-W.; Pyo, S.G.; Lee, D. The effect of calcium phosphate cement-silica composite materials on proliferation and differentiation of pre-osteoblast cells. *Mater. Lett.* **2013**, *109*, 302–305.

101. Van den Vreken, N.M.F.; de Canck, E.; Ide, M.; Lamote, K.; van der Voort, P.; Verbeeck, R.M.H. Calcium phosphate cements modified with pore expanded SBA-15 materials. *J. Mater. Chem.* **2012**, *22*, 14502–14509.

102. Hesaraki, S.; Alizadeh, M.; Borhan, S.; Pourbaghi-Masouleh, M. Polymerizable nanoparticulate silica-reinforced calcium phosphate bone cement. *J. Biomed. Mater. Res. B Appl. Biomater.* **2012**, *100*, 1627–1635.

103. Hamdan Alkhraisat, M.; Rueda, C.; Blanco Jerez, L.; Marino, F.T.; Torres, J.; Gbureck, U.; Lopez Cabarcos, E. Effect of silica gel on the cohesion, properties and biological performance of brushite cement. *Acta Biomater.* **2010**, *6*, 257–265.

104. Andersson, J.; Areva, S.; Spliethoff, B.; Linden, M. Sol-gel synthesis of a multifunctional, hierarchically porous silica/apatite composite. *Biomaterials* **2005**, *26*, 6827–6835.

105. Heinemann, S.; Heinemann, C.; Bernhardt, R.; Reinstorf, A.; Nies, B.; Meyer, M.; Worch, H.; Hanke, T. Bioactive silica-collagen composite xerogels modified by calcium phosphate phases with adjustable mechanical properties for bone replacement. *Acta Biomater.* **2009**, *5*, 1979–1990.

106. Sousa, A.; Souza, K.C.; Sousa, E.M.B. Mesoporous silica/apatite nanocomposite: Special synthesis route to control local drug delivery. *Acta Biomater.* **2008**, *4*, 671–679.

107. The International Organization for Standardization (ISO). *International Standard ISO 5833 Implants for Surgery—Acrylic Resin Cements*; ISO: Geneva, Switzerland, 2002.

108. Wilke, H.-J.; Mehnert, U.; Claes, L.E.; Bierschneider, M.M.; Jaksche, H.; Boszczyk, B.M. Biomechanical evaluation of vertebroplasty and kyphoplasty with polymethyl methacrylate or calcium phosphate cement under cyclic loading. *Spine* **2006**, *31*, 2934–2941.

109. Lewis, G.; Schwardt, J.D.; Slater, T.A.; Janna, S. Evaluation of a synthetic vertebral body augmentation model for rapid and reliable cyclic compression life testing of materials for balloon kyphoplasty. *J. Biomed. Mater. Res. B Appl. Biomater.* **2008**, *87*, 179–188.

Mechanical Properties and Cytocompatibility Improvement of Vertebroplasty PMMA Bone Cements by Incorporating Mineralized Collagen

Hong-Jiang Jiang, Jin Xu, Zhi-Ye Qiu, Xin-Long Ma, Zi-Qiang Zhang, Xun-Xiang Tan, Yun Cui and Fu-Zhai Cui

Abstract: Polymethyl methacrylate (PMMA) bone cement is a commonly used bone adhesive and filling material in percutaneous vertebroplasty and percutaneous kyphoplasty surgeries. However, PMMA bone cements have been reported to cause some severe complications, such as secondary fracture of adjacent vertebral bodies, and loosening or even dislodgement of the set PMMA bone cement, due to the over-high elastic modulus and poor osteointegration ability of the PMMA. In this study, mineralized collagen (MC) with biomimetic microstructure and good osteogenic activity was added to commercially available PMMA bone cement products, in order to improve both the mechanical properties and the cytocompatibility. As the compressive strength of the modified bone cements remained well, the compressive elastic modulus could be significantly down-regulated by the MC, so as to reduce the pressure on the adjacent vertebral bodies. Meanwhile, the adhesion and proliferation of pre-osteoblasts on the modified bone cements were improved compared with cells on those unmodified, such result is beneficial for a good osteointegration formation between the bone cement and the host bone tissue in clinical applications. Moreover, the modification of the PMMA bone cements by adding MC did not significantly influence the injectability and processing times of the cement.

Reprinted from *Materials*. Cite as: Jiang, H.-J.; Xu, J.; Qiu, Z.-Y.; Ma, X.-L.; Zhang, Z.-Q.; Tan, X.-X.; Cui, Y.; Cui, F.-Z. Mechanical Properties and Cytocompatibility Improvement of Vertebroplasty PMMA Bone Cements by Incorporating Mineralized Collagen. *Materials* **2015**, *8*, 2616-2634.

1. Introduction

Vertebral compression fractures (VCF) are one of the most common fractures for the elders with osteoporosis. In the United States, it was reported that about 25% of postmenopausal women suffered from VCF, and such morbidity rate was estimated to be 40% for those women over 80 years old [1]. With the current accelerated trend of the aging of the world population, the occurrence of VCF will continue increasing. Besides osteoporosis, VCF can also be induced by other disease, such as osteogenesis imperfecta [2], spinal tumors [3], and so on.

Percutaneous vertebroplasty (PVP) and percutaneous kyphoplasty (PKP) are the major applications of the polymethyl methacrylate (PMMA) bone cement in the treatment of VCF. In either PVP or PKP, the bone cement is injected into the vertebral body for the augmentation of the fractured vertebral body. The immediate effect and safety of the PMMA bone cements used in PVP and PKP have been deeply investigated and verified by long-term clinical practices. However, existing commercially available PMMA bone cement products for PVP and PKP have been reported to cause some complications, mainly includes secondary fractures of the adjacent vertebral bodies, and loosening or even dislodgement of the set PMMA bone cement, due to the high elastic modulus and bioinert of the PMMA.

The compressive elastic modulus of normal human vertebral body is 50–800 MPa [4–6], while the PMMA bone cements form hard solid body with an elastic modulus of 2000–3700 MPa [7,8], which is much higher than that of normal human vertebral body. The vertebral body filled with PMMA bone cement has a significantly higher stiffness than the adjacent segments, and the resulting stress concentration will easily cause secondary fracture on adjacent vertebral bodies and endplate near the surgical segment [9,10]. The incidence of the secondary fracture of the adjacent bodies after PVP and PKP was reported as high as 7%–20% [11], which is 4.62 times than those occurred on other segments [12].

On the other hand, PMMA is a bioinert material that neither form chemical bonding, nor form osteointegration with the bone tissue at the implant site [13], resulting in obvious interface and weak combination strength between the bone cement and the host bone. Micro motion cannot be avoided under such weak combination in daily activities, and small wear debris produced by the micro motion would cause osteolysis and further aseptic loosening or even dislodgement of the bone cement implant [14,15].

A new PVP or PKP surgery, or even more are necessary for the treatment of the secondary fracture on the adjacent vertebral body, which increase pain and economic burden of the patient. For serious loosening or dislodgement of the bone cement, further revision surgery is inevitable. Therefore, the modification of PMMA bone cement for the treatment of VCF is important and extremely urgent for clinical applications. Many approaches were tried to improve mechanical properties and/or biocompatibility of the PMMA bone cement by, for example, adding biocompatible hydroxyapatite (HA) powder, or partially modifying methyl methacrylate (MMA) monomer. However, ideal results were not achieved by previous reported modification studies, since the compressive strength decreased too much to meet the requirement of corresponding standard (ISO 5833-2002), or the compressive elastic modulus increased rather than decreased, or the injectability was limited and is not available in the use of PVP or PKP.

Mineralized collagen (MC) is a biomimetic biomaterial with the same chemical composition and hierarchical structures to natural bone tissue. The MC is usually prepared by an *in vitro* biomimetic mineralization process that is similar to the formation of natural bone tissue [16,17]. Within the MC, the organic type-I collagen is orderly arranged with the inorganic nano-sized HA [16]. Many laboratory studies and clinical practices have demonstrated that the MC could be used to fill bone defects and is able to promote new bone formation at the bone defect sites [18,19].

In this study, MC particles were added to commercially available PMMA bone cement products to improve both the mechanical properties and the cytocompatibility. The modification parameters, including MC particle size range and additive percentage were investigated for each PMMA bone cement. Injectability, mechanical properties, maximum temperature and setting time were tested to determine the modification availability and effectiveness. Cell experiments were performed to evaluate cytocompatibility improvement of the modification by observing adhesion and quantifying proliferation of pre-osteoblasts on the modified bone cements.

2. Materials and Methods

2.1. PMMA Bone Cement Products

Three commercially available PMMA bone cement products for PVP and PKP were purchased. The three products were Osteopal® V (Heraeus Medical GmbH, Hanau, Germany), Mendec® Spine (Tecres S. P. A., Verona, Italy) and Spineplex™ (Stryker Instruments, Kalamazoo, MI, USA). All these three bone cements were certified by medical administration of many countries and regions, and have been used in clinics for many years.

2.2. Preparation of MC Particles

MC particles used for the modification of the PMMA bone cements were made from a commercially available artificial bone graft "BonGold" produced by Beijing Allgens Medical Science and Technology Co., Ltd. (Beijing, China). The MC bone grafts were prepared by following main steps described in [20]. Briefly, water-soluble calcium salt solution and phosphate salt solution were added into acidic collagen solution to form MC deposition by adjusting pH value and temperature of the reaction system. This step is a biomineralization process, which was similar to the mineralization process of the natural bone tissue that the HA crystal nucleation and growth were directed by collagen molecular templates. The deposition was then collected by centrifugation and freeze-dried to obtain MC bone graft product.

The MC bone graft was ground into small particles and screened out 4 groups with different particle sizes by sieving. The particle size range for each group was: <200 μm, 200–300 μm, 300–400 μm, and 400–500 μm, respectively. Since the inner diameter of bone filler device for delivering bone cement in PVP and PKP are usually 2.5–4.0 mm, MC particles less than 500 μm were used in this modification study.

2.3. Addition Methods of the MC

MC particles with different addition amounts and size ranges were added into the bone cements for the modification. In our preliminary experiments, too much MC addition (>20wt % of the powder part of the bone cement) would lead to hard stirring of the bone cement and losing injectability. Therefore, 4 addition amount groups, 5 wt%, 10 wt%, 15 wt%, and 20 wt% of the powder part of the bone cement, were studied for each particle size range.

In the modification process, powder and liquid parts of the bone cement were firstly mixed for 30 s to form a uniform flowing phase, and MC particles were then added into with rapid stirring for 30 s to ensure homogeneous distribution within the bone cement. There were two adding methods for the MC particles. One is direct addition of a certain amount of MC particles, the other is partial replacement of the powder part of the bone cement by equivalent amount of MC particles. Specifically, in the replacement method, a portion of the powder part of the bone cement was firstly removed, and then the MC particles equivalent to the removed bone cement powder in weight would be added into the mixed bone cement. The direct addition is preferred since such operation is convenient for clinical use.

2.4. Injectability of the Modified Bone Cements

A bone filler device with an inner diameter of 2.8 mm (Shanghai Kinetic Co., Ltd., Shanghai, China) was used to investigate the injectability of the MC modified PMMA bone cements. The uniformly mixed bone cement was extracted into a 20 mL syringe, injected into the bone filler device, and then pushed out to determine whether the modified bone cement was injectable or not.

2.5. Mechanical Property Tests

Mechanical properties of the MC modified PMMA bone cements were tested by using a universal materials testing machine (Instron-5880, Instron, Norwood, MA, USA) according to annex E and F of ISO 5833-2002. Cylindrical specimens with 6 diameter and 12 mm height were prepared for compressive strength and compressive modulus tests, and flat plate specimens with 75 mm length, 10 mm width and 3.3 mm depth were prepared for four-point bending strength and bending modulus tests.

The compressive strength, bending strength and bending modulus for each specimen were calculated according to related expressions provided by ISO 5833-2002. The compressive modulus for each specimen was calculated as the slope of the linear region of the stress-strain curve, which was derived from the displacement-load curve recorded by the testing machine, the height and the diameter of the specimen.

2.6. Maximum Temperature and Setting Time Tests

Maximum temperature and setting time of the MC modified PMMA bone cement were tested and recorded as described by annex C of ISO 5833-2002. Briefly, approximately 25 g immediately mixed bone cement was filled into a polytetrafluoroethylene (PTFE) mold, and the temperature was measured via a thermocouple and an electronic converting device having an accuracy of ±0.1 °C. The maximum temperature would be directly recorded by the electronic converting device, and the setting time was determined as the time corresponding to the average value of the maximum and the ambient temperature [21]. The best modification solution screened by above mechanical property tests was tested for each PMMA bone cement product, and the two parameters of each unmodified original product were also tested as the control. The tested were performed at 23 °C and relative humidity of 50%.

2.7. Processing Times Tests

Processing times are of importance for clinical operation of the bone cements by a surgeon. The processing times consisted of four phases, including mixing, waiting, application, and setting.

In this study, processing times were tested for each bone cement before and after the modification to investigate the influence of the MC addition on the operation properties of the bone cements.

The measurement principles for the four phases were as follows:

Mixing time: time for completely mixing of the powder part and liquid part of the bone cement, as well as the MC particles;

Waiting time: time from the bone cement being extracted in to the syringe to being suitable for the injection;

Application time: time from the bone cement being applicable to being hard to inject;

Setting time: time from the injection of the bone cement to it become hardened.

2.8. In Vitro Cytocompatibility Evaluation

Cytocompatibility improvement of the MC modified bone cement were evaluated by culturing pre-osteoblasts on modified and unmodified Osteopal® V and Mendec® Spine bone cements. The use of these two bone cements was because that they contained different contrast agents, Osteopal® V contained ZrO_2 and Mendec® Spine contained $BaSO_4$ in their respect powder part. A clonal osteogenic cell line derived from newborn mouse calvarias, MC3T3-E1 (purchased from Cell Bank of Chinese Academy of Sciences, Shanghai, China), was used in this cytocompatibility evaluation. The cells were cultured in Dulbecco's Modified Eagle Medium (DMEM) with 10% fetal bovine serum (FBS), 100 U/mL penicillin and 0.1 mg/mL streptomycin at 37 °C in an incubator with 5% CO_2.

To prepare bone cement samples for cell culturing, the modified and unmodified bone cements were injected into respective 5 wells in a 96-well plate with 0.1 mL per well, immediately after all the components were fully mixed together. After setting for 24 h, cells were seeded on the set bone cements by adding 100 μL cell suspension into each well at a concentration of 1×10^5 cells/mL. Wells without bone cement were seeded with cells as the control group. Four such 96-well plates were maintained at 37 °C in an incubator with 5% CO_2, and the culture medium was replaced by fresh medium 1, 3, 5 and 7 days after the cell seeding.

Cell proliferation on both modified and unmodified PMMA bone cements were tested by cell counting kit-8 (CCK-8, Dojindo, Japan) at the 1st, 3rd, 5th and 7th day after cell seeding. At each time point, one 96-well plate was randomly selected after refreshing culturing medium, and 10 μL of CCK-8 solution was added into each well. After 2 h incubation at 37 °C, 100 μL solution of each well was transferred to another 96-well plate. Optical density (OD) values at 450 nm of all the wells were measured by a microplate reader (Bio-Rad, Model 680, Hercules, CA, USA).

Cytocompatibility improvement of the modified bone cement was also studied by observing cell attachment on the bone cements before and after the modification. The bone cement samples used for SEM observation of cell attachment were discs with 10 mm diameter and 2 mm thickness. The cell attachment was observed by scanning electron microscopy (SEM; FEI Quanta 200, Hillsboro, OR, USA) 48 h after cell seeding. Samples for the SEM observation were prepared as follows: bone cement samples with the cells were washed with phosphate buffer saline (PBS) to remove any non-adherent cells, and fixed in 2.5% glutaraldehyde in PBS for 24 h; the samples were then dehydrated in ascending series of ethanol solution from 50% to 100% and stored in frozen tert-butyl alcohol (TBA); followed by thoroughly freeze-drying, cell samples were sputter-coated with nano gold particles and observed by SEM.

2.9. Statistical Methods

The results were compared using standard analysis of Student's t-test and expressed as means $\pm$ SD. $p < 0.05$ was considered statistically significant.

3. Results

3.1. Injectability of the Modified Bone Cements

Tables 1–3 list the injectability of the MC modified bone cements. The symbol "○" refers to injectable, and "×" refers to uninjectable. The expression "100/x" means the direct addition method, and "(100 − x)/x" means the replacement method.

Table 1. The injectability of the MC modified Osteopal® V bone cement.

Particle size	Powder part of the bone cement/MC particle (w/w)								
(μm)	100/0	100/5	100/10	100/15	100/20	95/5	90/10	85/15	80/20
<200	○	○	×	×	×	○	×	×	×
200–300	○	○	×	×	×	○	○	○	×
300–400	○	○	○	×	×	○	○	○	○
400–500	○	○	○	×	×	○	○	○	×

Table 2. The injectability of the MC modified Mendec® Spine bone cement.

Particle size	Powder part of the bone cement/MC particle (w/w)								
(μm)	100/0	100/5	100/10	100/15	100/20	95/5	90/10	85/15	80/20
<200	○	○	×	×	×	○	×	×	×
200–300	○	○	○	○	×	○	○	○	○
300–400	○	○	○	○	×	○	○	○	○
400–500	○	○	○	○	×	○	○	○	○

Table 3. The injectability of the MC modified Spineplex™ bone cement.

Particle size	Powder part of the bone cement/MC particle (w/w)								
(μm)	100/0	100/5	100/10	100/15	100/20	95/5	90/10	85/15	80/20
<200	○	○	×	×	×	○	×	×	×
200–300	○	○	○	×	×	○	○	○	×
300–400	○	○	○	○	×	○	○	○	×
400–500	○	○	○	○	×	○	○	○	×

The results show that either small particles or high MC addition amount largely affected the injectability of the bone cement. For Osteopal® V bone cement, the equivalent placement method less influenced the injectability than the direct addition method. Once a MC modified bone cement was extracted into the syringe, it can easily be injected into the bone filler device and then be pushed out.

3.2. The Appearance of the Modified Bone Cements

Figure 1 shows the appearance of the unmodified and MC modified Spineplex™ bone cement. MC particles were homogeneously dispersed in the polymerized PMMA without obvious aggregation or vacancy, indicating that the MC particles were mixed well within the bone cement during flowing phase and MC was compatible with the PMMA material. The homogeneity of the MC modified bone cement ensures uniform mechanical properties throughout the bone cement, thus avoiding stress concentration in clinical applications.

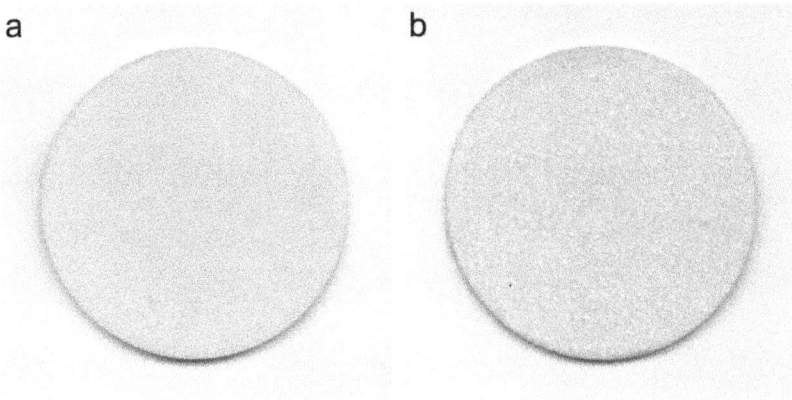

Figure 1. Appearance of the (**a**) unmodified PMMA bone cement and (**b**) MC modified PMMA bone cement.

3.3. Mechanical Properties of the Modified Bone Cements

3.3.1. Mechanical Properties of the Modified Osteopal® V Bone Cement

For Osteopal® V bone cement, partial replacement of the powder part of the bone cement by equivalent 400–500 μm MC particles kept injectable. Therefore, mechanical properties of different amount of 400–500 μm MC modified bone cement were tested, so as to screen out the best modification resolution. The compressive strength and modulus are shown in Figure 2.

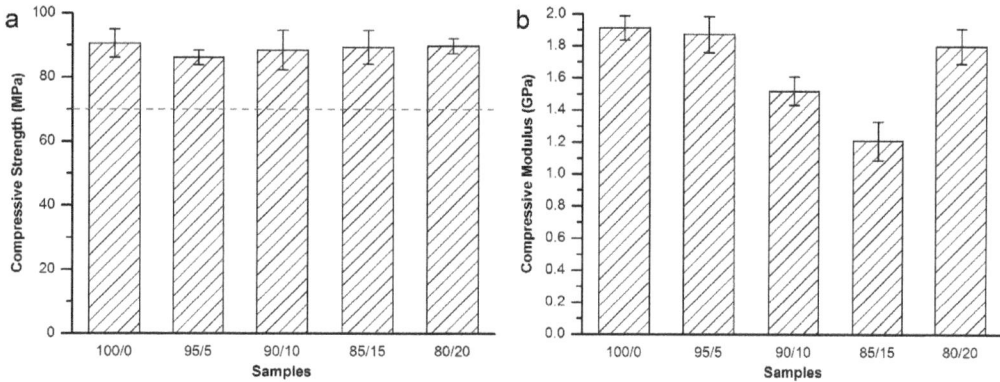

Figure 2. (a) Compressive strength and (b) compressive modulus of the MC modified Osteopal® V bone cement.

As shown in Figure 2a, the addition of the MC particles did not affected the compressive strength of the Osteopal® V bone cement. There were no significant differences between the control and each experimental group, or among experimental groups. The compressive strength for each group was higher than the 70 MPa specified by ISO 5833-2002 (red dash line in Figure 2a), thus meeting the requirement of clinical applications. Figure 2b demonstrates that replacement of the bone cement powder part by 10 wt% or 15 wt% could obtain significant down-regulation effects. 90/10 group down-regulated 20.6% and 85/15 group down-regulated 36.8%. There were statistical differences between 85/15 group and each of the other groups. Other experimental groups achieved very small down-regulation effects.

In order to investigate the effects of the particle size range on the compressive mechanical properties of the modified bone cements, and obtain the best modification result, nearby factors and levels of above experimental groups were further tested. MC particles with 200–300 μm and 300–400 μm were used to prepared 90/10 and 85/15 groups, respectively. The compressive strength and modulus are shown in Figure 3.

As shown in Figure 3a, the addition of MC particles with the particle size of either 200–300 μm or 300–400 μm did not affect the compressive strength of the set bone cements (red dash line in Figure 3a). Figure 3b demonstrated that the replacement of the bone cement powder part by 10wt % of 400–500 μm MC particles could obtain a 16.4% down-regulation effect on the compressive modulus, which was statistically different from the control (100/0) group or (85/15, 400–500) group. However, the modification results were much inferior to the 90/10 and 85/15 groups shown in Figure 2b. Therefore, equivalent replacement of Osteopal® V bone cement powder part by 15 wt% MC particles with 300–400 μm particle size achieved the best modification result for the compressive modulus of the bone cement.

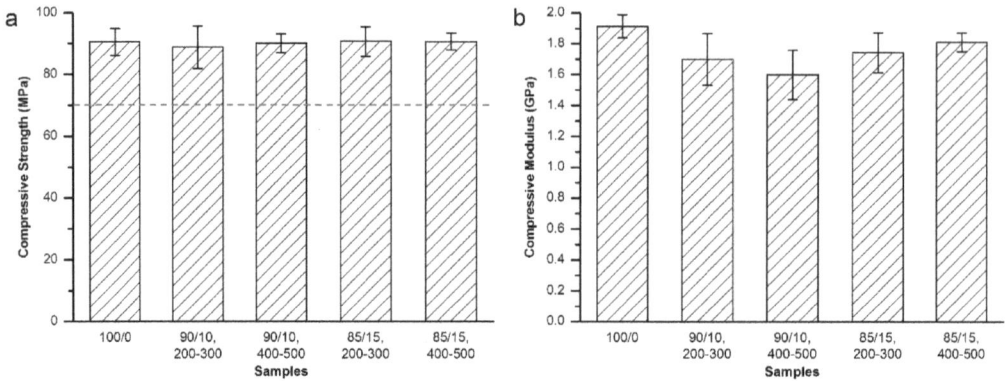

Figure 3. (**a**) Compressive strength and (**b**) compressive modulus of 200–300 μm and 300–400 μm MC particles modified Osteopal® V.

Then, bending strength and modulus were tested for the Osteopal® V bone cement specimens modified by equivalent replacement of the bone cement powder part by 10 wt% and 15 wt% MC particles with 300–400 μm particle size. As shown in Figure 4, it can be seen from the Figure 4 that both of the bending strength and bonding modulus decreased with the partial replacement of the powder part by MC particles. However, both of the bending strength and bending modulus were in conformity with related requirements in ISO 5833-2002 (red dash lines in Figure 4).

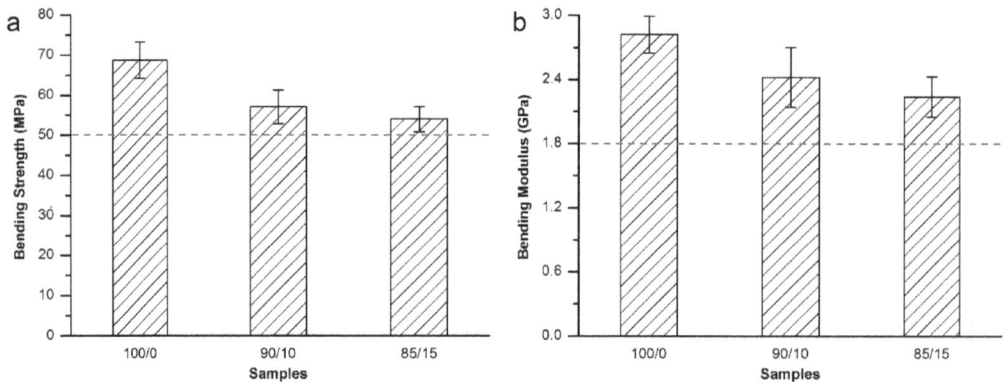

Figure 4. (**a**) Bending strength and (**b**) bending modulus of the MC modified Osteopal® V bone cement.

As a result, equivalent replacement of the bone cement powder part by 15 wt% MC particles with 300–400 μm particle size was considered to be the best resolution for the modification of the Osteopal® V bone cement. The mechanical properties met the requirement of the standard and the clinical applications after the modification.

In light of the modification study on Osteopal® V bone cement, 10 wt%–15 wt% were found to obtain better modification effects than other addition amounts. Moreover, the more MC contained within the bone cement, the better modification effects achieve for cytocompatibility improvement.

Therefore, 15 wt% addition amount of MC particles was considered prior to other amounts, and different particle size ranges were investigated on the premise of injectability.

3.3.2. Mechanical Properties of the Modified Mendec® Spine Bone Cement

MC particles with the size ranges of 200–300 μm, 300–400 μm and 400–500 μm were used for the modification of Mendec® Spine bone cement. The addition amount was 15 wt% for each group, and both direct addition and equivalent replacement methods were investigated. The compressive strength and modulus are shown in Figure 5.

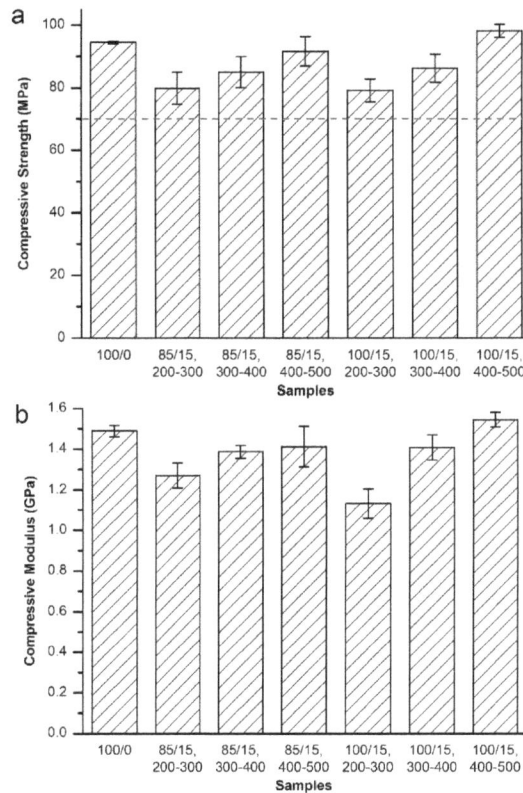

Figure 5. (**a**) Compressive strength and (**b**) compressive modulus of the MC modified Mendec® Spine bone cement.

As shown in Figure 5a, the compressive strength of the MC modified Mendec® Spine bone cement met the requirement of ISO 5833-2002 (red dash line in Figure 5a). Wherein, the direct addition of 200–300 μm MC particles achieved the best effect that the compressive modulus decreased by 24.0% (Figure 5b), and was statistically different from each of the other groups. Although the equivalent replacement using the same particle size range also obtained obvious down-regulatory effect, the direct addition would be more convenient.

Figure 6 shows the bending strength and modulus of the Mendec® Spine bone cement modified by 200–300 µm MC particles. Both specimens that modified by equivalent replacement and direct addition methods using were tested. The results show the bending strength and modulus slightly decreased by MC addition but met the requirement of ISO 5833-2002 (red dash lines in Figure 6).

As a result, direct addition of 15 wt% MC particles with 200–300 µm particle size was considered to be the best resolution for the modification of the Mendec® Spine bone cement.

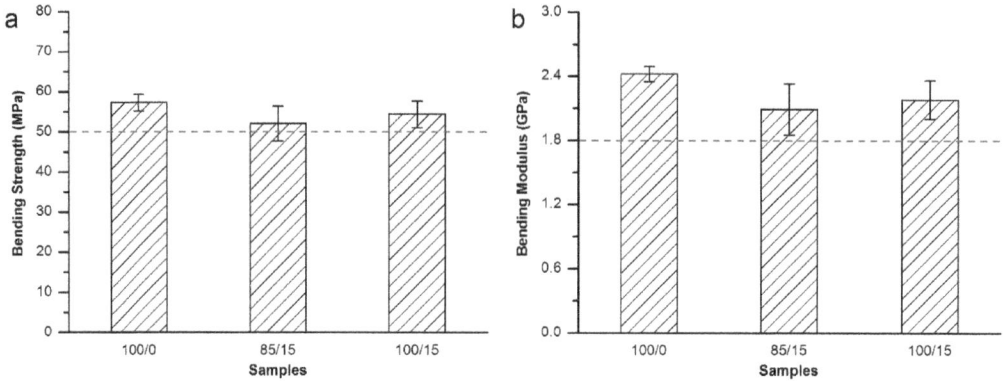

Figure 6. (**a**) Bending strength and (**b**) bending modulus of the MC modified Mendec® Spine bone cement.

3.3.3. Mechanical Properties of the Modified Spineplex™ Bone Cement

Similar to the study process of the Mendec® Spine bone cement, six experimental groups including three particle size ranges and two addition methods were investigated. The compressive strength and modulus are shown in Figure 7.

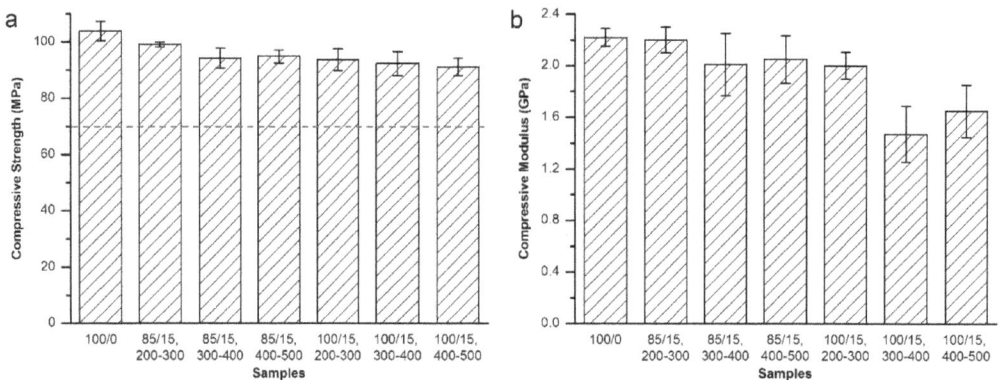

Figure 7. (**a**) Compressive strength and (**b**) compressive modulus of the MC modified Spineplex™ bone cement.

As shown in Figure 7a, the compressive strength of the Spineplex™ bone cement modified by MC particles slightly decreased, and met the requirement of ISO 5833-2002 (red dash line in

Figure 7a). Figure 7b shows direct addition of MC with 300–400 µm particle size obtained the best modification effect that the compressive modulus was down-regulated by 33.8%, and was statistically different from each of the other groups, except the (100/15, 400–500) group, which also obtained obvious down-regulation effect in comparison with the control group.

Figure 8 shows the bending strength and modulus of the Spineplex™ bone cement modified by direct addition of the MC particles. Specimens modified by 300–400 µm and 400–500 µm MC particles were tested. The results show the bending strength and modulus decreased a little after the modification but met the requirement of ISO 5833-2002 (red dash line in Figure 8).

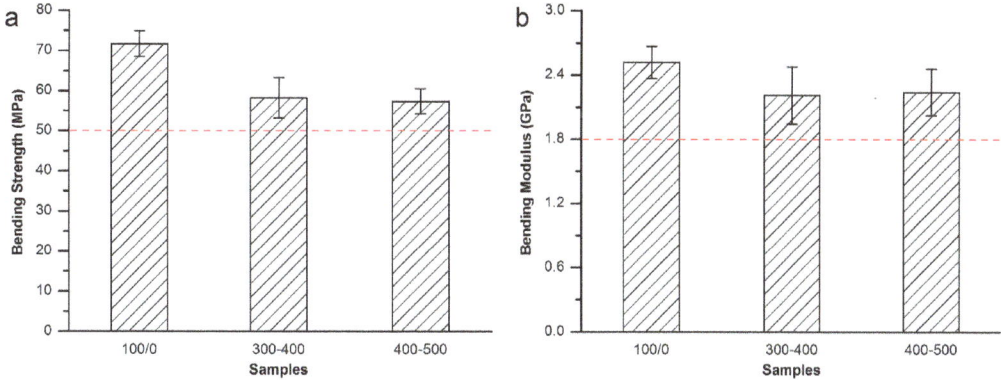

Figure 8. (**a**) Bending strength and (**b**) bending modulus of the MC modified Spineplex™ bone cement.

As a result, direct addition of 15 wt% MC particles with 300–400 µm particle size was considered to be the best resolution for the modification of the Spineplex™ bone cement.

3.4. Maximum Temperature and Setting Time

The maximum temperature comparisons between the unmodified bone cements and their perspective optimal modification group are shown in Figure 9.

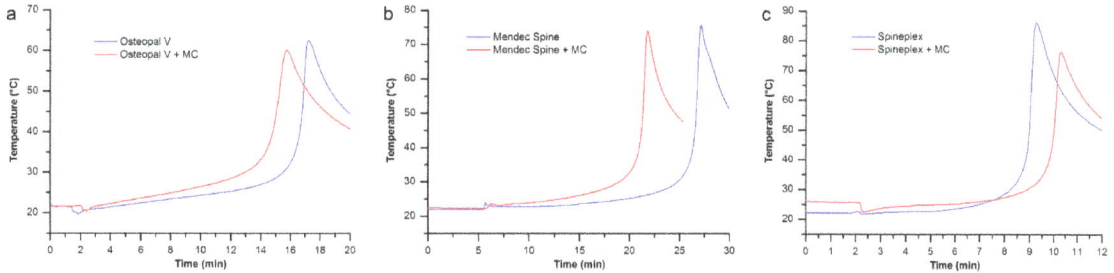

Figure 9. The maximum temperature comparisons between the unmodified and modified bone cements: (**a**) Osteopal® V bone cement; (**b**) Mendec® Spine bone cement; and (**c**) Spineplex™ bone cement.

For each bone cement product investigated in this study, the maximum temperature of the modified bone cement decreased compared to its original product. Because the added MC particles absorbed a portion of heat generated by the polymerization of the PMMA bone cements. The lower maximum temperature is beneficial for clinical applications, since such low temperature could reduce damage on tissues near the bone cement caused by the heat of polymerization.

Table 4 lists the setting time of the unmodified and modified bone cements. The addition of MC particles took effects on these bone cements. Wherein, the setting time shortened for Osteopal® V and Mendec® Spine bone cements after the modification, while the setting time of Spineplex™ became longer. In regard to specific setting time for each bone cement, Osteopal® V and Mendec® Spine bone cements had overlong setting time, while Spineplex™ bone cement set too fast. A change in setting time for all bone cements, by some modification, would make it more convenient for clinical use by a surgeon.

Table 4. Setting time of the original bone cement products and modified bone cements.

Bone cements	Osteopal® V	Mendec® Spine	Spineplex™
Original product	16'44"	26'36"	9'02"
Modified by MC particles	14'51"	21'18"	10'01"

3.5. Processing Times for the Modified Bone Cements

Processing times for each bone cement, before and after the modification, are shown in Figure 10. The processing times for each bone cement varied a little after the modification by MC particles, and the variation was 0.5–1 min for each phase. Such small variation in the processing times makes no changes to the operating habits of surgeons.

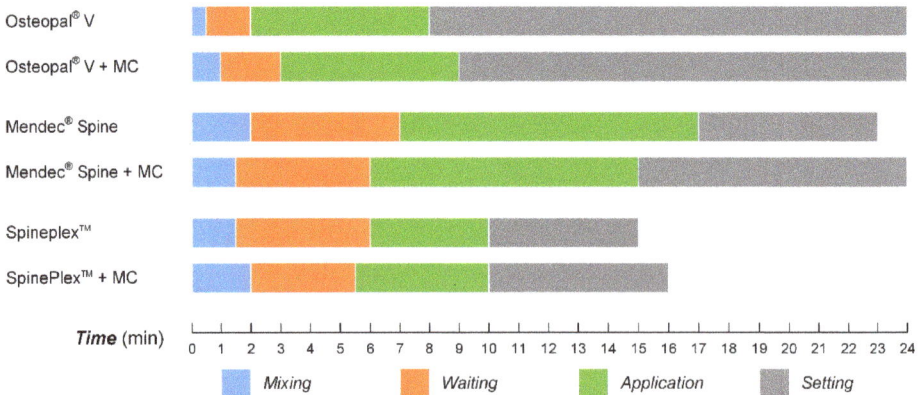

Figure 10. Processing times for the bone cements before and after the modification by MC.

Figure 11 takes Osteopal® V bone cement as an example to show operation process and processing times of the MC modified PMMA bone cement.

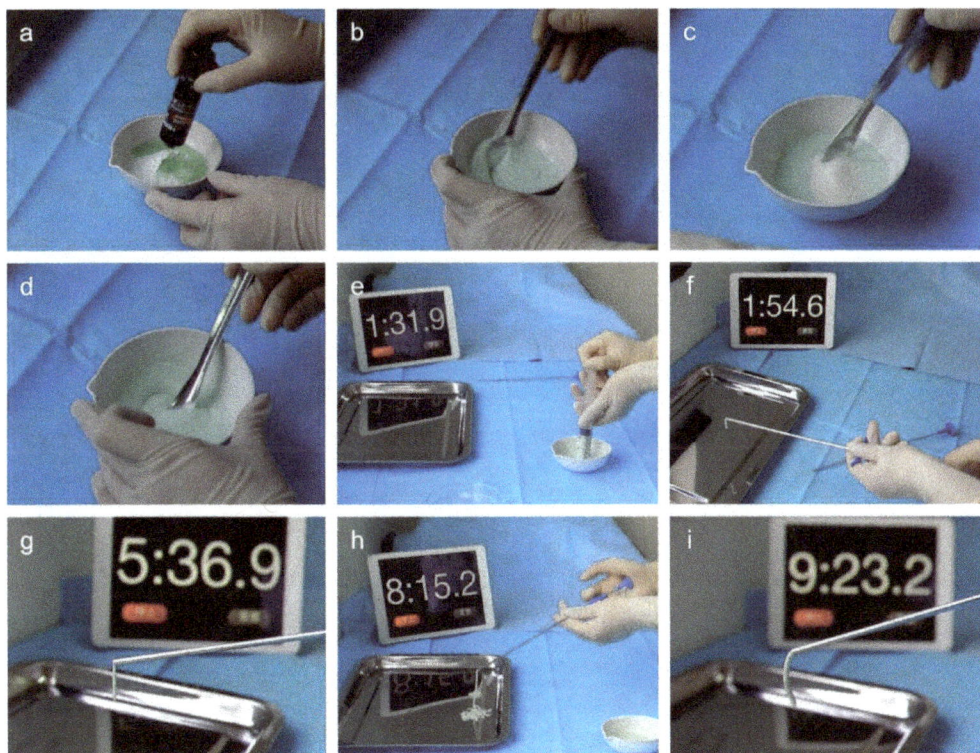

Figure 11. Operation process and processing times of the MC modified Osteopal® V bone cement: (**a**) mixing powder and liquid parts of the bone cement; (**b**) uniformly mixing powder and liquid parts; (**c**) addition of MC particles; (**d**) uniformly mixing all components; (**e**) extracting flowing bone cement by a syringe; (**f**) injection of the bone cement into a bone filler device; (**g**) earlier stage of the bone cement; (**h**) middle stage of the bone cement; and (**i**) later stage of the bone cement.

3.6. Cytocompatibility Improvement of the Modified Bone Cements

Cytocompatibility improvement of the MC modified bone cements were evaluated by proliferation quantification and attachment observation of MC3T3-E1 cells on the unmodified and MC modified bone cements. The proliferation of the cells on the bone cements are shown in Figure 12.

For both Osteopal® V and Mendec® Spine bone cements, cells proliferated well on each bone cement. Cell count on the MC modified bone cement was significantly higher than that on the unmodified original bone cement, with regard to either Osteopal® V or Mendec® Spine bone cement. At day 5 and 7, there were statistically significant differences between the MC incorporated group and the modified group, as well as between the MC incorporated group and the blank control. Cell counts for the unmodified group and the blank control group were closed without statistical differences at day 5 and 7, for both PMMA bone cement products, since pure PMMA bone cements and well-plates were all bioinert materials that had no effect on cell proliferation. The result indicated that the modification by using MC largely improved cytocompatibility of the PMMA bone cements, and the contrast agent, ZrO_2 or $BaSO_4$, did not affect such improvement effects.

Figure 13 Shows cell morphology on the Osteopal® V bone cement before and after the modification. Figure 13b,d are the amplification of the center areas (noted by dash boxes) of Figure 13a,c, respectively.

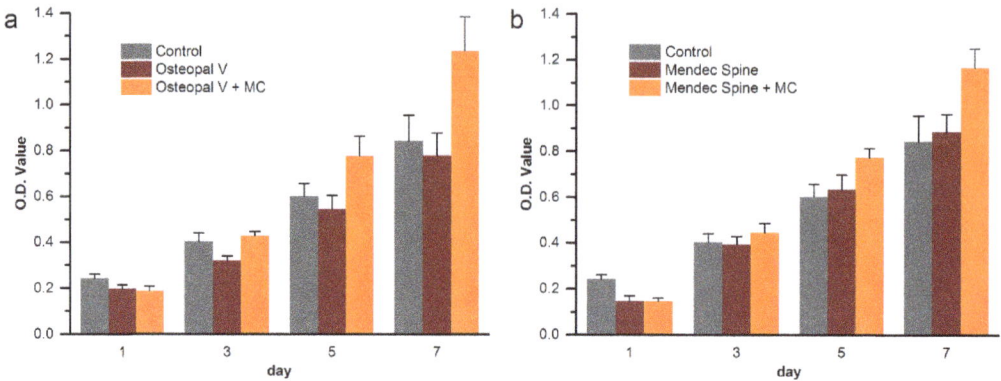

Figure 12. Cell proliferation on (**a**) Osteopal® V and (**b**) Mendec® Spine bone cements.

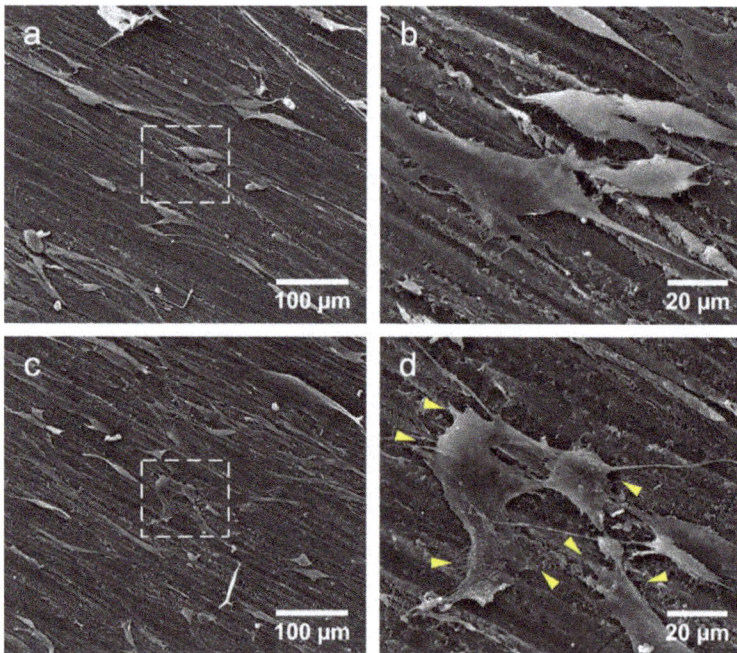

Figure 13. Cell observations on the bone cements before and after the MC modification by SEM: (**a**) cells on unmodified bone cement; (**b**) amplification of the center of Figure 13a; (**c**) cells on MC modified bone cement: and (**d**) amplification of the center of Figure 13c.

As shown in Figure 13, cells grew well on the bone cements and filopodia stretched out to anchor the cells on the bone cements. By comparing cells on the unmodified and modified bone cements, there were no differences on cell counts, which were in conformity with histograms shown in Figure 12. From the detail of the cell morphology shown in Figure 13b,d, it can be seen that a large number of filopodia stretched out from the cells on the MC modified bone cement (noted by yellow arrows), while the cells on the unmodified bone cement had less filopodia. The cytocompatibility of the MC modified PMMA bone cement was better than that of the unmodified original bone cement, and cell adhesion would be improved by such modification. The results indicates that the modification of PMMA bone cements by addition of MC could improve its cytocompatibility, which is beneficial for the formation of good osteointegration between the bone cement and the host bone in clinical applications.

4. Discussion

The spine is the load bearing structure in the human skeleton, and vertebral bodies are the basic structural units. For upright walking human beings, the major direction of the loading on the vertebral body is compressive force in the vertical direction, including compressive force from above lower endplate and support force from bottom upper endplate. Therefore, the compressive strength and modulus are key mechanical factors for those bone cements used for PVP and PKP. Overhigh compressive modulus of the bone cement produces overhigh stiffness of the PVP or PKP treated segment, which resulting in stress concentration at the segment, and would easily cause secondary fracture on adjacent vertebral bodies and endplate near the surgical segment [9,10].

Bioinert is another disadvantage of the PMMA bone cement, since osteocytes cannot grow into the bioinert PMMA, it is unable to form stable osteointegration between the bone cement and the host bone at the implant site [13–15]. As described by the analysis in the introduction section, aseptic loosening or even dislodgement of the bone cement are very dangerous for patients, as a free hard block may press on the spinal nerve to produce hazardous results [22].

Many efforts were made to improve the mechanical properties and biocompatibility of the PMMA bone cements for PVP and PKP. In light of above-mentioned disadvantages, these studies were focused on down-regulation of the compressive modulus, as well as improvement of the biocompatibility of the PMMA bone cement.

As the main inorganic component of natural bone tissue, HA was popular in the modification studies on PMMA bone cements. Many studies used HA and element-doped HA, such as strontium-doped HA to modify the PMMA bone cement. However, in some studies, the addition of HA largely decreased compressive strength of the bone cement that cannot meet the requirement of ISO 5833-2002 [23]; in some other studies, the compressive modulus even largely increased after the addition of HA [24]. Moreover, the addition of HA into PMMA bone cement did not exhibit improved biocompatibility [25].

Introduction of a biodegradable component was another modification idea. For example, chitosan and sodium hyaluronate were studied to form porous structure by degradation [26,27]. However, the compressive strength of the bone cement also decreased with the degradation of the biodegradable component, and became much lower than the lower limit specified by ISO 5833-2002 [26,27].

Modification of MMA monomer was tried by some researchers to down-regulate the compressive modulus of the PMMA bone cements. For example, *N*-methyl-pyrrolidone monomer and linoleic acid were, respectively, used to partially replace the MMA monomer in the polymerization of the PMMA bone cement. However, with the down-regulation of the compressive modulus, the compressive strength also decreased to be much lower than the requirement of ISO 5833-2002 [28,29].

In summary, previous studies on the modification of PMMA bone cement did not obtain a perfect solution that both down-regulated compressive modulus without affecting the compressive strength, and improved biocompatibility of the PMMA bone cement. In this study, a biomimetic material MC with good biocompatibility and osteogenic activity was used for the modification of the PMMA bone cement. MC was compatible with the PMMA and could be homogeneously dispersed within the PMMA bone cement. The dispersed MC particles broke integrality of the polymerized bone cement, and were able to regulate mechanical properties by verifying addition amounts, particle size range, and addition method of the MC particles. Through a series of experiments, both of mechanical properties and cytocompatibilities of three commonly used PMMA bone cements for PVP and PKP were successfully improved by addition of different MC particles with different addition methods. However, related mechanical properties regulation mechanisms need further investigations, and the clinical outcomes of the modification need long-term clinical observations.

5. Conclusions

Biomimetic MC with good osteogenic activity was added to commercially available PMMA bone cement products to improve both the mechanical properties and the cytocompatibility in this study. As the compressive strength of the modified bone cements remained well, the compressive elastic modulus were significantly down-regulated by the MC. Meanwhile, the adhesion and proliferation of pre-osteoblasts on the modified bone cements were improved compared with cells on those unmodified. The results are beneficial for both reducing the pressure on the adjacent vertebral bodies, and the osteointegration formation between the bone cement and the host bone tissue in clinical applications. Moreover, the modification of the PMMA bone cements by adding MC did not much influence the injectability and processing times of the cement. As a result, improvement of PMMA bone cements by incorporating MC particles is an effective and easy-to-operate clinical approach for improving the quality of the surgery and reducing complications after PVP and PKP.

Acknowledgments

This work was in part supported by the National Basic Research Program of China funded by the Ministry of Science and Technology of China (2011CB606205), and the National Natural Science Fund funded by the National Natural Science Foundation of China (21371106, 51402167).

Author Contributions

Hong-Jiang Jiang conceived the study and drafted the manuscript. Jin Xu and Yun Cui performed the experiments and analyzed the data. Zhi-Ye Qiu performed the experiments and drafted the manuscript. Xin-Long Ma participated in the experiment design and analyzed the data. Zi-Qiang Zhang analyzed the data. Xun-Xiang Tan participated in the experiment design. Fu-Zhai Cui conceived the study and participated in the experiment design. All the authors read and approved the final manuscript.

Conflicts of Interest

The authors declare no conflict of interest.

References

1. Old, J.L.; Calvert, M. Vertebral compression fractures in the elderly. *Am. Fam. Physician* **2004**, *69*, 111–116.
2. Furstenberg, C.H.; Grieser, T.; Wiedenhofer, B.; Gerner, H.J.; Putz, C.M. The role of kyphoplasty in the management of osteogenesis imperfecta: Risk or benefit? *Eur. Spine J.* **2010**, *19* (Suppl. 2), S144–S148.
3. Gu, Y.F.; Li, Y.D.; Wu, C.G.; Sun, Z.K.; He, C.J. Safety and efficacy of percutaneous vertebroplasty and interventional tumor removal for metastatic spinal tumors and malignant vertebral compression fractures. *AJR Am. J. Roentgenol.* **2014**, *202*, W298–W305.
4. Banse, X.; Sims, T.J.; Bailey, A.J. Mechanical properties of adult vertebral cancellous bone: Correlation with collagen intermolecular cross-links. *J. Bone Miner. Res.* **2002**, *17*, 1621–1628.
5. Hou, F.J.; Lang, S.M.; Hoshaw, S.J.; Reimann, D.A.; Fyhrie, D.P. Human vertebral body apparent and hard tissue stiffness. *J. Biomech.* **1998**, *31*, 1009–1015.
6. Morgan, E.F.; Bayraktar, H.H.; Keaveny, T.M. Trabecular bone modulus-density relationships depend on anatomic site. *J. Biomech.* **2003**, *36*, 897–904.
7. Kurtz, S.M.; Villarraga, M.L.; Zhao, K.; Edidin, A.A. Static and fatigue mechanical behavior of bone cement with elevated barium sulfate content for treatment of vertebral compression fractures. *Biomaterials* **2005**, *26*, 3699–3712.
8. Jasper, L.E.; Deramond, H.; Mathis, J.M.; Belkoff, S.M. Material properties of various cements for use with vertebroplasty. *J. Mater. Sci. Mater. Med.* **2002**, *13*, 1–5.
9. Grados, F.; Depriester, C.; Cayrolle, G.; Hardy, N.; Deramond, H.; Fardellone, P. Long-term observations of vertebral osteoporotic fractures treated by percutaneous vertebroplasty. *Rheumatology (Oxford)* **2000**, *39*, 1410–1414.
10. Trout, A.T.; Kallmes, D.F.; Layton, K.F.; Thielen, K.R.; Hentz, J.G. Vertebral endplate fractures: An indicator of the abnormal forces generated in the spine after vertebroplasty. *J. Bone Miner. Res.* **2006**, *21*, 1797–1802.

11. Burton, A.W.; Mendoza, T.; Gebhardt, R.; Hamid, B.; Nouri, K.; Perez-Toro, M.; Ting, J.; Koyyalagunta, D. Vertebral compression fracture treatment with vertebroplasty and kyphoplasty: Experience in 407 patients with 1156 fractures in a tertiary cancer center. *Pain Med.* **2011**, *12*, 1750–1757.

12. Trout, A.T.; Kallmes, D.F.; Kaufmann, T.J. New fractures after vertebroplasty: Adjacent fractures occur significantly sooner. *AJNR Am. J. Neuroradiol.* **2006**, *27*, 217–223.

13. Sugino, A.; Miyazaki, T.; Kawachi, G.; Kikuta, K.; Ohtsuki, C. Relationship between apatite-forming ability and mechanical properties of bioactive PMMA-based bone cement modified with calcium salts and alkoxysilane. *J. Mater. Sci. Mater. Med.* **2008**, *19*, 1399–1405.

14. Portigliatti-Barbos, M.; Rossi, P.; Salvadori, L.; Carando, S.; Gallinaro, M. Bone-cement interface: A histological study of aseptic loosening in twelve prosthetic implants. *Ital. J. Orthop. Traumatol.* **1986**, *12*, 499–505.

15. Mann, K.A.; Miller, M.A.; Cleary, R.J.; Janssen, D.; Verdonschot, N. Experimental micromechanics of the cement-bone interface. *J. Orthop. Res.* **2008**, *26*, 872–879.

16. Cui, F.Z.; Li, Y.; Ge, J. Self-assembly of mineralized collagen composites. *Mater. Sci. Eng.* **2007**, *57*, 1–27.

17. Zhang, W.; Liao, S.S.; Cui, F.Z. Hierarchical Self-Assembly of Nano-Fibrils in Mineralized Collagen. *Chem. Mater.* **2003**, *15*, 3221–3226.

18. Liao, S.S.; Cui, F.Z. *In vitro* and *in vivo* degradation of mineralized collagen-based composite scaffold: Nanohydroxyapatite/collagen/poly(*L*-lactide). *Tissue Eng.* **2004**, *10*, 73–80.

19. Liao, S.S.; Guan, K.; Cui, F.Z.; Shi, S.S.; Sun, T.S. Lumbar spinal fusion with a mineralized collagen matrix and rhBMP-2 in a rabbit model. *Spine (Phila Pa 1976)* **2003**, *28*, 1954–1960.

20. Liao, S.S.; Cui, F.Z.; Zhang, W.; Feng, Q.L. Hierarchically biomimetic bone scaffold materials: Nano-HA/collagen/PLA composite. *J. Biomed. Mater. Res. B Appl. Biomater.* **2004**, *69*, 158–165.

21. International Organization for Standardization. *Implants for Surgery—Acrylic Resin Cements*; ISO 5833:2002(E); International Organization for Standardization: Geneva, Switzerland, 2002.

22. Tsai, T.T.; Chen, W.J.; Lai, P.L.; Chen, L.H.; Niu, C.C.; Fu, T.S.; Wong, C.B. Polymethylmethacrylate cement dislodgment following percutaneous vertebroplasty: A case report. *Spine (Phila Pa 1976)* **2003**, *28*, E457–E460.

23. Lam, W.; Pan, H.B.; Fong, M.K.; Cheung, W.S.; Wong, K.L.; Li, Z.Y.; Luk, K.D.; Chan, W.K.; Wong, C.T.; Yang, C.; Lu, W.W. *In vitro* characterization of low modulus linoleic acid coated strontium-substituted hydroxyapatite containing PMMA bone cement. *J. Biomed. Mater. Res. B Appl. Biomater.* **2011**, *96*, 76–83.

24. Hernandez, L.; Gurruchaga, M.; Goni, I. Injectable acrylic bone cements for vertebroplasty based on a radiopaque hydroxyapatite. Formulation and rheological behaviour. *J. Mater. Sci. Mater. Med.* **2009**, *20*, 89–97.

25. Hernandez, L.; Parra, J.; Vazquez, B.; Bravo, A.L.; Collia, F.; Goni, I.; Gurruchaga, M.; San Roman, J. Injectable acrylic bone cements for vertebroplasty based on a radiopaque hydroxyapatite. Bioactivity and biocompatibility. *J. Biomed. Mater. Res. B Appl. Biomater.* **2009**, *88*, 103–114.

26. Kim, S.B.; Kim, Y.J.; Yoon, T.L.; Park, S.A.; Cho, I.H.; Kim, E.J.; Kim, I.A.; Shin, J.W. The characteristics of a hydroxyapatite-chitosan-PMMA bone cement. *Biomaterials* **2004**, *25*, 5715–5723.

27. Boger, A.; Bohner, M.; Heini, P.; Verrier, S.; Schneider, E. Properties of an injectable low modulus PMMA bone cement for osteoporotic bone. *J. Biomed. Mater. Res. B Appl. Biomater.* **2008**, *86*, 474–482.

28. Boger, A.; Wheeler, K.; Montali, A.; Gruskin, E. NMP-modified PMMA bone cement with adapted mechanical and hardening properties for the use in cancellous bone augmentation. *J. Biomed. Mater. Res. B Appl. Biomater.* **2009**, *90*, 760–766.

29. Lopez, A.; Mestres, G.; Karlsson Ott, M.; Engqvist, H.; Ferguson, S.J.; Persson, C.; Helgason, B. Compressive mechanical properties and cytocompatibility of bone-compliant, linoleic acid-modified bone cement in a bovine model. *J .Mech. Behav. Biomed. Mater.* **2014**, *32*, 245–256.

Microstructure and Deformation of Coronary Stents from CoCr-Alloys with Different Designs

Sabine Weiss and Bojan Mitevski

Abstract: Coronary heart disease is still one of the most common sources for death in western industrial countries. Since 1986, a metal vessel scaffold (stent) has been inserted to prevent the vessel wall from collapsing. Most of these coronary stents are made from CrNiMo-steel (316L). Due to its austenitic structure, the material shows a good combination of strength, ductility, corrosion resistance, and biocompatibility. However, this material has some disadvantages like its non-MRI compatibility and its poor fluoroscopic visibility. Other typically used materials are the Co-Base alloys L-605 and F-562 which are MRI compatible as well as radiopaque. Another interesting fact is their excellent radial strength and therefore the ability to produce extra thin struts with increased strength. However, because of a strut diameter much less than 100 μm, the cross section consists of about 5 to 10 crystal grains (oligo-crystalline). Thus, very few or even just one grain can be responsible for the success or failure of the whole stent. To investigate the relation between microstructure, mechanical factors and stent design, commercially available Cobalt-Chromium stents were investigated with focus on distinct inhomogeneous plastic deformation due to crimping and dilation. A characteristic, material related deformation behavior with predominantly primary slip was identified to be responsible for the special properties of CoCr stents.

Reprinted from *Materials*. Cite as: Weiss, S.; Mitevski, B. Microstructure and Deformation of Coronary Stents from CoCr-Alloys with Different Designs. *Materials* **2015**, *8*, 2467-2479.

1. Introduction

Most commercially available coronary stents are made of 316L type CrNiMo-steels (e.g., DIN EN 1.4441) [1]. Due to its austenitic structure, the material shows a good combination of strength, ductility, corrosion resistance, and biocompatibility [2–4]. However, this material has some disadvantages like its non-MRI compatibility and its poor fluoroscopic visibility [5]. Other typically used materials are the Co-Base alloys L-605 and F-562 which are MRI compatible [6] as well as radiopaque [7]. Another interesting fact is their excellent radial strength, flexibility and deliverability and therefore the ability to produce extra thin struts with increased strength [5,8]. Stent strut thickness plays an important role in vascular injury and consequent neointimal proliferation. Clinical studies revealed that the thin-strut stents are significantly associated with reduced early restenosis [9–11]. A positive influence of thin strut stents on long-term luminal response with the results that neointimal atherosclerotic change occurred more frequently for patients with thick-strut stents and that the incidence of late in stent restenosis was significantly lower in the thin-strut group of patients was observed, too [7,12,13]. However, thin strut stents especially from CoCr-alloys are not indisputable. In recent investigations significantly higher acute elastic recoil was observed for thin strut CoCr-stents [14]. The authors compared 316L steel stents with a strut thickness of 135 μm, PtCr stents and CoCr stents, both with strut thicknesses of 81 μm. Low acute recoil was found for 316L and PtCr in contrast to much higher recoil for CoCr. Because of the same strut thickness of the PtCr stents and the CoCo stents they assume, that acute stent recoil is more correlated with stent design and materials than with stent strut thickness. Another current study used nanoindentation to analyze the local mechanical properties of stents. They could identify regions of high hardness and related them to the gradient of plastic strains generated during the deployment process and the associated stress. These regions were correlated with the locations where macroscopic fractures have been observed in both mechanically tested and human explanted stents because of a reduction in plastic work ratio [15].

Despite all these investigations there is only limited knowledge available about the correlations between plastic deformation, strut thickness, design, material and recoil. Furthermore, due to the small dimensions of stents, the material has an oligo- crystalline structure (only a few grains distributed over the cross section of a stent strut). These structures can in fact neither be described as multi-crystalline materials, nor be treated as single crystals. The result is an orientation dependent inhomogeneous deformation behavior [16,17]. With regard to the importance of the orientation parameter, the Electron BackScatter Diffraction (EBSD) technique has been used to compare the crystallographic orientation of the grains in CoCr stents with different designs after deformation due to dilation. By means of transmission electron microscopy (TEM) the deformation mechanisms can be investigated. For this study, the microstructural alterations after plastic deformation of three different stent designs (Multi Link Vision™, GUIDANT Corporation, Santa Clara, CA, USA; Coroflex® Blue, B. Braun Melsungen AG, Melsungen, Germany and Driver Medtronic Inc., Minneapolis, MN, USA) were investigated.

2. Experimental

2.1. Material

For this purpose, commercially available cobalt chromium based L-605 and F-562 coronary artery stents of three different designs (Multi Link Vision L-605, Coroflex® Blue L-605 and Driver F-562) with characteristic strut thicknesses were compared. Multi Link Vision and Driver are among the market leading cobalt chromium stents. Coroflex® Blue was chosen because of its very low strut thickness. The stent designs are depicted in Figure 1.

All three designs are laser cut slotted tube stents with typical v-shape structure. The Driver design exclusively consists of regular v-shapes with 91 μm strut thickness. In the Coroflex® Blue with only 60 μm strut thickness one of the two flanks is curved and in the Multi Link Vision design with 81 μm strut thickness stabilization struts with loops connecting two lines of the v-structure are added.

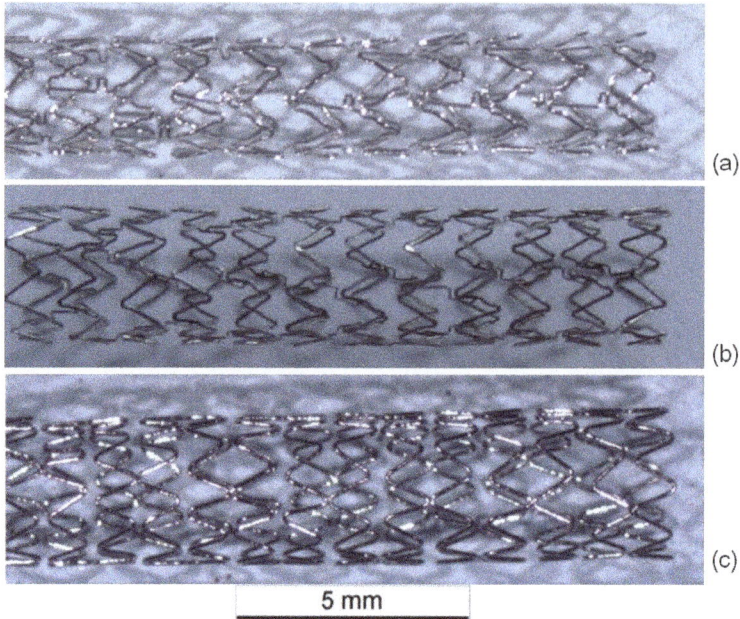

Figure 1. (**a**) Multi Link Vision (L-605, 81 μm); (**b**) Coroflex® Blue (L-605, 60 μm); (**c**) Medtronic Driver (F-562, 91 μm).

2.2. Preparation

All stents were dilated in air according to the descriptions in the manufacturers' instructions. Microstructure characterization of dilated stents was carried out by means of scanning electron microscopy (SEM) as well as transmission electron microscopy (TEM). Furthermore, single grain orientation determination by means of Electron BackScatter Diffraction (EBSD) was performed, in order to reveal microstructural alterations during deformation. For some stents no further metallographic preparation was necessary to obtain EBSD pattern, because the final stent production step is electrochemical polishing. Stents with a coated surface were electrochemically polished for 10 s in an electrolyte composed of sulfuric acid and phosphorous acid to reach the bare metal surface.

2.3. SEM-Imaging and EBSD-Measurement

For measurement some geometrical difficulties had to be taken into account; Because of the tube profiles of the stents significant decrease in pattern quality is observed depending on the position on the sample surface. For a tube with 3 mm diameter, very close to the top of the tube in the secondary-electron (SE)-image, a small part of the upper region of the EBSD-pattern is shaded and appears dark. Best pattern quality is obtained in a distance of about 50 μm from the top; at about 100 μm from the top the shaded area in the lower part of the pattern has increased substantially and at a distance of about 200 μm (from the top) no orientation determination was possible any more. The measurable distance depends on the radius R of the tube (or stent) and can be calculated as:

$$u = R \left(1 - \cos \alpha \right) \tag{1}$$

α corresponds to 90° minus the minimum tilting angle at which pattern can be reached (for traditional EBSD equipment α is between 65° and 75°). Further information on this field is published elsewhere (by the authors) [18].

For SEM images a TESCAN MIRA\\ (TESCAN, Brno, Czech Republic) was used. An SEM Gemini 1530 (Zeiss, Oberkochen, Germany) with an EBSD system Crystal (Oxford Instruments, Wiesbaden, Germany) and calculation software Channel 5 from HKL-Technology (Hobro, Denmark) as well as an EBSD system from EDAX (Tilburg, The Netherlands) with OIM™ 6.0 Software were used. EBSD analyzes were carried out by two different people using two different EBSD systems without knowledge about the results of the respective partner to support the repeatability of the results. For both an accelerating voltage of 20 kV and a working distance of 15 mm were applied to obtain single grain orientations as well as orientation maps. The orientation differences are presented by means of Kernel misorientation maps.

2.4. TEM-Imaging

For the investigation of stents by means of TEM, parts of interest were cut out with focused ion beam technique for direct observation or with special preparation scissors for further manual preparation. For manual preparation these parts were inserted into a grid net. Stent and net were glued together using a suitable adhesive (M-Bond 610, Gatan GmbH, Munich, Germany). Afterwards it was ground down to a thickness of 80 μm and further thinned from both sides using a dimple grinder (Model 656, Gatan GmbH, Munich, Germany) and a nitrogen cooled ion mill (Model Pips II cool; Gatan GmbH). TEM investigations were performed with a Philips EM 400 TEM (FEI, Eindhoven, the Netherlands) as well as a Zeiss EM 912 (Zeiss, Oberkochen, Germany) both using an accelerating voltage of 120 kV. For TEM imaging the results of three different investigators as well as FIB or manually prepared specimens were analyzed. The images presented in Section 3 are representative for the materials microstructure of the corresponding CoCr stents.

3. Results and Discussion

After dilation, stents of all three designs exhibit regions of high deformation with characteristic deformation structures like slip traces on the surface. In contrast to this, before dilation, a shiny polished surface was apparent. In Figure 2, representative SEM images of the inner part of stent bows of a Driver Stent are depicted. Slip traces (fine parallel lines) extrusions and intrusions of grains are clearly visible. This occurrence is generated by heavy plastic deformation and is accompanied with orientation gradients within the deformed grains. These so called misorientations are correlated to the amount of plastic deformation in the material.

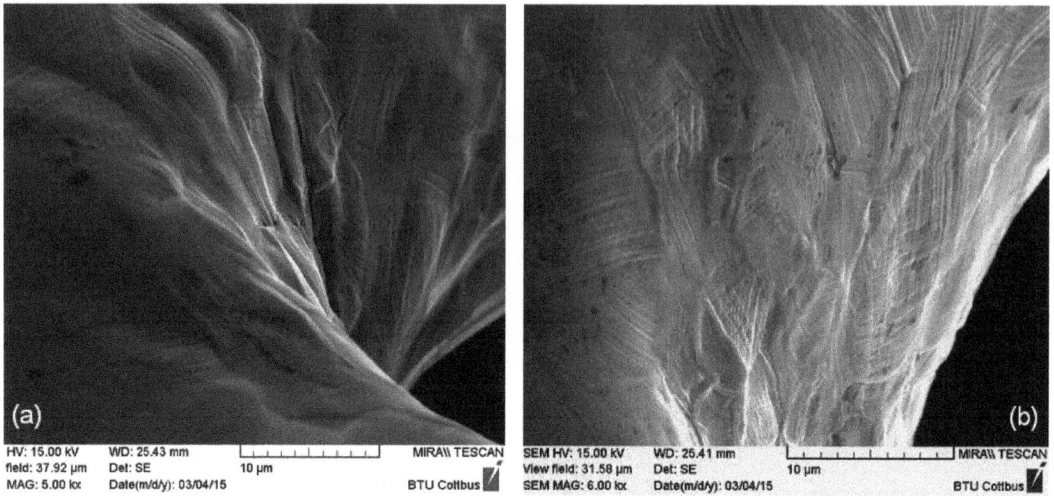

Figure 2. (a) Inner part of a stent bow and (b) passage between bow and strut of a Driver Stent (F-562) with slip traces on the surface as well as extrusions and intrusions of grains.

Orientation mappings of several stent bows of all three types of stents were measured. A typical grain orientation determination at a highly deformed stent bow of a Driver Stent after dilation is shown in Figure 3a. The orientation data is represented parallel to the longitudinal direction of the stent. The coloring of the mapping is in accordance to the legend in Figure 3a. As available from the mapping the material has an oligo-crystalline structure with only few grains distributed over the strut diameter. In contrast to the material prior to load [19] where completely homogeneous orientation mappings without any orientation gradient were measured, regions with large misorientations occur, identifiable from the iridescent colors within the grains. The high deformation becomes more obvious in the corresponding local misorientation map (Figure 3b). Low misorientations occur in blue and increasing misorientations are represented by green and yellow colors. A concentration of large orientation differences in the inner and outer region of the stent bow as well as close to the grain boundaries becomes obvious.

In Figure 4a, a comparison of representative local misorientation mappings of a stent bow and a stent strut of a Coroflex® Blue Stent is depicted. As available from the extended areas of green and yellow colors in the stent bow and in contrast to that predominantly blue coloring in the stent strut the much lower deformation in the strut than in the bow becomes obvious. This is in good agreement with the locations of high misorientations corresponding to high plastic deformation as well as high hardness measured by Kapnisis [15]. In his investigations of the hardness by means of nanoindentation he identified regions of high hardness in the stent bows and correlated them to high plastic deformation.

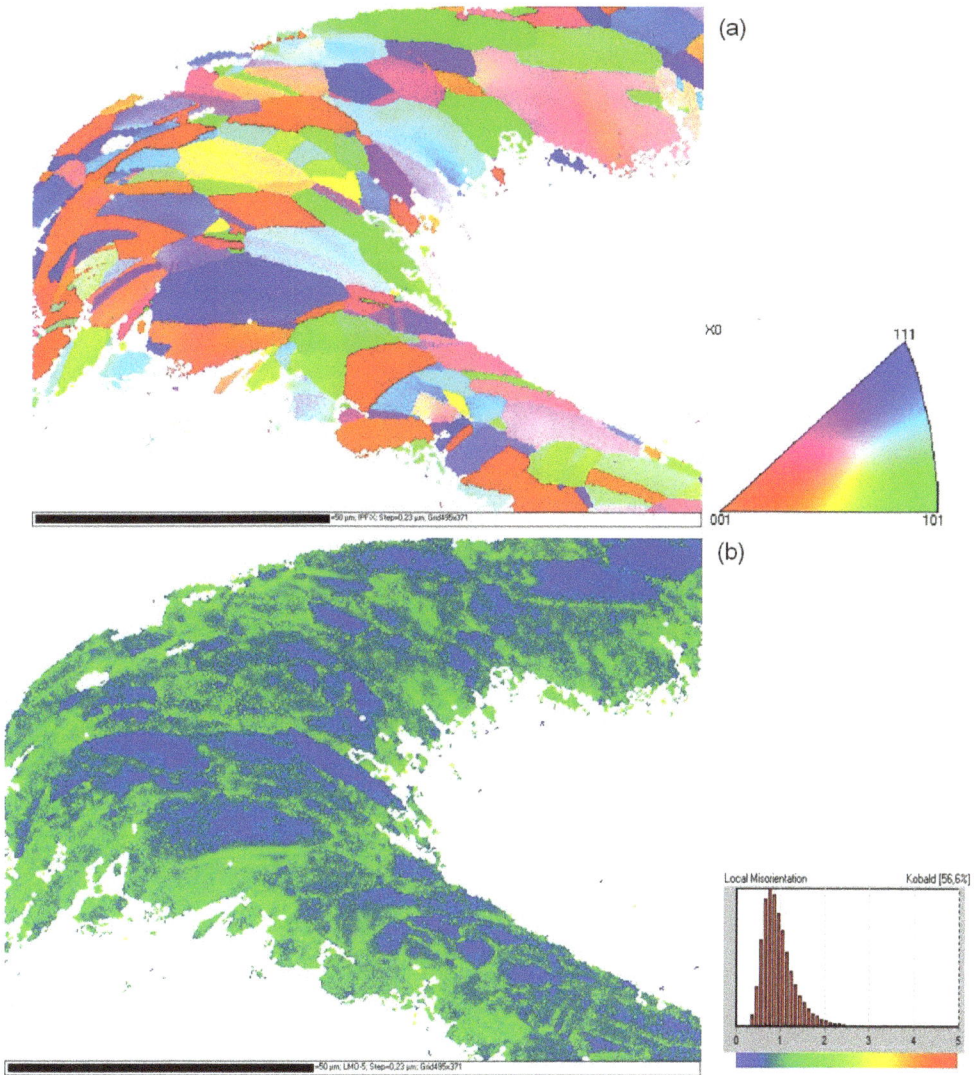

Figure 3. (a) Inverse pole figure orientation mapping of an F-562 stent bow (Driver) after dilation; **(b)** Kernel misorientation mapping of this stent bow. The colorings of the mappings are according to the given legends.

For comparison of the microstructures of the different stent designs in Figure 5 representative orientation mappings of bow segments of each stent type are given. All three designs show microstructures typical for CoCr alloys with relatively straight grain boundaries and lots of twins inside the grains but without preferred orientation. Regions with large misorientations occur in all three designs especially in the inner parts of the bows as available from more or less iridescent colors in the lower right parts of all mappings (Figure 5a–c). Excluding the twins the average grain diameters as well as the average number of grains per strut diameter have been determined. The lowest strut size to grain size ratio, thus, the strongest oligocrystallinity was measured for the

Coroflex® Blue, the largest for the Medtronic Driver. The influence of this ratio will be discussed later. It is difficult to present a quantitative comparison of the amount of misorientation by means of orientation mappings. Therefore the orientation gradient inside the grains along a line from one grain boundary to the opposite one was determined for a large number of grains (300 to 500 grains) separately for each stent design. One characteristic orientation gradient curve as well as the distribution of all orientation gradients of each stent design classified in four main crystallographic orientation groups is shown in Figure 6a,b, respectively.

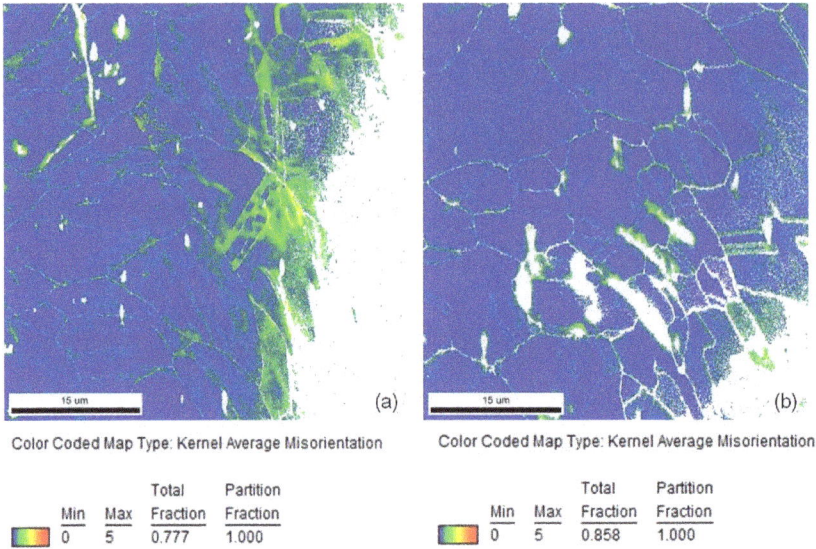

Color Coded Map Type: Kernel Average Misorientation Color Coded Map Type: Kernel Average Misorientation

	Min	Max	Total Fraction	Partition Fraction
	0	5	0.777	1.000

	Min	Max	Total Fraction	Partition Fraction
	0	5	0.858	1.000

Figure 4. Comparison of local misorientation mappings of (**a**) a stent bow and (**b**) a stent strut of a Coroflex® Blue Stent (coloring according to the legend).

A comparison between the three different designs indicates a significantly different behavior. The largest orientation gradients up to more than four degrees are observed in the Coroflex® Blue stent design, the smallest, not more than 2.5 degrees, in the Multilink Vision. These results indicate that larger plastic deformation occurs in the bows of the Coroflex® Blue stent design than in the Multilink Vision design. An interesting correlation with the results of Schmidt [20] is found: In his experiments he measured the elastic recoil of different stent designs and found the largest recoil for the Coroflex® Blue stent design, medium recoil for the Driver, and lowest recoil for the Multilink Vision design. Therefore, one factor of this design dependence of the recoil is possibly attributed to the amount of plastic deformation in the material.

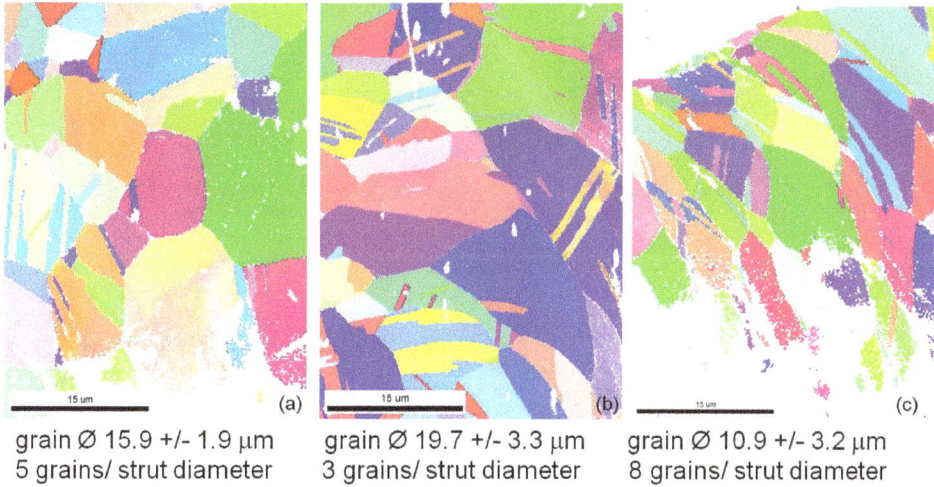

grain Ø 15.9 +/- 1.9 µm   grain Ø 19.7 +/- 3.3 µm   grain Ø 10.9 +/- 3.2 µm
5 grains/ strut diameter   3 grains/ strut diameter   8 grains/ strut diameter

Figure 5. Inverse pole figure orientation mappings of (**a**) Multi Link Vision; (**b**) Coroflex® Blue; and (**c**) Medtronic Driver stent. The colorings of the mappings are according to the legend in Figure 3a.

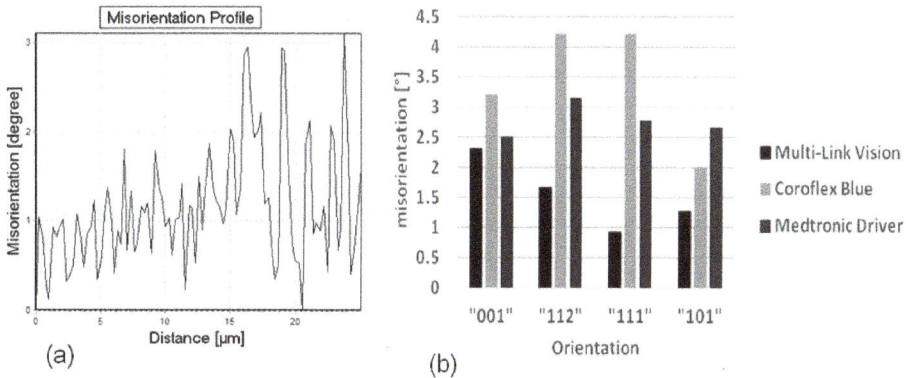

Figure 6. (**a**) Representative misorientation profile and (**b**) Maximum misorientation for different stent designs as a function of the grain orientation.

Furthermore the misorientation is found to be more or less dependent on the grain orientation (Figure 6b). In the Coroflex® Blue and Driver stent designs the highest misorientations are measured in grains with {112}- and {111}- orientations. In the Multilink Vision design the highest misorientations are found in the {001} and {112} grains. This orientation dependence of plastic deformation is found to be more pronounced in the Coroflex® Blue and in the Multilink Vision design and less pronounced in the Driver design. The characteristic strut thickness of Coroflex® Blue is only 60 µm, combined with a relatively large grain size strong oligocrystallinity is obtained for this design. The Multilink Vision with 81 µm strut thickness and medium grain size shows oligocrystalline behavior, too. In contrast, the Medtronic Driver shows only low orientation dependency, thus a nearly polycrystalline behavior occurs. This result is in good correlation to the assumption that oligocrystalline deformation behavior begins when the grain size reaches the construction size. Up to now it is still not clear at which grain size to strut size ratio the homogenous deformation changes into an orientation dominated deformation but this value can be localized between 6 and 10 grains per diameter and seems to be floating. Nevertheless, the current results indicate that the influence of the individual grains on the total deformation decreases with increasing strut thickness and decreasing grain size. Hence, further investigations are necessary for a statistical validation of these results.

Sliding Behavior of the Material

As available from Figure 6a there is not only a high misorientation gradient within the deformed grains but also considerable fluctuations between neighboring measuring points, for example 1.5 degrees along a distance of only one micrometer. A possible explication is the typical deformation behavior of cobalt chromium alloys, the primary planar slip. Due to their low stacking fault energy slip is concentrated on only a few slip systems. There is nearly no possibility for the dislocations to deviate from their slip plane by cross slip or climbing. Figures 7 to 9 present the microstructures of highly deformed areas of stent bows of all three designs. In all three designs characteristic deformation structures consisting of parallely-oriented and crossing deformation bands become visible. In addition, nanocarbides can be identified in the Multilink Vision microstructure (Figure 7b). A deformation structure like this can be an explanation for the considerable orientation fluctuations between neighboring measuring points observed in the orientation gradient distribution in Figure 6a. In Figures 8b and 9b two deformation bands are visible. The width of these bands, about 500 nanometers, is in exact correlation with the spacing of the oscillating orientation gradient. A deformation induced phase transformation from fcc alpha cobalt to hcp beta cobalt is possible.

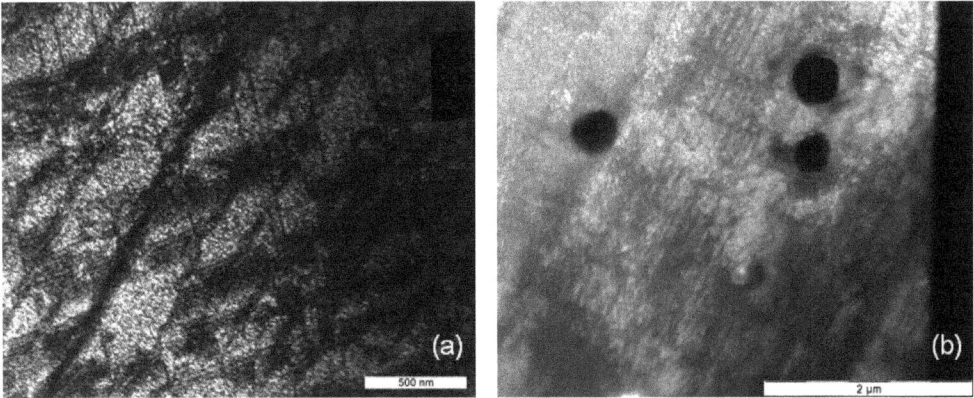

Figure 7. (**a**) Deformation structure and (**b**) nanocarbides in an L605 stent bow (Multilink Vision).

Under planar slip dislocations, twins, stacking faults, and beta cobalt are generated and remain mobile, but only on their discrete sliding planes (see Figures 7a, 8a and 9a, straight lines crossing under distinct angles). Thus, this material shows such distinct localization of sliding under plastic deformation and fatigue, and is known for cyclic softening. However, such sliding behavior may be one reason for the higher tendency for elastic recoil in CoCr-alloy stents than in 316L stents. Even if the strength increases within the shear zone, the dislocations remain mobile but are concentrated on few slip systems with only few obstruction of dislocation movement due to immobile dislocations. Therefore back sliding of dislocations on their slip system can occur resulting in a decline of the plastic deformation introduced during dilation. As a result the "elastic" recoil is probably attributed to a plastic reaction of the material, too.

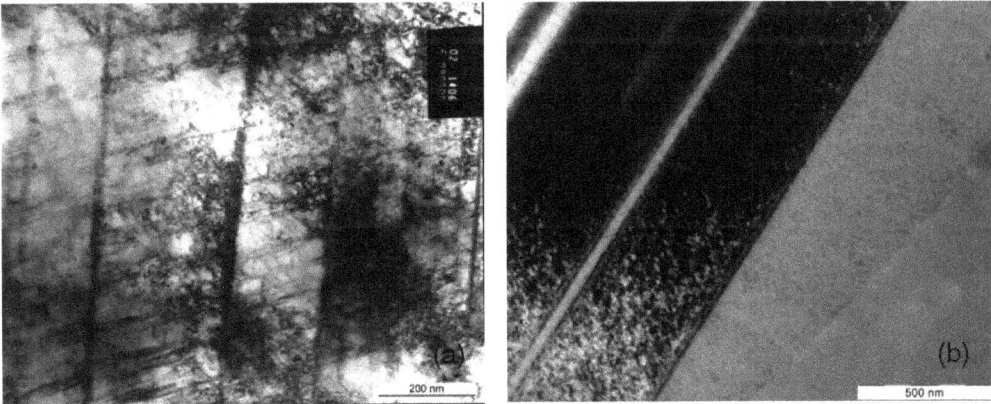

Figure 8. (**a**) Deformation structure and (**b**) deformation bands in an L605 stent bow (Coroflex® Blue).

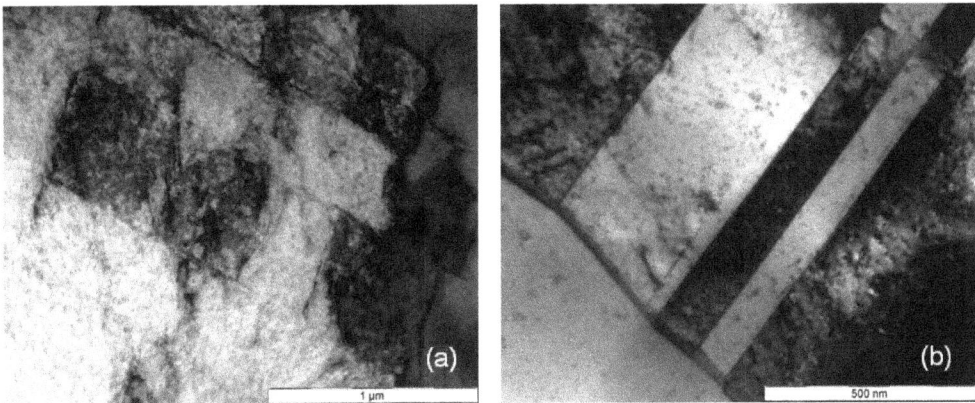

Figure 9. (**a**) Deformation structure and (**b**) deformation bands in an F-562 stent bow (Medtronic Driver).

In summary, it appears that in those metals with predominantly primary slip mechanical instabilities, e.g., like shear bands, crack initiation and propagation take place at distinctly higher stress levels, but are concentrated on few slip planes which can be an explanation for the high radial stiffness of CoCr stents. The result of slip concentration is higher dislocation mobility and therefore a higher flexibility of the material, advantageous for delivery and dilation of the stent but combined with the negative effect of larger recoil. In the current study no macroscopic fractures have been observed and no cyclic load was investigated. However, materials with strongly localized slip like

CoCr alloys tend to a moderate cyclic softening. Therefore, the potential risk of fracture can be classified as low. This estimation is in good agreement with the results of Kapnisis [15]. He compared stainless steel and CoCr stents with similar geometry with regard to their fracture behavior after 100 million load cycles. According to him, strut fracture could be attributed to fretting wear in overlapping regions in case of multiple stenting. For application of single CoCr stents in straight or low curved arteries an extremely low risk of fracture is expected, even lower than that of the market leading stainless steel stents.

4. Conclusions

The present study gives a more comprehensive understanding of the influence of the stent design on the structure property relationship under monotonic deformations, as a basis for ongoing development of new designs for stent optimization. Therefore, stents of different designs produced from the cobalt chromium alloys L-605 and F-562 were investigated. Microstructure characterization by means of scanning electron microscopy, as well as transmission electron microscopy and single grain orientation determination, show the microstructure and microtexture evolution during deformation. The comparison reveals differences in the amount of plastic deformation between the three designs. Larger plastic deformation occurs in the bows of the Coroflex® Blue stent design than in the Multilink Vision design. Because of the small grain size/sample size ratio, orientation dependent deformation behavior occurs, which is more pronounced the thinner the strut thickness is.

The investigation of microstructure indicates that in all three CoCr alloy stent designs typical deformation behavior of primary planar slip takes place. This deformation behavior can be attributed not only to the high radial stiffness of CoCr stents but also to a high flexibility of the material, nonetheless combined with the negative effect of larger recoil.

As a first result, the advantageous properties of CoCr alloy stents predominate for clinical use, but the research is still ongoing. Further experiments, for example with regard to the grain size to strut size relationship of the stents or a modification of the dilation process are in progress and will be published in the near future.

Acknowledgments

Thanks to Sarah Radau for the performance of some of her measurements. The authors would like to thank Birgit Gleising from ICAN, Duisburg for her assistance with specimen preparation and the staff of the chair "materials science" of Ruhr University Bochum for FIB preparation. In addition we are indebted to the "Deutsche Forschungsgemeinschaft" (DFG) for financial support under contract WE 2671/5-1.

Author Contributions

Please prepare a short, one paragraph statement giving the individual contribution of each co-author to the reported research and the writing of the paper.

Conflicts of Interest

The authors declare no conflict of interest.

References

1. *Nichtaktive chirurgische Implantate Besondere Anforderungen an Herz- und Gefäßimplantate—Spezielle Anforderungen an Arterienstents*; DIN EN 14299; Beuth Verlag: Berlin, Germany, 2004.
2. Polak, J.; Hajek, M.; Obrtlik, K. Stress-strain response of austenitic steel 316L in cyclic straining. *Metal. Mater.* **1993**, *31*, 198–205.
3. Bannard, J.-E.; O-Malley, P. The effect of ECM on the fatigue life of Co27Cr5Mo3Ni and stainless steel 316L. ISEM 7. In Proceedings of the 7th International Symposia on Electromachining, IFS (Conferences)/CIRP, Birmingham, UK, 12–14 April 1983; pp. 443–452.
4. Windelband, B.; Schinke, B.; Munz, D. Determination of strain components in low-cycle multiaxial fatigue tests on tubes. *Int. J. Fatigue* **1996**, *162*, 47–53.
5. Mani, G.; Feldman, M.D.; Patel, D.; Agrawal, C.M. Coronary stents: A materials perspective. *Biomaterials* **2007**, *28*, 1689–1710.
6. Klocke, A.; Kemper, J.; Schulze, D.; Adam, G.; Kahl-Nieke, B. Magnetic field interactions of orthodontic wires during magnetic resonance imaging (MRI) at 1.5T. *J. Orofac. Orthop.* **2005**, *66*, 279–287.
7. Kereiakes, D.J.; Cox, D.A.; Hermiller, J.B.; Midei, M.G.; Bachinsky, W.B.; Nukta, E.D.; Leon, M.B.; Fink, S.; Marin, L.; Lansky, A.J. Usefulness of a cobalt chromium coronary stent alloy. *Am. J. Cardiol.* **2003**, *92*, 463–466.
8. Grogan, J.A.; Leen, S.B.; McHugh, P.E. Comparing coronary stent material performance on a common geometric platform through simulated bench testing. *J. Mech. Behav. Biomed. Mater.* **2012**, *12*, 129–138.
9. Kastrati, A.; Mehilli, J.; Dirschinger, J.; Dotzer, F.; Schühlen, H.; Neumann, F.-J.; Fleckenstein, M.; Pfafferott, C.; Seyfarth, M.; Schömig, A. Intracoronary stenting angiographic results: Strut thickness effect on restenosis outcome (ISAR-STEREO) trial. *Circulation* **2001**, *103*, 2816–2821.
10. Pache, J.; Kastrati, A.; Mehilli, J.; Schühlen, H.; Dotzer, F.; Hausleiter, J.; Fleckenstein, M.; Neumann, F.-J.; Sattelberger, U.; Schmitt, C.; *et al.* Intracoronary stenting angiographic results: Strut thickness effect on restenosis outcome (ISAR-STEREO-2) trial. *J. Am. Coll. Cardiol.* **2003**, *41*, 1283–1288.
11. Briguori, C.; Sarais, C.; Pagnotta, P.; Liistro, F.; Montorfano, M.; Chieffo, A.; Sgura, F.; Corvaja, N.; Albiero, R.; Stankovic, G.; *et al.* In-stent restenosis in small coronary arteries: Impact of strut thickness. *J. Am. Coll. Cardiol.* **2002**, *40*, 403–409.
12. Rittersma, S.Z.H.; de Winter, R.J.; Koch, K.T.; Bax, M.; Schotborgh, C.E.; Mulder, K.J.; Tijssen, J.G.P.; Piek, J.J. Impact of strut thickness on late luminal loss after coronary artery stent placement. *Am. J. Cardiol.* **2004**, *93*, 477–480.

13. Kitabata, H.; Kubo, T.; Komukai, K.; Ishibashi, K.; Tanimoto, T.; Ino, Y.; Takarada, S.; Ozaki, Y.; Kashiwagi, M.; Orii, M.; *et al.* Effect of strut thickness on neointimal atherosclerotic change over an extended follow-up period (≥4 years) after bare-metal stent implantation: Intracoronary optical coherence tomography examination. *Am. Heart J.* **2012**, *163*, 608–616.

14. Ota, T.; Ishii, H.; Sumi, T.; Okada, T.; Murakami, H.; Suzuki, S.; Kada, K.; Tsuboi, N.; Murohara, T. Impact of coronary stent designs on acute stent recoil. *J. Cardiol.* **2014**, *64*, 347–352.

15. Kapnisis, K.; Constantinides, G.; Georgiou, H.; Cristea, D.; Gabor, C.; Munteanu, D.; Brott, B.; Anderson, P.; Lemons, J.; Anayiotos, A. Multi-scale mechanical investigation of stainless steel and cobalt-chromium stents. *J. Mech. Behav. Biomed. Mater.* **2014**, *40*, 240–251.

16. Weiß, S.; Schnauber, T.; Fischer, A. Microstructure characterization of thin structures after deformation. In *Materials and Devices for Smart Systems III*; Su, J., Wang, L.-P., Furuya, Y., Trolier-McKinstry, S., Leng, J., Eds.; Cambridge University Press: Warrendale, PA, USA, 2009; pp. V11–V19.

17. Weiss, S.; Meißner, A.; Fischer, A. Microstructural changes within similar coronary stents produced from two different austenitic steels. *J. Mech. Behav. Biomed. Mater.* **2009**, *2*, 210–216.

18. Weiß, S.; Klement, U. Orientation determination on non planar surfaces. In Proceedings of the Channel Users Meeting, Ribe, Denmark, 9–11 June 2004; pp. 62–66.

19. Weiß, S.; Meißner, A. Ermüdung und mikrostruktur von koronaren stents. *Mater. Werkst.* **2006**, *37*, 755–761.

20. Schmidt, W.; Behrens, P.; Schmitz, K.P. Biomechanical aspects of potential stent malapposition at coronary stent implantation. In *IFMBE Proceedings 25/VI*; Dössel, O., Schlegel, W.C., Eds.; Springer Verlag: Berlin, Heidelberg, Germany; New York, NY, USA, 2009; pp. 136–139.

Additively Manufactured Open-Cell Porous Biomaterials Made from Six Different Space-Filling Unit Cells: The Mechanical and Morphological Properties

Seyed Mohammad Ahmadi, Saber Amin Yavari, Ruebn Wauthle, Behdad Pouran, Jan Schrooten, Harrie Weinans and Amir A. Zadpoor

Abstract: It is known that the mechanical properties of bone-mimicking porous biomaterials are a function of the morphological properties of the porous structure, including the configuration and size of the repeating unit cell from which they are made. However, the literature on this topic is limited, primarily because of the challenge in fabricating porous biomaterials with arbitrarily complex morphological designs. In the present work, we studied the relationship between relative density (RD) of porous Ti6Al4V EFI alloy and five compressive properties of the material, namely elastic gradient or modulus (E_{s20-70}), first maximum stress, plateau stress, yield stress, and energy absorption. Porous structures with different RD and six different unit cell configurations (cubic (C), diamond (D), truncated cube (TC), truncated cuboctahedron (TCO), rhombic dodecahedron (RD), and rhombicuboctahedron (RCO)) were fabricated using selective laser melting. Each of the compressive properties increased with increase in RD, the relationship being of a power law type. Clear trends were seen in the influence of unit cell configuration and porosity on each of the compressive properties. For example, in terms of E_{s20-70}, the structures may be divided into two groups: those that are stiff (comprising those made using C, TC, TCO, and RCO unit cell) and those that are compliant (comprising those made using D and RD unit cell).

Reprinted from *Materials*. Cite as: Ahmadi, S.M.; Yavari, S.A.; Wauthle, R.; Pouran, B.; Schrooten, J.; Weinans, H.; Zadpoor, A.A. Additively Manufactured Open-Cell Porous Biomaterials Made from Six Different Space-Filling Unit Cells: The Mechanical and Morphological Properties. *Materials* **2015**, *8*, 1871-1896.

1. Introduction

In orthopaedic surgery, cellular structures are used as three-dimensional porous biomaterials that try to mimic the structure and function of bone [1]. The porous biomaterial could be used either as a bone substitute or a cell-seeded scaffold used as a part of a tissue engineering approach. In either case, the porous biomaterial should be designed such that its mechanical properties match those of bone, while considering the other factors that maximize bone ingrowth. For example, the permeability of the porous structures used in bone tissue engineering could influence cell migration and mass transport and should be carefully designed [2,3]. During the last two decades, several design principles have been proposed for the design of bone tissue engineering scaffolds that consider the mechanical properties, biocompatibility, biodegradability, and bio-functionality of the scaffold biomaterials [4–9].

In this study, we focused on the compressive properties of porous titanium biomaterials aimed for application in orthopaedic surgery. Solid titanium alloys are often very stiff, exceeding the mechanical properties of bone by up to one order of magnitude [10,11]. The mismatch between the mechanical properties of bone and those of the biomaterial could hinder bone ingrowth, result in stress shielding, bone resorption, and ultimately cause loosening of orthopaedic implants [12–15]. Creating porous structures out of bulk materials, however, results in much lower stiffness values that are comparable with those of bone [10,16,17]. Traditionally, various techniques have been used for fabrication of porous biomaterials including space-holder method, hot isostatic pressing, gel casting, and chemical vapor deposition/infiltration [18–21]. Recently, additive manufacturing techniques have been introduced for manufacturing of porous biomaterials and have several advantages over conventional techniques including their ability to create arbitrarily complex 3D structures, highly accurate and predictable porous structure, and wide materials selection [22–25]. Two widely used AM methods are selective laser melting [26–30] and electron beam melting [31–34]. In addition to favorable mechanical properties, highly porous biomaterials have a large pore space that could be used for controlled release of growth factors [35] as well as huge surface area that could be treated using chemical and electrochemical techniques for obtaining desired bio-functional properties [36–39].

The mechanical properties of additively manufactured porous biomaterials are highly dependent on the type of unit cell from which they are made [40–45]. Optimizing the mechanical properties of porous biomaterials for different applications may require combining various types and dimensions of unit cells in one single porous structure. It is therefore important to have a good understanding of the relationship between the type and dimensions of unit cell and the resulting mechanical properties of the porous structure [46]. Many different types of unit cells are available. However, data on the mechanical properties of porous structures from many different unit cell configurations are limited.

In the present work, we used six different unit cell configurations, namely, cubic, diamond, truncated cube, truncated cuboctahedron, rhombic dodecahedron, and rhombicuboctahedron are considered in the current study. For each of these configurations, we used selective laser melting to manufacture porous structures with different porosities. Micro-CT imaging and compression testing were performed to determine the morphological and mechanical properties of the porous materials and to study the relationship between these parameters.

2. Materials and Methods

2.1. Materials and Manufacturing

Selective laser melting (SLM) method (Layerwise NV, Leuven, Belgium) was used for processing of Ti6Al4V-ELI powder (according to ASTM B348, grade 23) on top of a solid titanium substrate and in an inert atmosphere. Porous titanium structures were thereby manufactured based on six different unit cells configurations, namely, cubic, diamond, truncated cube, rhombicuboctahedron, rhombic dodecahedron, and truncated cuboctahedron (Figure 1). The details of the laser process technique were reported in our previous studies [10,16,40,47,48]. For each unit cell, different porosities were achieved by changing the strut thickness and pore size (Table 1). Cylindrical specimens with the length of 15 mm, diameter of 10 mm and unit cell size of 1.5 mm were manufactured for static compression testing (Figure 2). After fabrication, electro discharge machining (EDM) was used to remove the specimens from the substrate.

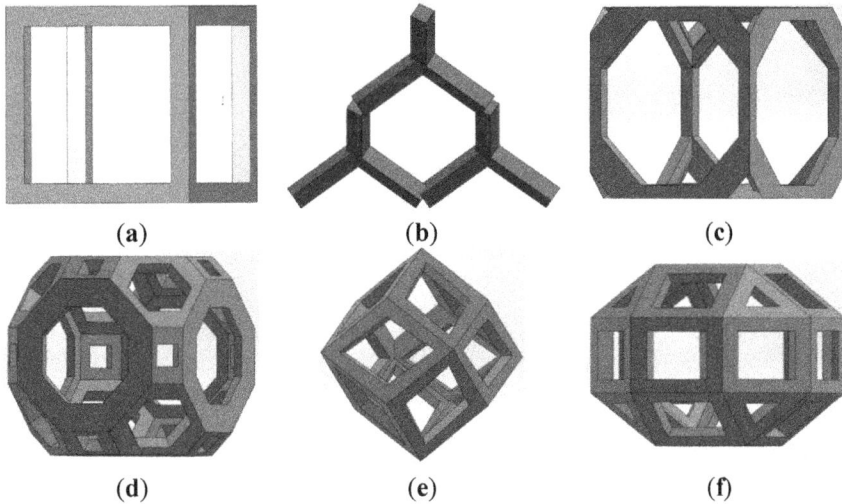

Figure 1. Schematic drawings of the unit cells used in the porous structure: (**a**) Cubic; (**b**) Diamond; (**c**) Truncated cube; (**d**) Truncated cuboctahedron; (**e**) Rhombic dodecahedron; (**f**) Rhombicuboctahedron.

2.2. Morphological Characterization

For morphological characterization, we scanned the titanium scaffolds using a micro-CT (Quantum FX, Perkin Elmer, Waltham, MA, USA). The scans were made under tube voltage of 90 kV, tube current of 180 μA, scan time of 3 min, and resolution of 42 μm. The 3D images of the porous structures were automatically reconstructed using the in-built software of the micro-CT. The reconstructed images were then transferred to the Caliper Analyze 11.0 (provided by the manufacturer) to align the geometry along the major axis of the specimens and to acquire 2D slices. The 2D slices contained transverse views of the scaffolds, *i.e.*, circular cross-sections. The 2D slices were then imported into the ImageJ 1.47v (http://imagej.nih.gov/ij/) in order to create region of interests (ROIs) and segment the titanium volume using the optimal thresholding algorithm available in the boneJ [49] plugin of ImageJ 1.47v (16 bit images). Segmented images were then exported to the boneJ plugin to calculate the ratio of the void volume to the 3D ROI volume that was ultimately reported as the structure relative density of the porous structures.

Table 1. Morphological properties of the porous structures used.

	Strut Diameter (µm)		Pore Size (µm)	
	Nominal (Design)	µCT (SD)	Nominal (Design)	µCT (SD)
Cubic (C)				
C-1	348	451 (147)	1452	1413 (366)
C-2	540	654 (190)	1260	1139 (359)
C-3	612	693 (200)	1188	1155 (354)
C-4	720	823 (230)	1080	1020 (311)
Diamond (D)				
D-1	277	240 (46)	923	958 (144)
D-2	450	416 (65)	750	780 (141)
D-3	520	482 (70)	680	719 (130)
D-4	600	564 (76)	600	641 (137)
Truncated Cube (TC)				
TC-1	180	331 (76)	1720	1625 (398)
TC-2	240	363 (80)	1660	1615 (392)
TC-3	304	395 (88)	1596	1593 (382)
TC-4	380	463 (126)	1520	1535 (370)
TC-5	460	568 (183)	1440	1497 (360)
TC-6	530	620 (200)	1370	1426 (357)
Truncated Cubeoctahedron (TCO)				
TCO-1	324	350 (60)	876	862 (349)
TCO-2	460	416 (64)	1040	1142 (383)
TCO-3	520	452 (65)	980	1098 (386)
TCO-4	577	482 (70)	923	1079 (391)
TCO-5	621	516 (82)	862	1065 (361)
TCO-6	693	564 (76)	807	1049 (383)
Rhombicdodecahdron (RD)				
RD-1	250	246 (53)	1250	1299 (449)
RD-2	310	305 (97)	1190	1224 (455)
RD-3	370	440 (126)	1130	1168 (364)
RD-4	430	461 (163)	1070	1305 (554)
RD-5	490	430 (122)	1010	920 (300)
RD-6	550	506 (144)	950	1058 (356)
Rhombic Cubeoctahedron (RCO)				
RCO-1	380	348 (59)	820	877 (355)
RCO-2	410	369 (59)	790	847 (349)
RCO-3	440	486(113)	760	1089 (402)
RCO-4	470	437 (61)	730	754 (359)
RCO-5	500	539 (120)	700	1043 (401)
RCO-6	530	438 (61)	670	794 (368)

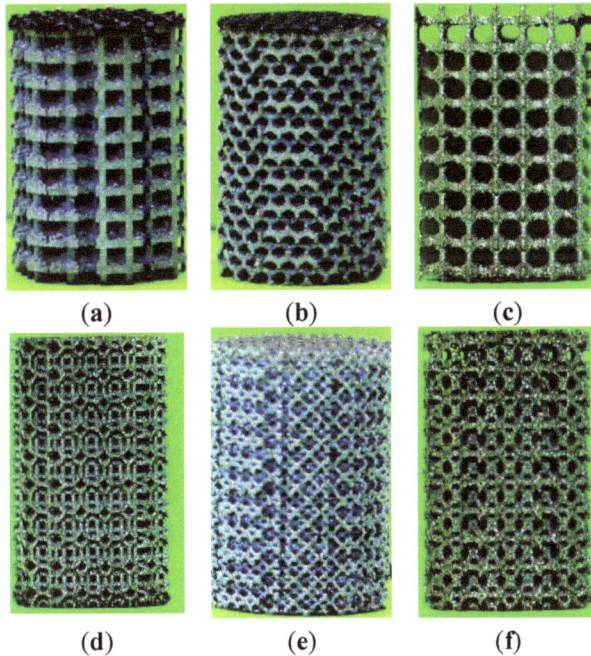

Figure 2. Sample specimens from the porous structures based on different types of unit cells: (**a**) Cubic; (**b**) Diamond; (**c**) Truncated cube; (**d**) Truncated cuboctahedron; (**e**) Rhombic dodecahedron; (**f**) Rhombicuboctahedron.

In addition, the Archimedes technique and dry weighing were used for determining the structure relative density of the specimens (Table 2) using five specimens from each porous type of porous structure, except for the case of rhombic dodecahedron unit cells that only 2 samples were used for measurement of the Archimedes porosity values. In both cases, an OHAUS Pioneer balance was used for weight measurements that were performed in normal atmospheric conditions in room temperature. As for the dry weighing, the weight of the porous specimens was divided by the theoretical weight of the corresponding solid specimens assuming a theoretical density of 4.42 g/cm^3 for Ti6Al4V-ELI [50]. In the Archimedes technique, the specimens were weighed both in dry conditions and in pure ethanol to determine the actual and macro volume and calculating overall porosity of the porous structures.

2.3. Compressive Testing

The mechanical properties of the porous structures were obtained by static compression test using a static test machine (INSTRON 5985, 100 kN load cell) by applying a constant deformation rate of 1.8 mm/min. The compression test was carried out in accordance with ISO standard 13314:2011 [51] which refers to mechanical testing of porous and cellular metals. The tests were continued until 60% strain was applied to the specimens. Five specimens were tested for every variation of the porous structures. The stress-strain curves were obtained and the mean and standard deviation of each of five compressive properties were determined. According to the above-mentioned standard, the elastic gradient ($E\sigma_{20-70}$) was calculated as the gradient of the elastic straight line between two stress values, namely σ_{70} and σ_{20}. The first maximum compressive strength (σ_{max}) that corresponds to the first local maximum in the stress-strain curve was also calculated. The plateau stress (σ_y) was defined according to the same standard as the arithmetical mean of the stresses between 20% and 40% compressive strain and was calculated for all specimens. [40,51]. Energy absorption, which is defined as the energy required for deforming a specimen to a strain (ε), was calculated from the area under the strain-stress curve up to 50% strain [52,53].

In order to analyze the compressive properties of porous structures more systematically, power laws relating structure relative density (the weight per unit volume of a material, including voids that exist in the tested material" as defined in ASTM D1895) to different compressive properties were fitted to the measured experimental data:

$$X = a\rho^b \tag{1}$$

where X is any of the above-mentioned compressive properties measured for the porous structures and ρ is structure relative density. The parameters a and b are dependent on the type of the unit cell.

2.4. Correlational Analysis

MATLAB and Simulink R2014a, The MathWorks Inc., Natick, MA, USA, and Microsoft Excel, Microsoft Corporation, Redmond, WA, USA, were used to determine the correlation between the compressive properties of specimens and relevant density. Closeness of the data to the fitted regression line was measured by coefficient of determination.

3. Results

The structure relative density of each unit cell configuration is presented in Table 2. The trends observed in the stress strain curves of the specimens with different types of unit cells were quite different (Figures 3–8). There were also differences in the shape of stress-strain curves of specimens with the same type of unit cell configuration but different relative density (RD) (Figures 3–8). In many cases, the typical stress-strain response of porous alloy was observed including the initial linear response that was followed by a plateau region and the subsequent fluctuations of the stress-strain curve (Figures 3–8). The final part of the stress-strain curves was often associated with stiffening of the porous structure (Figures 3–8). In general, the level of fluctuations following the plateau region tended to decrease as the structure relative density of the porous structures increased (Figures 3–8). However, this was not, the case for porous structures based on the truncated cube unit cell (Figure 8).

Table 2. Summary of the structure relative density results (in %).

	Structure Relative Density (%)			
	CAD File	Dry Weighing (SD)	Archimedes (SD)	μCT
Cubic (C)				
C-1	10	11 (0.1)	12 (0.1)	13
C-2	22	21 (0.2)	22 (0.2)	24
C-3	27	26 (0.2)	26 (0.2)	28
C-4	35	34 (0.1)	34 (0.2)	37
Diamond (D)				
D-1	11	11 (0.1)	11 (0.2)	11
D-2	21	20 (0.2)	21 (0.1)	21
D-3	28	26 (0.4)	27 (0.3)	28
D-4	37	34 (0.3)	35 (0.4)	36
Truncated cube (TC)				
TC-1	6	7 (0.1)	7(0.1)	9
TC-2	9	9 (0.1)	9 (0.1)	11
TC-3	12	12 (0.1)	12 (0.1)	12
TC-4	16	14 (0.2)	15 (0.2)	14
TC-5	21	17 (0.2)	18 (0.1)	17
TC-6	24	20 (0.2)	20 (0.2)	20
Truncated Cubeoctahedron (TCO)				
TCO-1	18	20 (0.4)	20 (0.4)	19
TCO-2	21	23 (0.2)	23 (0.2)	21
TCO-3	26	25 (0.5)	25 (0.5)	23
TCO-4	31	28 (0.2)	28 (0.3)	28
TCO-5	34	31 (0.3)	31 (0.3)	32
TCO-6	36	34 (0.2)	35 (0.3)	36
Rhombicdodecahdron (RD)				
RD-1	10	11 (0.3)	11 (0.4)	11
RD-2	15	17 (0.2)	17 (0.1)	16
RD-3	20	23 (0.2)	23 (0.1)	22
RD-4	25	27 (0.1)	27 (0.2)	27
RD-5	29	28 (0.3)	28 (0.3)	28
RD-6	34	33 (0.3)	33 (0.2)	32
Rhombic Cubeoctahedron (RCO)				
RCO-1	16	18 (0.2)	18 (0.2)	18
RCO-2	18	21 (0.2)	21 (0.2)	21
RCO-3	21	23 (0.3)	23 (0.3)	24
RCO-4	26	25 (0.3)	26 (0.4)	25
RCO-5	31	29 (0.4)	29 (0.4)	27
RCO-6	36	32 (0.3)	33 (0.5)	31

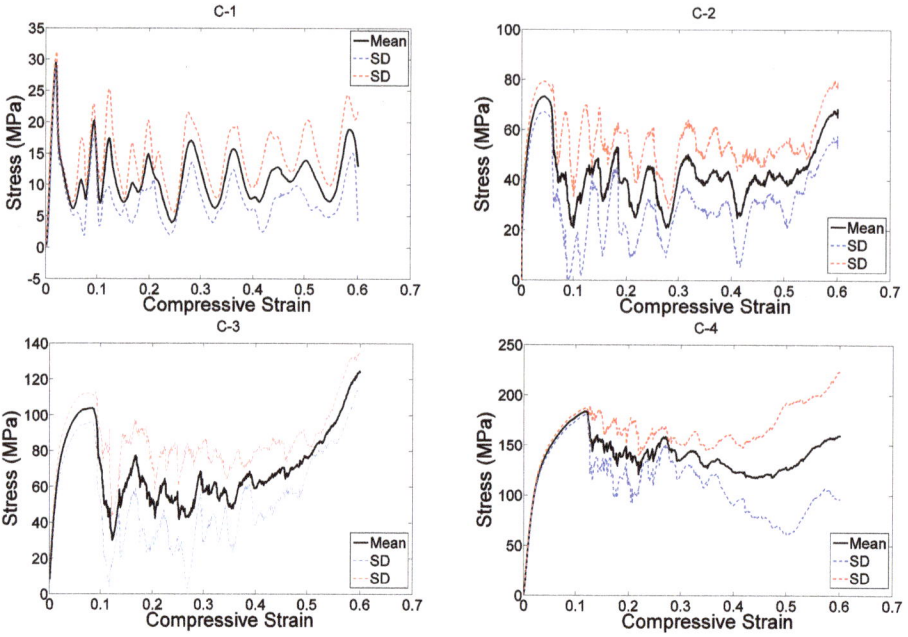

Figure 3. Compressive stress-*versus*-compressive strain curves for specimens based on the cube unit cell and with different porosities (see Table 2).

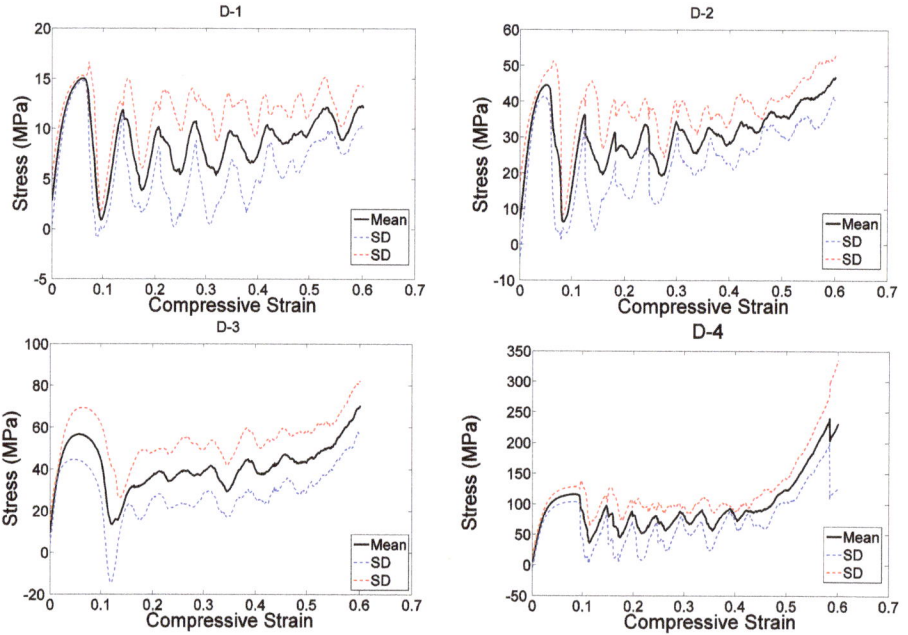

Figure 4. Stress-strain curves for specimens based on the diamond unit cell and with different porosities (see Table 2).

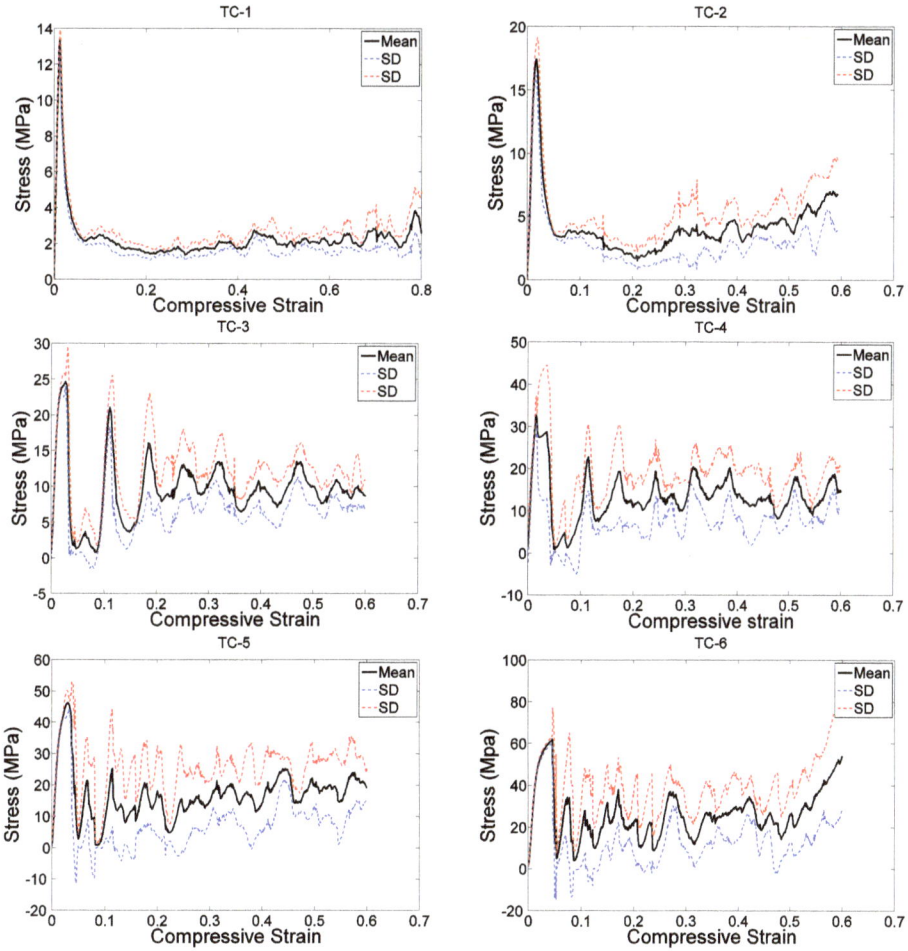

Figure 5. Compressive stress-*versus*-compressive strain curves for specimens based on the truncated cube unit cell and with different porosities (see Table 2).

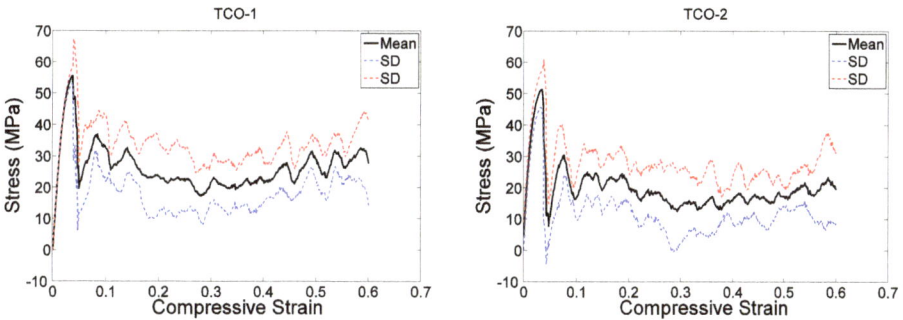

Figure 6. *Cont.*

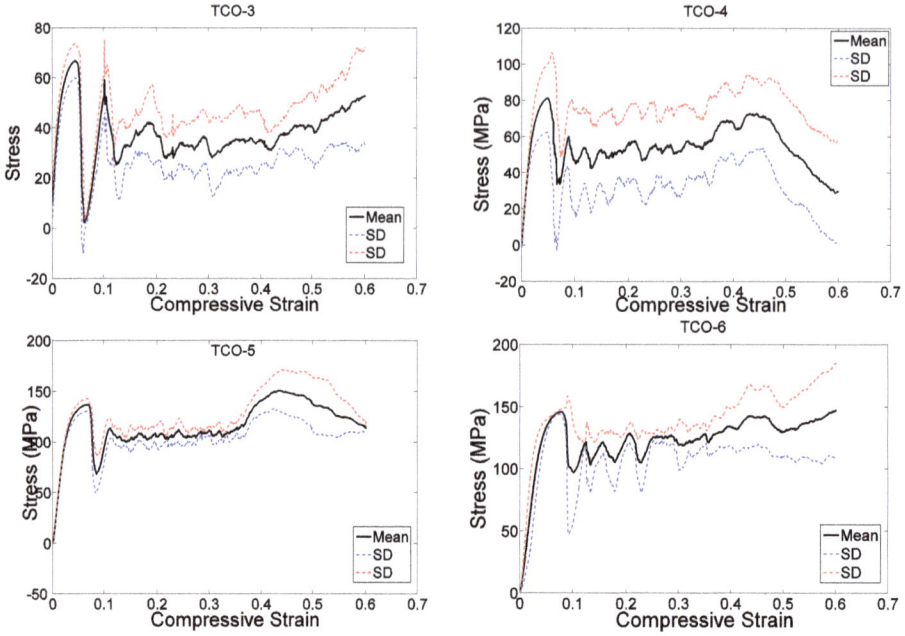

Figure 6. Compressive stress-*versus*-compressive strain curves for specimens based on the truncated cuboctahedron unit cell and with different porosities (see Table 2).

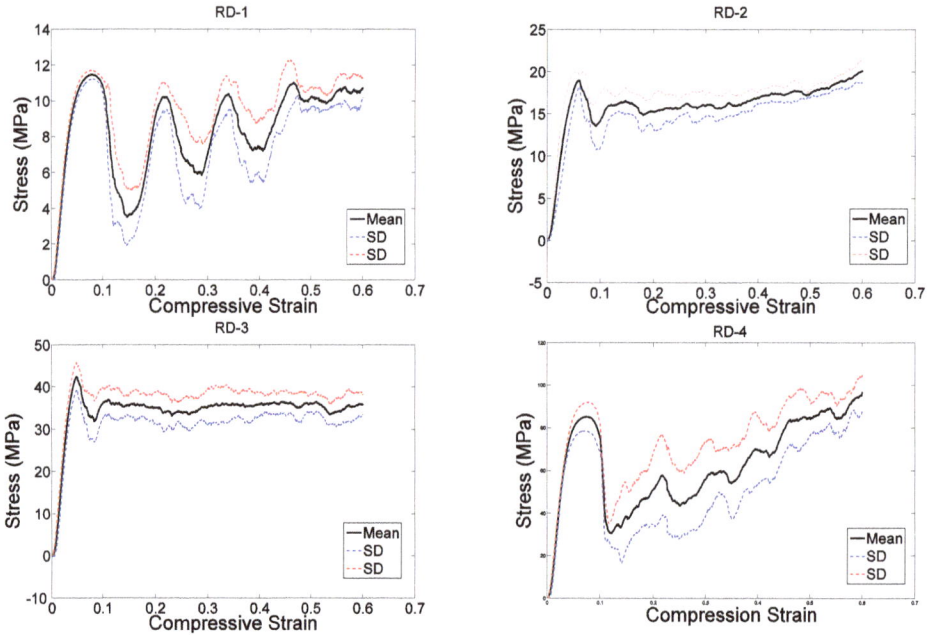

Figure 7. *Cont.*

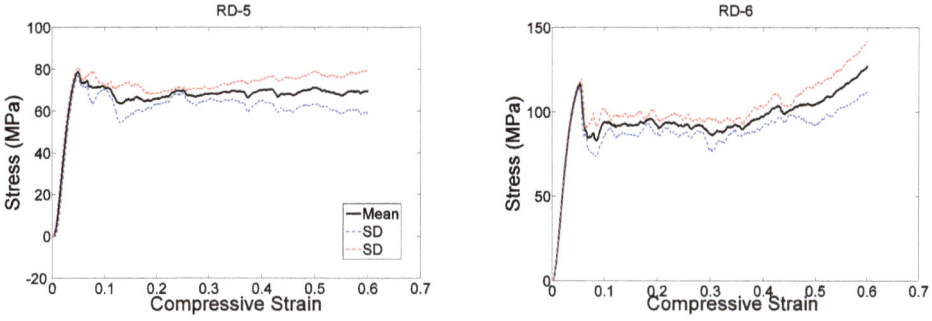

Figure 7. Compressive stress-*versus*-compressive strain for specimens based on the rhombic dodecahedron unit cell and with different porosities (see Table 2).

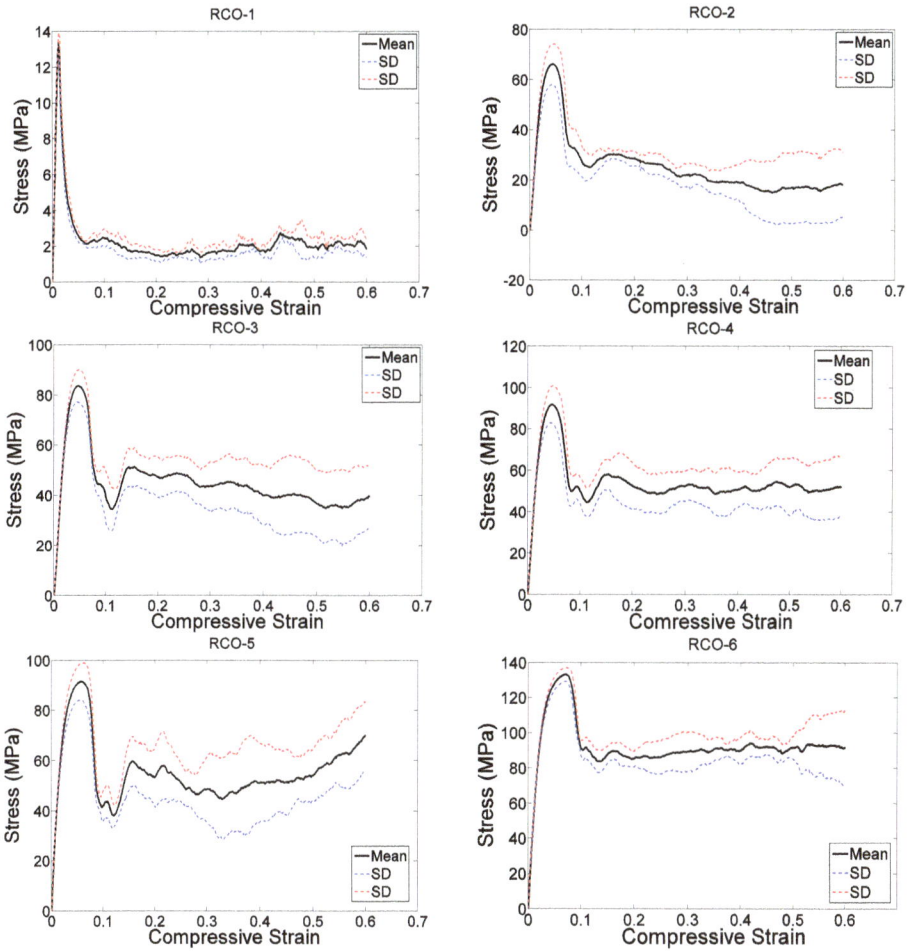

Figure 8. Compressive stress-*versus*-compressive strain curves for specimens based on the rhombicuboctahedron unit cell and with different porosities (see Table 2).

As expected, each of the compressive properties increased with increase in structure relative density (Figures 9–13). The exponent of the power law fitted to the experimental data points (Figures 9–13) varied between 0.93 and 2.34 for the elastic gradient (Figure 9), between 1.28 and 2.15 for the first maximum stress (Figure 10), between 1.75 and 3.5 for the plateau stress (Figure 11), between 1.21 and 2.31 for the yield stress (Figure 12), and between 2.18 and 73 for energy absorption (Figure 13).

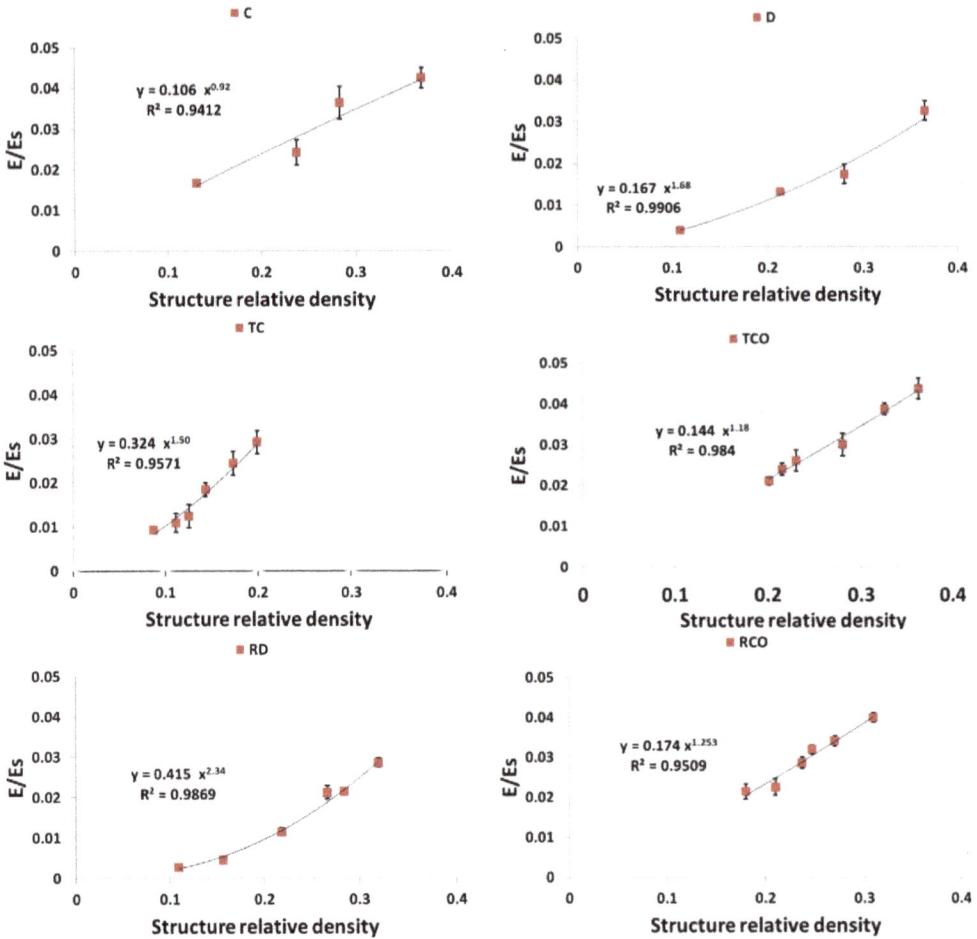

Figure 9. Summary of the elastic gradient results for porous structures based on different types of unit cell configurations (cubic (C); diamond (D); truncatedcube (TC); truncated cuboctahedron (TCO); rhombic dodecahedron (RD); rhombicuboctahedron (RCO)) and different structure relative densities (see Table 2) (E_s indicates the elastic gradient of the structure if it was solid).

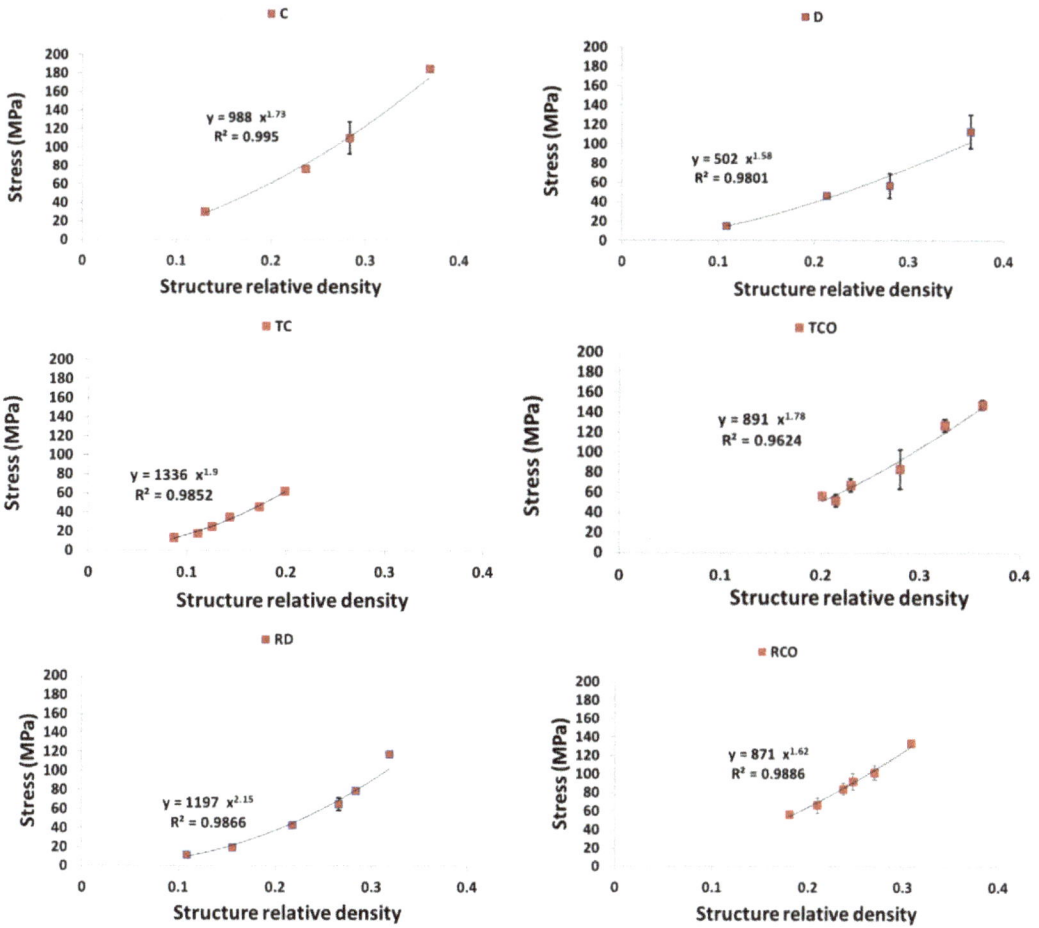

Figure 10. Summary of the first maximum stress results for porous structures based on different types of unit cell configurations (cubic (C); diamond (D); truncated cube (TC); truncated cuboctahedron (TCO); rhombic dodecahedron (RD); rhombicuboctahedron (RCO)) and different structure relative densities (see Table 2).

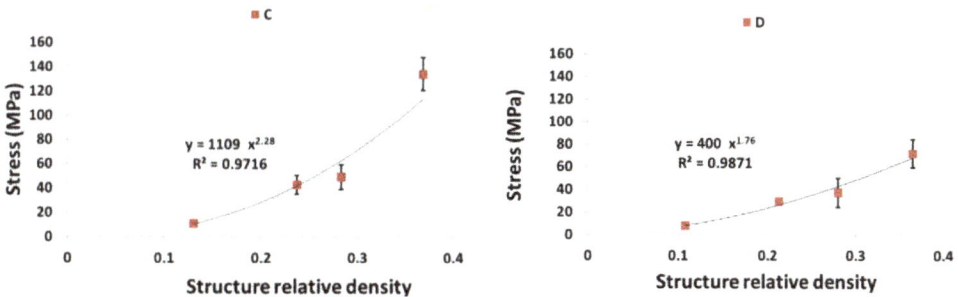

Figure 11. *Cont.*

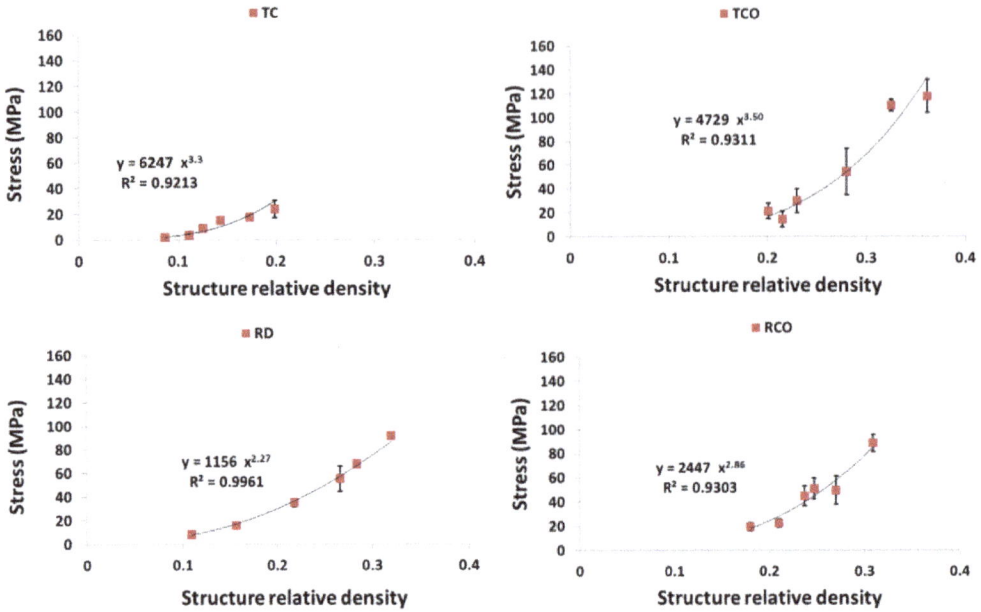

Figure 11. Summary of the plateau stress results for porous structures based on different types of unit cell configurations (cubic (C); diamond (D); truncated cube (TC); truncated cuboctahedron (TCO); rhombic dodecahedron (RD); rhombicuboctahedron (RCO)) and different structure relative densities (see Table 2).

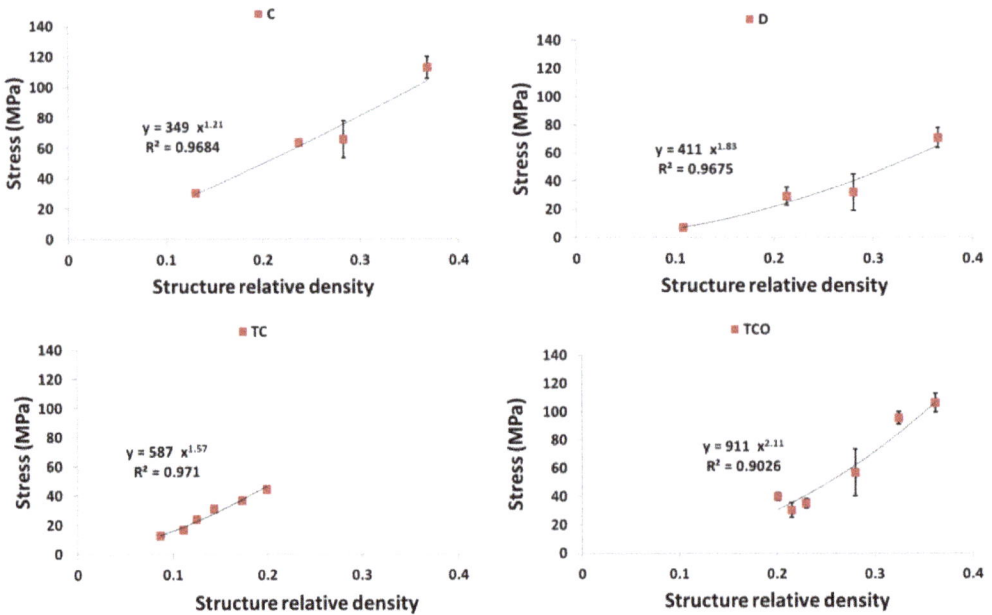

Figure 12. *Cont.*

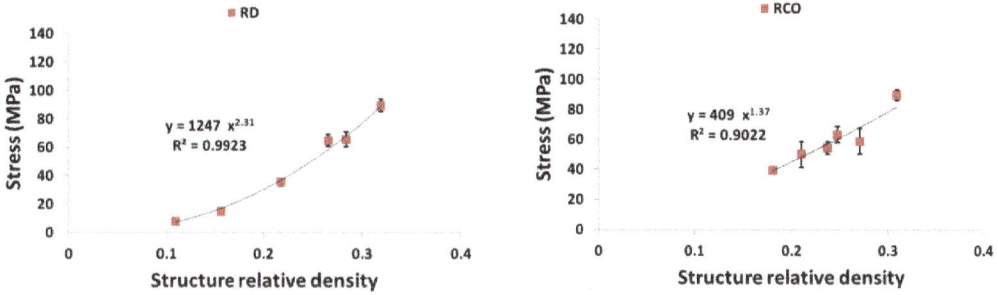

Figure 12. Summary of the yield stress results for porous structures based on different types of unit cell configurations (cubic (C); diamond (D); truncated cube (TC); truncated cuboctahedron (TCO); rhombic dodecahedron (RD); rhombicuboctahedron (RCO)) and different structure relative densities (see Table 2).

Among all the unit cells studied here, the structure with the diamond unit cell was the most compliant, especially at RD > 0.15, whereas the stiffest structure was that having a truncated cube unit cell, especially when RD > 0.30 (Figure 9). When RD was small (RD < 0.2) the structures may be divided into two groups, with those in the first group (truncated cube, truncated cuboctahedron, rhombicuboctahedron, and cube unit cells) having larger stiffness than those in the second group (diamond and rhombic dodecahedron unit cells) (Figures 9 and 14a).

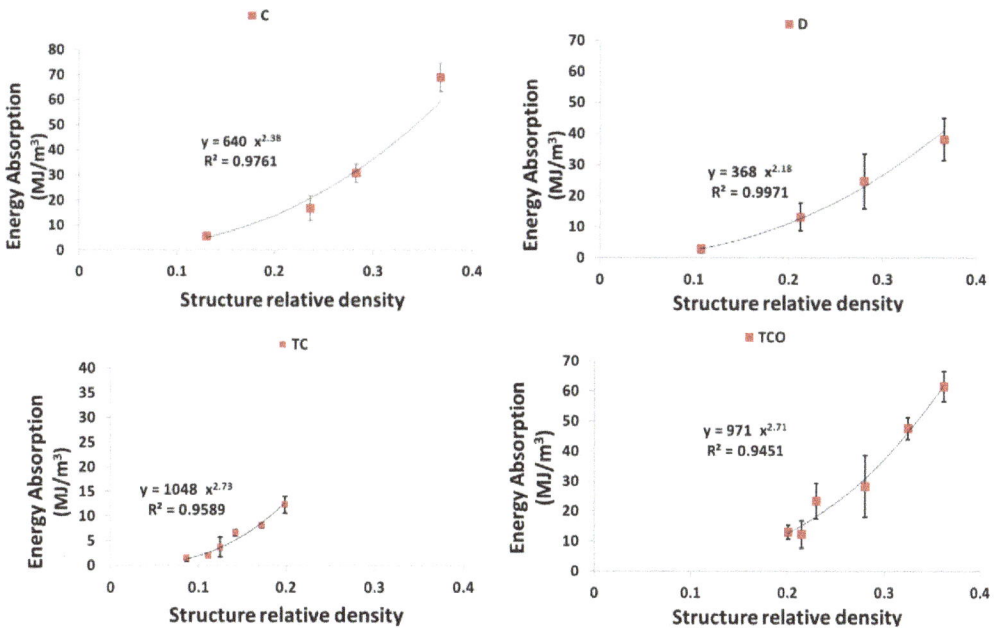

Figure 13. *Cont.*

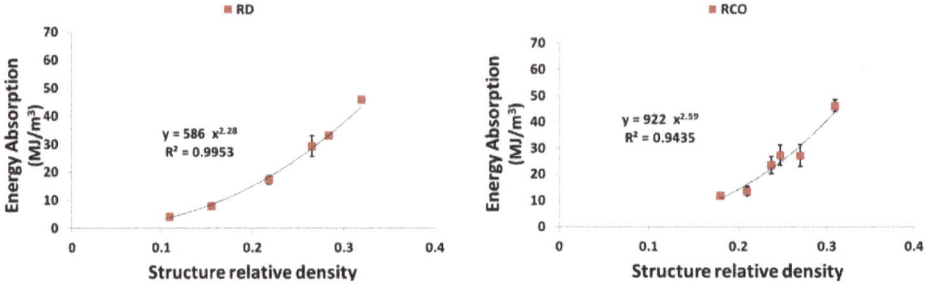

Figure 13. Summary of the energy absorption results for porous structures based on different types of unit cell configurations (cubic (C); diamond (D); truncated cube (TC); truncated cuboctahedron (TCO); rhombic dodecahedron (RD); rhombicuboctahedron (RCO)) and different structure relative densities (see Table 2).

With regard to σmax, there is also separation of the structures into two groups. When RD < 0.2, the structures with the highest and lowest value of this compressive property were built using rhombiccuboctahedron and rhombic dodecahedron unit cells, respectively (Figure 10). However, when RD > 0.2, the structures with the highest and lowest value of this compressive property were built using the truncated cube and diamond unit cells, respectively (Figures 10 and 14b). When RD < 0.2, there is no difference in plateau stress between the different structures, but, when RD > 0.2, the highest and lowest value of this compressive property were built using the truncated cube and diamond unit cells, respectively (Figures 11 and 14c). The four remaining unit cells are relatively close in terms of the plateau stress values they exhibit (Figures 11 and 14c).

Regarding σy, structures with the diamond unit cell show the lowest value throughout the RD range (Figures 12 and 14d). The one group comprising structures having the truncated cube rhombicuboctahedron, and cube and cube and the other group comprising structures having truncated cuboctahedron and rhombic dodecahedron, When RD < 0.2, the former group has clearly higher yield stress values as compared to the latter group, but, when RD > 0.2, the results for the two groups overlapped (Figures 12 and 14d). When RD < 0.2, Energy absorption (EA) for the structures with different unit cell configurations are practically the same, but, at higher RD, EA of structure with diamond unit cell is much lower than that of a structure with any other type of unit cell configuration (Figures 13 and 14e).

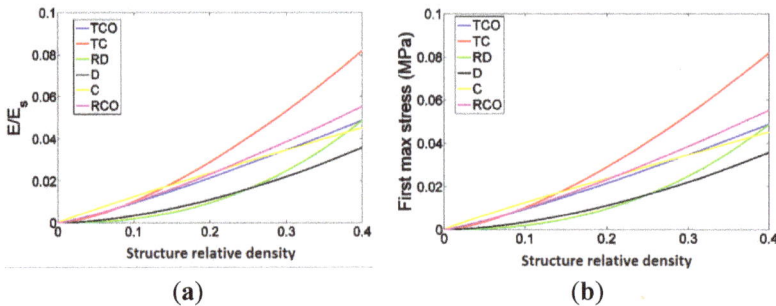

(a)   (b)

Figure 14. *Cont.*

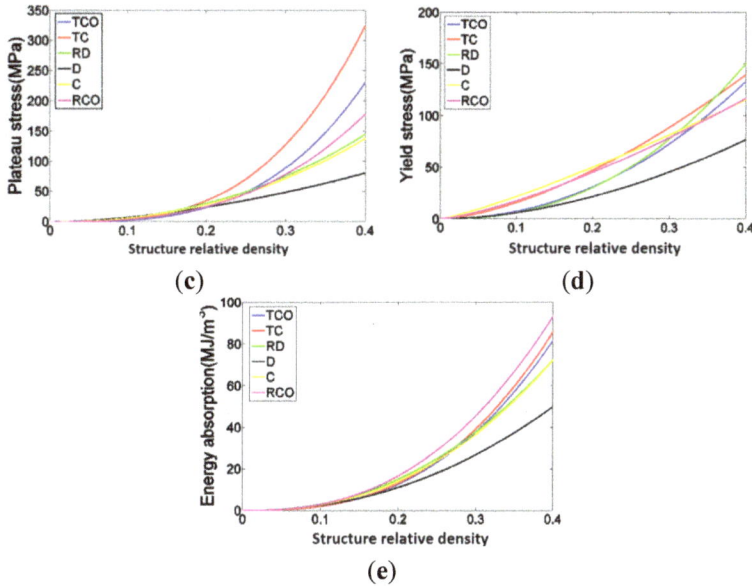

Figure 14. Comparison between the mechanical properties measured for different types of porous structures based on the six different unit cells studied here including (**a**) Elastic gradient; (**b**) First maximum stress. (**c**) Plateau stress; (**d**) Yield stress; (**e**) Energy absorption. In these figures, the power laws fitted to the experimental data points, and not the experimental data points themselves, are compared with each other.

The ratio of plateau stress to yield stress was more or less constant and close to one for the diamond and rhombic dodecahedron unit cells (Figure 15a). For the other types of unit cells, the ratio of plateau stress to yield stress remarkably increased with the relative density (Figure 15a). As for the ratio of plateau stress to first maximum stress, it was relatively stable for diamond, rhombic dodecahedron, and rhombicuboctahedron (Figure 15b). For the three remaining types of unit cells, the ratio of plateau stress to first maximum stress drastically increased with the relative density Figure 15b).

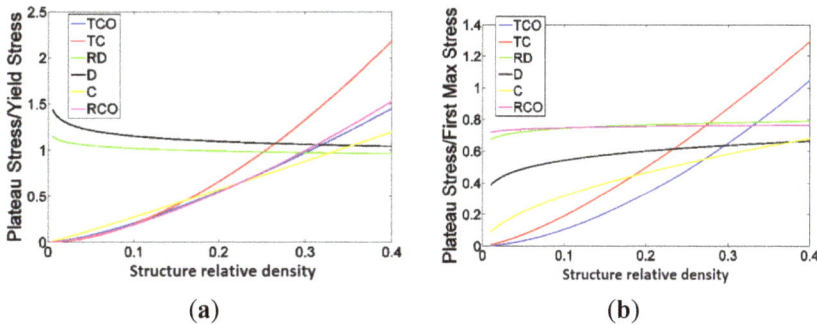

Figure 15. (**a**) The ratio of plateau stress to yield stress as well as (**b**) the ratio of plateau stress to first maximum stress for different types of unit cells. In these figures,

the power laws fitted to the experimental data points, and not the experimental data points themselves, are compared with each other.

4. Discussion

The results of this study clearly show the difference between the porous structures made using different types of unit cells. Not only do the mechanical properties of the porous structures differ drastically between the various unit cells studied here, the deformation and failure mechanisms change as well particularly at the plateau region as well as in the succeeding regions of the stress-strain curves. These different failure mechanisms are reflected in the different shapes of stress-strain curves.

4.1. Comparison between the Different Types of Unit Cells

Since all other parameters are kept constant during the manufacturing of the specimens, the only factor that differentiates the different classes of porous structures from each other is the geometry of unit cell. For example, it was observed that the unit cells that include vertical struts, exhibit a different failure mechanism as compared to the other unit cells. In the unit cells with vertical struts, failure of one (vertical) strut usually resulted in the collapse of the entire unit cell, causing a sudden drop of the measured force to values close to zero. Once one unit cell, that is often the weakest link in the remaining porous structure, has collapsed, the other unit cells take over the force-carrying function of the missing unit cell and the force increases again. This will continue until the next weakest link in the remaining porous structure has collapsed and the force drops to near-zero values again. The presence of vertical struts could not, however, explain all the cases where force repeatedly dropped to near-zero values. An important exception was the diamond unit cell. In this unit cell, the geometry of the unit cell is such that the failure of one strut could easily cause the collapse of the entire unit cell, as the shape of the unit cell is relatively simple and the different struts provide only limited support to each other. This could be also found back in all of the compressive properties measured for the diamond unit cell. Comparatively speaking, the diamond unit cell showed the lowest values of the compressive properties for the entire range of apparent densities. There are only two exceptions, elastic gradient and first maximum stress, where rhombic dodecahedron shows slightly lower compressive properties for the lowest values of the structure relative density.

The stiffness of the porous structures made from different types of unit cells is probably the most important property of these structures when they are used as bone-mimicking biomaterials. The elastic gradient is the best indicator of the stiffness of the porous structure, among all the compressive properties presented here. For small apparent densities, *i.e.*, < 0.15, one could speak of two groups of unit cells, namely strong unit cells and weak unit cells. The strong unit cells group includes truncated cube, truncated cuboctahedron, rhombicuboctahedron, and cube, while the weak unit cell group includes diamond and rhombic dodecahedron. Within each of the groups, there is not much difference between the different types of unit cells for small structure relative density values, meaning that they are interchangeable from mechanical viewpoint. The other considerations such as permeability [3,9] could therefore play more important role when deciding which of those unit cells is used in bone regeneration applications. For larger structure relative density values, *i.e.*, >0.15, the truncated cube unit cell shows remarkably higher stiffness values and could therefore be used in the applications where high stiffness values are required. Since cube and truncated cube are relatively similar unit cells, it is remarkable that such small variation in the geometry of the cubic unit cells results in such improvement in the stiffness values for relatively large apparent densities. One could explain this by noting that in the cube unit cell force transmission occurs at a few junction points that are also prone to stress concentration. Truncated cube replaces the single junction of the cubic unit cell with a supporting structure that could better distribute and transmit the forces. This improves the stiffness of the porous structure particularly for higher apparent densities where the thick struts at the truncation region of the truncated cube unit cell are particularly closed-pack and support the porous structure very efficiently.

4.2. Ratio of Plateau Stress to Yield Stress

One of the important findings of the current study is the point that the relationship between the plateau stress and yield stress is very different for different types of unit cells. In general, plateau stress has received more attention in the recent literature, partly because of the emphasis and explicit definition of the concept in the new ISO standard for the mechanical testing of metallic porous materials [51]. In comparison, there is less emphasis on the concept of yield (or compressive offset) stress in the standard, demoting it to the status of "optional information" in the standard test report [51]. As a consequence, a number of recent studies including our studies on porous structures made from the rhombic dodecahedron unit cell [16,40] and one study of the mechanical behavior of porous structures based on the diamond unit cell [10] have used the concept of plateau stress as a replacement for the yield stress. The results of the current study show that, interestingly, for both types of unit cells used in our previous studies, the plateau and yield stress are very close. Moreover, the ratio of plateau stress to yield stress is largely independent from the structure relative density. This justifies the use of plateau stress as a replacement for the yield stress for the porous structures based on those two types of unit cells. The results of this study, however, show that this is not necessarily the case for the other types of unit cells. Not only the plateau and yield stress are not close to each other for the other types of unit cells, their ratio could be very much dependent on the structure relative density. This is an important point in all future studies where one needs to choose a specific parameter for representing the elastic limit of additively manufactured porous structures based on the different types of unit cells.

4.3. Energy Absorption

Fracture toughness of bone is defined as the resistance to crack growth before the final fracture [54] and several studies on what can influence on fracture toughness of the human bone, cortical and trabecular [55–58] show the importance of this definition. Although tough bone resists more to fracture but it may have lower yield point and be considered weaker [59]. It is therefore important to select the right type of unit cell for bone-mimicking porous structure by comparing the energy absorption values of the porous structures based on the different types of unit cells with that of bone they are aimed to replace. This is an important design aspect has received less attention in the previous studies that look into the mechanical properties of bone-mimicking porous biomaterials and how they are related to those of bone.

4.4. Anisotropy

The mechanical properties of porous structures based on some of the unit cells included in the current study are anisotropic. In the current study, we only studied the mechanical properties of the porous structures in one direction (Figure 1). The mechanical properties of the porous structures may be therefore very different in the directions not tested in the current study. One needs to be careful when interpreting the results presented here, as they only pertain to specific directions of unit cells. The experiment required for characterizing the mechanical properties of the porous structures in all relevant directions is formidably large and expensive. A more feasible approach would be to develop analytical and computational models that are first validated against the experimental data presented here and could then be used for estimating the mechanical properties of the porous structures in all possible directions. In addition to the anisotropy caused by the geometry of the unit cells, the manufacturing process could also cause some directionality in the porous structure [44]. This directionality, which is dependent on the geometry of the unit cell, could also induce some additional anisotropy in the mechanical behaviour of the porous structures.

4.5. Applications in the Design of Implants and Tissue Engineering Scaffolds

The main application of the results presented in the current study is in the design of porous biomaterials used for bone substitution either as an implant or as a part of a bone tissue engineering scheme. The mechanical properties of the porous biomaterials are important from several viewpoints. First, one needs to ensure that there is a good match between the stiffness of porous biomaterial and those of the bone they replace. This could help in avoiding stress shielding. The elastic gradient values reported here for the different types of unit cells could be important in that context. Second, it is important to make sure that the porous biomaterials are capable of providing enough mechanical support and do not fail under the mechanical loading they are exposed to. The plateau stress as well as yield and first maximum stress values reported here could play important roles in that regard.

From a design viewpoint, one needs to ensure that the mechanical properties of the porous biomaterials are favorable for bone regeneration and ingrowth. That is because bone tissue formation is known to be largely driven by mechanical loading [60–63]. The results of the current study clearly show that, for the same structure relative density, the mechanical properties of bone-mimicking porous biomaterials are very much dependent on the morphology of the porous structure including the type of unit cell and the unit cell dimensions. On the other hand, the same morphological properties determine the other important properties of the porous biomaterials such as permeability and diffusivity [2,3,8,9]. The design of porous biomaterials for bone regeneration applications can therefore be defined as a multi-objective optimization problem. There are additional patient-specific aspects that need to be taken into account. It is therefore important to combine the computer models for optimal design of porous biomaterials with patient-specific finite element models of bones [64–66]. The complex and multi-objective nature of such an optimal design problem requires a high degree of flexibility in the design space. Studies such as the present study that help to establish the relationship between the morphological design and the different types of properties of porous biomaterials based on various types of unit cells are helpful in this context. That is because they enable the designers to use a larger library of unit cells for which the different types of properties including mechanical properties are known, thereby enlarging the design space for optimal design of bone substituting implants and tissue engineering scaffolds. Given the production flexibility offered by advanced additive manufacturing techniques such as selective laser melting, different types of unit cells could be combined in one single implant or scaffold so as to optimally distribute the properties within the entire implant or scaffold.

The results presented in this study are also valuable for corroboration of analytical and numerical models that are developed used for prediction of the mechanical properties of porous structures given their designed morphology. This type of experimental data is not currently available in the literature particularly for some of the unit cells studied here.

4.6. Future Research

In this study, all the manufacturing parameters such as building orientation and post processing of the samples [44] or laser power or energy density of the specimens processed by SLM [30] considered to be constant. Changing in any of these parameters will influence the results [42]. It is clear from the results of this study that the deformation and failure mechanisms of porous structures based on the considered unit cells are very different. Even though certain aspects of the deformation and failure mechanisms were studied in the current study, it was not the main focus of the paper. It is suggested that future studies should focus on the detailed deformation and failure mechanisms of additively manufactured porous biomaterials based on different types of unit cells. In particular, it would be useful to perform full-field strain measurement [67–70] during the mechanical testing of the structures, for example, using optical techniques such as digital image correlation (DIC). DIC has been previously used for measurement of strain in engineering [71–74] and biological materials [75–77] and is shown to be capable of capturing the detailed deformation and fracture mechanisms of both types of materials. For determining the mechanical properties only static compressive properties were determined in the present work. In future studies, other relevant mechanical properties, such as static bending strength [46], static torsional strength [46] and fatigue life [50], should be determined.

5. Conclusions

The relationship between morphological and mechanical properties of selective laser melted porous titanium alloy biomaterials based on six different types of space-filling unit cells were studied. It was observed that the mechanical behavior, mechanical properties, and failure mechanisms of the porous structures are highly dependent on the type and dimensions of the unit cells out of which the porous structures are made. As expected, compressive properties of all the porous structures increased with structure relative density. Moreover, for a given compressive property of a porous structure, the dependence on the structure relative density was of the power type. The exponent could be used for generalizing the relationships between structure relative density and the compressive properties of porous structures with different types of unit cells. When comparing the compressive properties of the porous structures based on the different types of unit cells, it was found that in many cases the comparative performance of the structures is different for low and high values of structure relative density with a separating structure relative density of 0.15–0.2. Among all unit cells, the diamond unit cell consistently showed lower compressive properties. Regarding the stiffness values, the unit cells were divided into a high stiffness group including truncated cube, truncated cuboctahedron, rhombicuboctahedron, and cube and a low stiffness group including diamond and rhombic dodecahedron. However, truncated cube showed remarkably higher stiffness than other members of its group for apparent densities exceeding 0.2. The results obtained in the present study revealed the relationship between the morphological and compressive properties of porous structures based on six different types of unit cells, many of which have been so far largely unexplored. Moreover, it could serves as a basis for validation of analytical and computational models developed for estimation of the mechanical properties of additively manufactured porous biomaterials.

Author Contributions

Seyed Mohammad Ahmadi, Saber Amin Yavari, Jan Schrooten, Harrie Weinans and Amir A. Zadpoor conceived the experiments; Seyed Mohammad Ahmadi and Saber Amin Yavari performed the experiments; Seyed Mohammad Ahmadi and Amir A. Zadpoor analyzed the data; Ruebn Wauthle, Behdad Pouran contributed materials and analysis tools; each of the authors reviewed the manuscript for intellectual content.

Conflicts of Interest

The authors declare no conflict of interest.

References

1. Butscher, A.; Bohner, M.; Holmann, S.; Gauckler, L.; Muller, R. Structural and material approaches for bone tissue engineering in powder based 3D printing. *Acta Biomater.* **2013**, *7*, 907–920.
2. Hollister, S.J. Porous scaffold design for tissue engineering. *Nat. Mater.* **2005**, *4*, 518–524.
3. Hollister, S.J. Scaffold design and manufacturing: From concept to clinic. *Adv. Mater.* **2009**, *21*, 3330–3342.
4. Hutmacher, D.W. Scaffold design and fabrication technologies for engineering tissues—State of the art and future perspectives. *J. Biomater. Sci. Polym. Ed.* **2001**, *12*, 107–124.
5. Cook, S.; Dalton, J. Biocompatibility and biofunctionality of implanted materials. *Alpha Omegan* **1991**, *85*, 41–47.

6. Gotman, I. Characteristics of metals used in implants. *J. Endourol.* **1997**, *11*, 383–389.

7. Goulet, R.W.; Goldstein, S.A.; Ciarelli, M.J.; Kuhn, J.L.; Brown, M.B.; Feldkamp, L.A. The relationship between the structural and orthogonal compressive properties of trabecular bone. *J. Biomech.* **1994**, *27*, 375–377.

8. Dias, M.R.; Guedes, J.M.; Flanagan, C.L.; Hollister, S.J.; Fernandes, P.R. Optimization of scaffold design for bone tissue engineering: A computational and experimental study. *Med. Eng. Phys.* **2014**, *36*, 448–457.

9. Dias, M.R.; Fernandes, P.R.; Guedes, J.M.; Hollister, S.J. Permeability analysis of scaffolds for bone tissue engineering. *J. Biomech.* **2012**, *45*, 938–944.

10. Ahmadi, S.M.; Campoli, G.; Yavari, S.A.; Sajadi, B.; Wauthle, R.; Schrooten, J.; Weinans, H.; Zadpoor, A.A. Mechanical behavior of regular open-cell porous biomaterials made of diamond lattice unit cells. *J. Mech. Behav. Biomed. Mater.* **2014**, *34*, 106–115.

11. Niinomi, M. Mechanical properties of biomedical titanium alloys. *Mater. Sci. Eng.* **1998**, *243*, 231–236.

12. Engh, C.; Bobyn, J.; Glassman, A. Porous-coated hip replacement. The factors governing bone ingrowth, stress shielding, and clinical results. *J. Bone Joint Surg. Br.* **1987**, *69*, 45–55.

13. Engh, C.A., Jr.; Young, A.M.; Engh, C.A.; Robert, H., Jr. Clinical consequences of stress shielding after porous-coated total hip arthroplasty. *Clin. Orthop. Relat. Res.* **2003**, *417*, 157–163.

14. Huiskes, R.; Weinans, H.; van Rietbergen, B. The relationship between stress shielding and bone resorption around total hip stems and the effects of flexible materials. *Clin. Orthop. Relat. Res.* **1992**, *274*, 124–134.

15. Nagels, J.; Stokdijk, M.; Rozing, P.M. Stress shielding and bone resorption in shoulder arthroplasty. *J. Shoulder Elbow Surg.* **2003**, *12*, 35–39.

16. Amin Yavari, S.; Ahmadi, S.M.; van der Stok, J.; Wauthle, R.; Riemslag, A.C.; Janssen, M.; Schrooten, J.; Weinans, H.; Zadpoor, A.A. Effects of bio-functionalizing surface treatments on the mechanical behavior of open porous titanium biomaterials. *J. Mech. Behav. Biomed. Mater.* **2014**, *36*, 109–119.

17. Van der Stok, J.; van der Jagt, O.P.; Yavari, S.A.; de Haas, M.F.P.; Waarsing, J.H.; Jahr, H.; van Lieshout, E.M.M.; Patka, P.; Verhaar, J.A.N.; Zadpoor, A.A.; *et al.* Selective laser melting-produced porous titanium scaffolds regenerate bone in critical size cortical bone defects. *J. Orthop. Res.* **2013**, *31*, 792–799.

18. Imwinkelried, T. Mechanical properties of open-pore titanium foam. *J. Biomed. Mater. Res. Part A* **2007**, *81*, 964–970.

19. Krishna, B.V.; Bose, S.; Bandyopadhyay, A. Low stiffness porous Ti structures for load-bearing implants. *Acta Biomater.* **2007**, *3*, 997–1006.

20. Torres, Y.; Pavón, J.J.; Rodríguez, J.A. Processing and characterization of porous titanium for implants by using NaCl as space holder. *J. Mater. Process. Technol.* **2012**, *212*, 1061–1069.

21. Yang, D.; Shao, H.; Guo, Z.; Lin, T.; Fan, L. Preparation and properties of biomedical porous titanium alloys by gelcasting. *Biomed. Mater.* **2011**, *6*, doi:10.1088/1748-6041/6/4/045010.

22. Bartolo, P.; Kruth, J.-P.; Silva, J.; Levy, G.; Malshe, A.; Rajurkar, K.; Mitsuishi, M.; Ciurana, J.; Leu, M. Biomedical production of implants by additive electro-chemical and physical processes. *CIRP Ann. Manuf. Technol.* **2012**, *61*, 635–655.

23. Gu, D.; Meiners, W.; Wissenbach, K.; Poprawe, R. Laser additive manufacturing of metallic components: Materials, processes and mechanisms. *Int. Mater. Rev.* **2012**, *57*, 133–164.

24. Mironov, V.; Trusk, T.; Kasyanov, V.; Little, S.; Swaja, R.; Markwald, R. Biofabrication: A 21st century manufacturing paradigm. *Biofabrication* **2009**, *1*, doi:10.1088/1758-5082/1/2/022001.

25. Murr, L.; Gaytan, S.M.; Medina, F.; Lopez, H.; Martinez, E.; Machado, B.I.; Hernandez, D.H.; Martinez, L.; Lopez, M.I.; Wicker, R.B.; *et al.* Next-generation biomedical implants using additive manufacturing of complex, cellular and functional mesh arrays. *Philos. Trans. R. Soc. A: Math. Phys. Eng. Sci.* **2010**, *368*, 1999–2032.

26. Louvis, E.; Fox, P.; Sutcliffe, C.J. Selective laser melting of aluminium components. *J. Mater. Process. Technol.* **2011**, *211*, 275–284.

27. Mullen, L.; Stamp, R.C.; Brooks, W.K.; Jones, E.; Sutcliffe, C.J. Selective Laser Melting: A regular unit cell approach for the manufacture of porous, titanium, bone in-growth constructs, suitable for orthopedic applications. *J. Biomed. Mater. Res. Part B: Appl. Biomater.* **2009**, *89*, 325–334.

28. Mullen, L.; Stamp, R.C.; Fox, P.; Jones, E.; Ngo, C.; Sutcliffe, C.J. Selective laser melting: A unit cell approach for the manufacture of porous, titanium, bone in-growth constructs, suitable for orthopedic applications. II. Randomized structures. *J. Biomed. Mater. Res. Part B: Appl. Biomater.* **2010**, *92*, 178–188.

29. Vandenbroucke, B.; Kruth, J. Selective laser melting of biocompatible metals for rapid manufacturing of medical parts. *Rapid Prototyp. J.* **2007**, *13*, 196–203.

30. Attar, H.; Calin, M.; Zhang, L.C.; Scudino, S.; Eckert, J. Manufacture by selective laser melting and mechanical behavior of commercially pure titanium. *Mater. Sci. Eng.: A* **2014**, *593*, 170–177.

31. Heinl, P.; Müller, L.; Körner, C.; Singer, R.F.; Müller, F.A. Cellular Ti–6Al–4V structures with interconnected macro porosity for bone implants fabricated by selective electron beam melting. *Acta Biomater.* **2008**, *4*, 1536–1544.

32. Li, X.; Wang, C.; Zhang, W.; Li, Y. Fabrication and characterization of porous Ti6Al4V parts for biomedical applications using electron beam melting process. *Mater. Lett.* **2009**, *63*, 403–405.

33. Parthasarathy, J.; Starly, B.; Raman, S.; Christensen, A. Mechanical evaluation of porous titanium (Ti6Al4V) structures with electron beam melting (EBM). *J. Mech. Behav. Biomed. Mater.* **2010**, *3*, 249–259.

34. Ponader, S.; von Wilmowsky, C.; Widenmayer, M.; Lutz, R.; Heinl, P.; Körner, C.; Singer, R.F.; Nkenke, E.; Neukam, F.W.; Schlegel, K.A. *In vivo* performance of selective electron beam-melted Ti-6Al-4V structures. *J. Biomed. Mater. Res. Part A* **2010**, *92*, 56–62.

35. Van der Stok, J.; Wang, H.; Amin, Y.S.; Siebelt, M.; Sandker, M.; Waarsing, J.H.; Verhaar, J.A.; Jahr, H.; Zadpoor, A.A.; Leeuwenburgh, S.C.; *et al.* Enhanced bone regeneration of cortical segmental bone defects using porous titanium scaffolds incorporated with colloidal gelatin gels for time-and dose-controlled delivery of dual growth factors. *Tissue Eng. Part A* **2013**, *19*, 2605–2614.

36. Amin Yavari, S.; Wauthle, R.; Böttger, A.J.; Schrooten, J.; Weinans, H.; Zadpoor, A.A. Crystal structure and nanotopographical features on the surface of heat-treated and anodized porous titanium biomaterials produced using selective laser melting. *Appl. Surf. Sci.* **2014**, *290*, 287–294.

37. Chen, X.-B.; Li, Y.C.; Du Plessis, J.; Hodgson, P.D.; Wen, C. Influence of calcium ion deposition on apatite-inducing ability of porous titanium for biomedical applications. *Acta Biomater.* **2009**, *5*, 1808–1820.

38. Liang, F.; Zhou, L.; Wang, K. Apatite formation on porous titanium by alkali and heat-treatment. *Surf. Coat. Technol.* **2003**, *165*, 133–139.

39. Lopez-Heredia, M.A.; Sohier, J.; Gaillard, C.; Quillard, S.; Dorget, M.; Layrolle, P. Rapid prototyped porous titanium coated with calcium phosphate as a scaffold for bone tissue engineering. *Biomaterials* **2008**, *29*, 2608–2615.

40. Campoli, G.; Borleffs, M.S.; Amin Yavari, S.; Wauthle, R.; Weinans, H.; Zadpoor, A.A. Mechanical properties of open-cell metallic biomaterials manufactured using additive manufacturing. *Mater. Des.* **2013**, *49*, 957–965.

41. Hazlehurst, K.B.; Wang, C.J.; Stanford, M. A numerical investigation into the influence of the properties of cobalt chrome cellular structures on the load transfer to the periprosthetic femur following total hip arthroplasty. *Med. Eng. Phys.* **2014**, *36*, 458–466.

42. Lewis, G. Properties of open-cell porous metals and alloys for orthopaedic applications. *J. Mater. Sci.: Mater. Med.* **2013**, *24*, 2293–2325.

43. Li, S.; Xu, Q.S.; Wang, Z.; Hou, W.T.; Hao, Y.L.; Yang, R.; Murr, L.E. Influence of cell shape on mechanical properties of Ti-6Al-4V meshes fabricated by electron beam melting method. *Acta Biomater.* **2014**, *10*, 4537–4547.

44. Wauthle, R.; Vrancken, B.; Beynaerts, B.; Jorissen, K.; Schrooten, J.; Kruth, J.-P.; van Humbeeck, J. Effects of build orientation and heat treatment on the microstructure and mechanical properties of selective laser melted Ti6Al4V lattice structures. *Addit. Manuf.* **2015**, *5*, 77–84.

45. Wieding, J.; Jonitz, A.; Bader, R. The effect of structural design on mechanical properties and cellular response of additive manufactured titanium scaffolds. *Materials* **2012**, *5*, 1336–1347.

46. Wieding, J.; Wolf, A.; Bader, R. Numerical optimization of open-porous bone scaffold structures to match the elastic properties of human cortical bone. *J. Mech. Behav. Biomed. Mater.* **2014**, *37*, 56–68.

47. Pyka, G.; Burakowski, A.; Kerckhofs, G.; Moesen, M.; van Bael, S.; Schrooten, J.; Wevers, M. Surface modification of Ti6Al4V open porous structures produced by additive manufacturing. *Adv. Eng. Mater.* **2012**, *14*, 363–370.

48. Van Bael, S.; Kerckhofs, G.; Moesen, M.; Pyka, G.; Schrooten, J.; Krutha, J.P. Micro-CT-based improvement of geometrical and mechanical controllability of selective laser melted Ti6Al4V porous structures. *Mater. Sci. Eng. A* **2011**, *528*, 7423–7431.

49. Doube, M.; Kłosowski, M.M.; Arganda-Carreras, I.; Cordelières, F.P.; Dougherty, R.P.; Jackson, J.S.; Schmid, B.; Hutchinson, J.R.; Shefelbine, S.J. BoneJ: Free and extensible bone image analysis in ImageJ. *Bone* **2010**, *47*, 1076–1079.

50. Amin Yavari, S.; Ahmadi, S.M.; Wauthle, R.; Pouran, B.; Schrooten, J.; Weinans, H.; Zadpoor, A.A. Relationship between unit cell type and porosity and the fatigue behavior of selective laser melted meta-biomaterials. *J. Mech. Behav. Biomed. Mater.* **2015**, *43*, 91–100.

51. International Organization for Standardization (ISO). *Mechanical Testing of Metals—Ductility Testing—Compression Test for Porous and Cellular Metals*; ISO: Genva, Switzerland, 2011; Volume ISO 13314:2011.

52. Kim, H.W.; Knowles, J.C.; Kim, H.E. Hydroxyapatite porous scaffold engineered with biological polymer hybrid coating for antibiotic Vancomycin release. *J. Mater. Sci.* **2005**, *16*, 189–195.

53. Kenesei, P.; Kádár, C.; Rajkovits, Z.; Lendvai, J. The influence of cell-size distribution on the plastic deformation in metal foams. *Scripta Mater.* **2004**, *50*, 295–300.

54. Yeni, Y.N.; Brown, C.U.; Wang, Z.; Norman, T.L. The influence of bone morphology on fracture toughness of the human femur and tibia. *Bone* **1997**, *21*, 453–459.

55. Garrison, J.G.; Gargac, J.A.; Niebur, G.L. Shear strength and toughness of trabecular bone are more sensitive to density than damage. *J. Biomech.* **2011**, *44*, 2747–2754.

56. Keaveny, T.M.; Wachtel, E.F.; Guo, X.E.; Hayes, W.C. Mechanical behavior of damaged trabecular bone. *J. Biomech.* **1994**, *27*, 1309–1318.

57. Moore, T.L.A.; Gibson, L.J. Fatigue Microdamage in Bovine, Trabecular Bone. *J. Biomech. Eng.* **2003**, *125*, 769–776.

58. Black, D.M.; Cummings, S.R.; Karpf, D.B.; Cauley, J.A.; Thompson, D.E.; Nevitt, M.C.; Bauer, D.C.; Genant, H.K.; Haskell, W.L.; Marcus, R.; *et al.* Randomised trial of effect of alendronate on risk of fracture in women with existing vertebral fractures. *Lancet* **1996**, *348*, 1535–1541.

59. Morgan, E.F.; Bouxsein, M. Biomechanics of bone and age-related fractures. In *Principles of Bone Biology*, 3rd ed.; Bilezikian, J.P., Raisz, L.G., Martin, J., Eds.; Elsevier: Amsterdam, The Netherlands, 2008; pp. 29–51.

60. Adachi, T.; Osako, Y.; Tanaka, M.; Hojo, M.; Hollister, S.J. Framework for optimal design of porous scaffold microstructure by computational simulation of bone regeneration. *Biomaterials* **2006**, *27*, 3964–3972.

61. Carter, D.R.; Beaupré, G.S.; Giori, N.J.; Helms, J.A. Mechanobiology of skeletal regeneration. *Clin. Orthop. Relat. Res.* **1998**, *355*, S41–S55.

62. Petite, H.; Viateau, V.; Bensaïd, W.; Meunier, A.; de Pollak, C.; Bourguignon, M.; Oudina, K.; Sedel, L.; Guillemin, G. Tissue-engineered bone regeneration. *Nat. Biotechnol.* **2000**, *18*, 959–963.

63. Zadpoor, A.A. Open forward and inverse problems in theoretical modeling of bone tissue adaptation. *J. Mech. Behav. Biomed. Mater.* **2013**, *27*, 249–261.

64. Harrysson, O.L.; Hosni, Y.A.; Nayfeh, J.F. Custom-designed orthopedic implants evaluated using finite element analysis of patient-specific computed tomography data: femoral-component case study. *BMC Musculoskelet. Disord.* **2007**, *8*, doi:10.1186/1471-2474-8-91.

65. Poelert, S.; Valstar, E.; Weinans, H.; Zadpoor, A.A. Patient-specific finite element modeling of bones. *Proc. Inst. Mech. Eng. Part H: J. Eng. Med.* **2013**, *227*, 464–478.

66. Schileo, E.; Taddei, F.; Malandrino, A.; Cristofolini, L.; Viceconti, M. Subject-specific finite element models can accurately predict strain levels in long bones. *J. Biomech.* **2007**, *40*, 2982–2989.

67. Lomov, S.V.; Boissec, P.; Deluycker, E.; Morestin, F.; Vanclooster, K.; Vandepitte, D.; Verpoest, I.; Willems, A. Full-field strain measurements in textile deformability studies. *Compos. Part A* **2008**, *39*, 1232–1244.

68. Pan, B.; Xie, H.; Guo, Z.; Hua, T. Full-field strain measurement using a two-dimensional Savitzky-Golay digital differentiator in digital image correlation. *Opt. Eng.* **2007**, *46*, doi:10.1117/1.2714926.

69. Schmidt, T.; Tyson, J.; Galanulis, K. Full-field dynamic displacement and strain measurement using advanced 3d image correlation photogrammetry: Part 1. *Exp. Tech.* **2003**, *27*, 47–50.

70. Zadpoor, A.A.; Sinke, J.; Benedictus, R. Experimental and numerical study of machined aluminum tailor-made blanks. *J. Mater. Process. Technol.* **2008**, *200*, 288–299.

71. Hild, F.; Roux, S. Digital image correlation: from displacement measurement to identification of elastic properties—A review. *Strain* **2006**, *42*, 69–80.

72. McCormick, N.; Lord, J. Digital image correlation. *Mater. Today* **2010**, *13*, 52–54.

73. Wattrisse, B.; Chrysochoos, A.; Muracciole, J.-M.; Némoz-Gaillard, M. Analysis of strain localization during tensile tests by digital image correlation. *Exp. Mech.* **2001**, *41*, 29–39.

74. Zadpoor, A.A.; Sinke, J.; Benedictus, R. Elastoplastic deformation of dissimilar-alloy adhesively-bonded tailor-made blanks. *Mater. Des.* **2010**, *31*, 4611–4620.

75. Thompson, M.; Schell, H.; Lienau, J.; Duda, G.N. Digital image correlation: A technique for determining local mechanical conditions within early bone callus. *Med. Eng. Phys.* **2007**, *29*, 820–823.

76. Verhulp, E.; Rietbergen, B.V.; Huiskes, R. A three-dimensional digital image correlation technique for strain measurements in microstructures. *J. Biomech.* **2004**, *37*, 1313–1320.

77. Zhang, D.; Arola, D.D. Applications of digital image correlation to biological tissues. *J. Biomed. Opt.* **2004**, *9*, 691–699.

Conductive Polymer Porous Film with Tunable Wettability and Adhesion

Yuqi Teng, Yuqi Zhang, Liping Heng, Xiangfu Meng, Qiaowen Yang and Lei Jiang

Abstract: A conductive polymer porous film with tunable wettability and adhesion was fabricated by the chloroform solution of poly(3-hexylthiophene) (P3HT) and [6,6]-phenyl-C61-butyricacid-methyl-ester (PCBM) via the freeze drying method. The porous film could be obtained from the solution of 0.8 wt%, whose pore diameters ranged from 50 nm to 500 nm. The hydrophobic porous surface with a water contact angle (CA) of 144.7° could be transferred into a hydrophilic surface with CA of 25° by applying a voltage. The water adhesive force on the porous film increased with the increase of the external voltage. The electro-controllable wettability and adhesion of the porous film have potential application in manipulating liquid collection and transportation.

Reprinted from *Materials*. Cite as: Teng, Y.; Zhang, Y.; Heng, L.; Meng, X.; Yang, Q.; Jiang, L. Conductive Polymer Porous Film with Tunable Wettability and Adhesion. *Materials* **2015**, *8*, 1817-1830.

1. Introduction

Macroporous materials with a high porous volume, specific surface area an tunable pore sizes, especially conductive polymer porous materials, have emerged as a hot topics due to their wide applications in gas sensing, adsorption, catalysis, porous electrodes, energy storage, tissue engineering and biomaterials [1–7]. Various fabrication techniques of porous materials have been developed, such as phase separation [8–10], emulsion templating [11,12], direct foaming [7,13], polymer foam replication [14,15], breath figures [16–18] and freeze drying [19–22]. Compared with other methods, freeze drying shows some advantages, such as large area preparation, with no need for further purification, and obtaining a number of pore morphologies and nanostructures by changing variables during freezing [22]. During the freeze drying process, the solution is frozen under a certain freezing temperature, followed by removing solvent by sublimation under vacuum, which leads to forming porous structures. Presently, many inorganic, polymer or composite porous materials have been fabricated by freeze drying, for example porous alumina [23,24], chitosan [25], glycosaminoglycan [26] and silylated nanocellulose sponges [27]. However, conductive polymer porous materials prepared by the freeze drying method have not yet drawn scientific attention.

The surface wettability and adhesive behaviors, as important properties of porous materials, have been paid more attention due to the desire for developing new functions, such as photoelectric conversion [28], photocatalysis [29], antireflection [30] and cell adhesion [31]. Many artificial surfaces with special wettability and adhesion have been prepared, for example vertical-aligned multiwalled carbon nanotubes [32], superhydrophobic polystyrene (PS) nanotube film [33] and an artificial biomimic polymer film duplicated by a rose petal surface [34]. In this field, our group [35] also reported a high-adhesive ordered porous structure surface fabricated by the breath figures method. The surface adhesive force of the as-prepared film can be effectively adjusted by changing the pore sizes. However, the wettability and adhesion regulation on the conductive polymer porous material surface by external voltage has generated fewer reports.

In this paper, a conductive polymer porous composite film was fabricated by the freeze drying method via using a blend system composed of poly(3-hexylthiophene) (P3HT) as an electron-donating polymer and [6,6]-phenyl-C61-butyricacid-methyl-ester (PCBM) as an electron-accepting fullerene. The chemical structures of P3HT and PCBM are shown in Scheme 1. Presently, The most extensive studies for the P3HT:PCBM blend system focus on its bulk heterojunction organic photovoltaics [36,37]. The mixing of P3HT and PCBM will not disrupt the crystalline P3HT domains, and PCBM can be dispersed well in disordered P3HT domains, which is prone to constructing a good interpenetrating network structure. Combined with the conductivity and formation of the network structure of P3HT:PCBM, we prepared an electro-responsive tunable wettability porous structure composed of P3HT and PCBM by freeze drying. Simultaneously, the surface adhesion forces of the porous film can also be controlled by electric stimuli. The surface wettability and water adhesive forces of the as-prepared films can be effectively controlled by an external electric field. The porous surface induced by external voltage showed relatively high adhesion for water, which will be very useful for manipulating liquid collection and transportation.

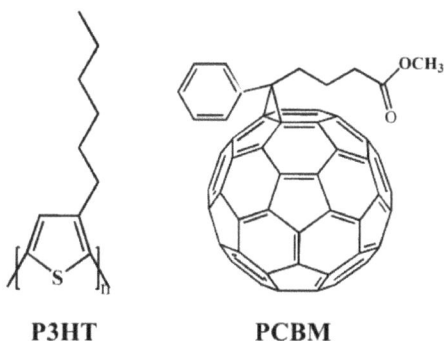

P3HT　　　　**PCBM**

Scheme 1. Chemical structures of poly(3-hexylthiophene) (P3HT) and [6,6]-phenyl-C61-butyricacid-methyl-ester (PCBM).

2. Experimental Section

2.1. Materials and Characterization

Commercially available P3HT (Rieke Metals, Inc., Lincoln, NE, USA) and PCBM (Fem Technology Co., *Groningen*, The Netherlands), were used directly without further treatment. Chloroform (Tianjin Hengxing Chemical Industry Co., Ltd., Tianjin, China) was used as the solvent to dissolve P3HT and PCBM and as the freezing vehicle.

The morphology of the polymer porous film was characterized by field-emission scanning electron microscopy (SEM, JEOL JSM-7500, Tokyo, Japan), after sputtering the samples with a thin layer of gold. The contact angle (CA) of the prepared porous film was measured on a CA system (JC2000C, Shanghai Zhongchen Technology Co. Ltd., Shanghai, China) at ambient temperature.

The water droplets (about 2 μL) were dropped onto the surface, and the contact angle average value of five measurements was performed at different positions on the same sample.

The adhesive forces were measured on a high-sensitivity microelectro-mechanical balance system (DCAT 11, Dataphysics, Goettingen, Germany) at different voltages. Typically, a water droplet of about 6 μL was hung on a clean copper cap connected to the microbalance. Then, the substrate was controlled to move toward the water droplet at a constant speed of 0.05 mm/s, until it made contact with the droplet, at which point the substrate was then moved in reverse direction and left the droplet. The distance between the break points recorded in the force-distance curve was taken as the maximum adhesion force. The adhesion values were the averages of 10 independent measurements. A variable-frequency power source of 60 HZ (Shanghai Ruijin Sci & Technol. Co. Ltd., Shanghai, China) was used to obtain the different external voltages.

2.2. Preparation of Conductive Polymer Porous Film

The conductive porous film of the P3HT:PCBM blend system was prepared by the freeze drying method, in which chloroform was used as the freezing vehicle. The same mass of P3HT and PCBM was dissolved in chloroform to prepare the conductive polymer solution with a mass concentration of 1% by stirring. The chloroform solutions of the P3HT:PCBM blend system, whose mass concentrations are 0.08%, 0.1%, 0.2%, 0.4%, 0.6% and 0.8%, respectively, were obtained by diluting the solution of 1%. The conductive polymer solution was dropped onto the surface of ITO substrate, followed by freezing in liquid nitrogen quickly. Then, the frozen polymer film on the ITO was lyophilized for 12 h at a temperature of −84 °C and a vacuum degree of 0.07 Pa. The schematic illustration of the preparation process is exhibited in Scheme 2. The polymer porous film was obtained. In addition, the prepared polymer solutions were spin-coated on ITO substrate and dried at ambient temperature and pressure to obtain the smooth conductive polymer film, which was used as control samples.

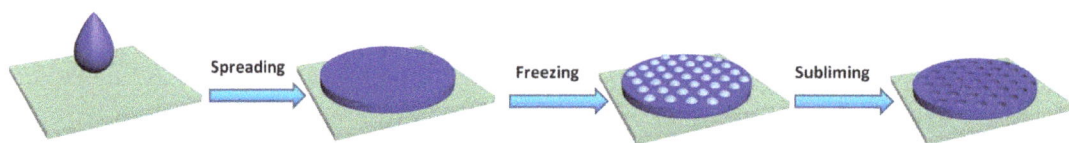

Scheme 2. Schematic illustration of preparing the conductive polymer porous film.

3. Results and Discussion

3.1. Preparation and Morphology of Conductive Polymer Porous Film

Freeze drying is a drying techniques based on sublimation. The material to be dried is frozen quickly at low temperature and then dried in a vacuum, in which the frozen water or other solvent molecules directly sublimate and escape as vapors [19–22]. In the P3HT:PCBM blend system, PCBM dispersed into the P3HT molecular chains, which made P3HT be in a disordered state. The π-π interaction between the molecular chains of P3HT with the conjugated system (Scheme 1) can form the multi-dimensional network structure, in which the shorter PCBM chain (Scheme 1) is attached to the network structure. As shown in Scheme 2, the chloroform solution of P3HT:PCBM dropped on the ITO substrate was frozen quickly below its freezing point in liquid nitrogen, which hindered the movement of the polymer chains and further mixture with the solvent. During the freezing process, the solvent crystallized, and its crystals grew. Then, polymer molecules were excluded from the frozen solvent until the sample was completely frozen, which induced the phase separation between the polymer and the solvent. After drying at low temperature and vacuum, the solvent sublimated, and porous structures are formed from the voids left by the removal of the solvent. Finally, the conductive polymer porous film with an interpenetrating network structure was obtained.

Figure 1 shows SEM images of the conductive polymer porous films prepared from different concentrations of the polymer solutions. When the mass concentration of the P3HT:PCBM blend system is between 0.08% and 0.4%, the obtained polymer films from the freeze drying method show non-uniform mesh-like structures (Figure 1a–d), in which nanoparticles with different sizes aggregate and are attached to the interconnected nanofibers. The interpenetrating nanofibers constructed larger pores whose diameter is more than 1 μm. However, from Figure 1a–d, we can observe that nanofibers in the films grow slowly and form a lamellar structure. When the concentration is 0.6%, the micrograph (Figure 1e) displays that most microstructures are lamellar and that non-uniform pores occur between the polymer layers. With increasing the mass concentration of the conductive polymer solution, an open pore microstructure with a high degree of interconnectivity formed during the freezing and lyophilization process, as shown in Figure 1f. The lamellar polymer layers arranged in different orientation and induced the formation of the macroporous structure. The range of the pore diameters is about 50–500 nm. The results demonstrate that the porosity and pore size distribution were affected by the solution's concentration. When the concentration is too low, the nanofiber structure caused the larger pores and could not form a stable porous surface. In contrast, if the concentration is too high, it is difficult to keep good fluidity, which will affect the growth of solvent crystals. Therefore, the porous structure can be adjusted by regulating the concentration of the conductive polymer solution.

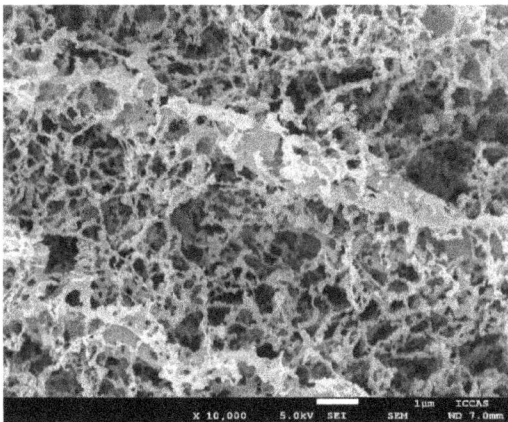

(a)

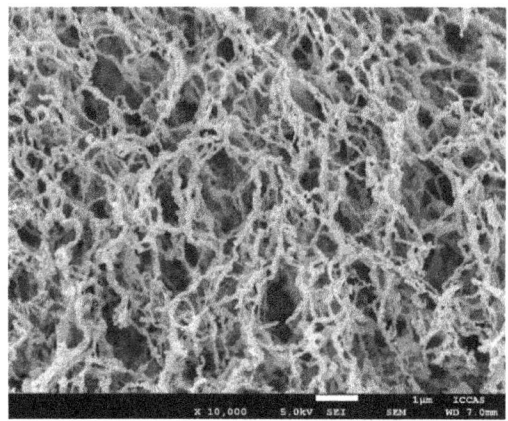

(b)

Figure 1. *Cont.*

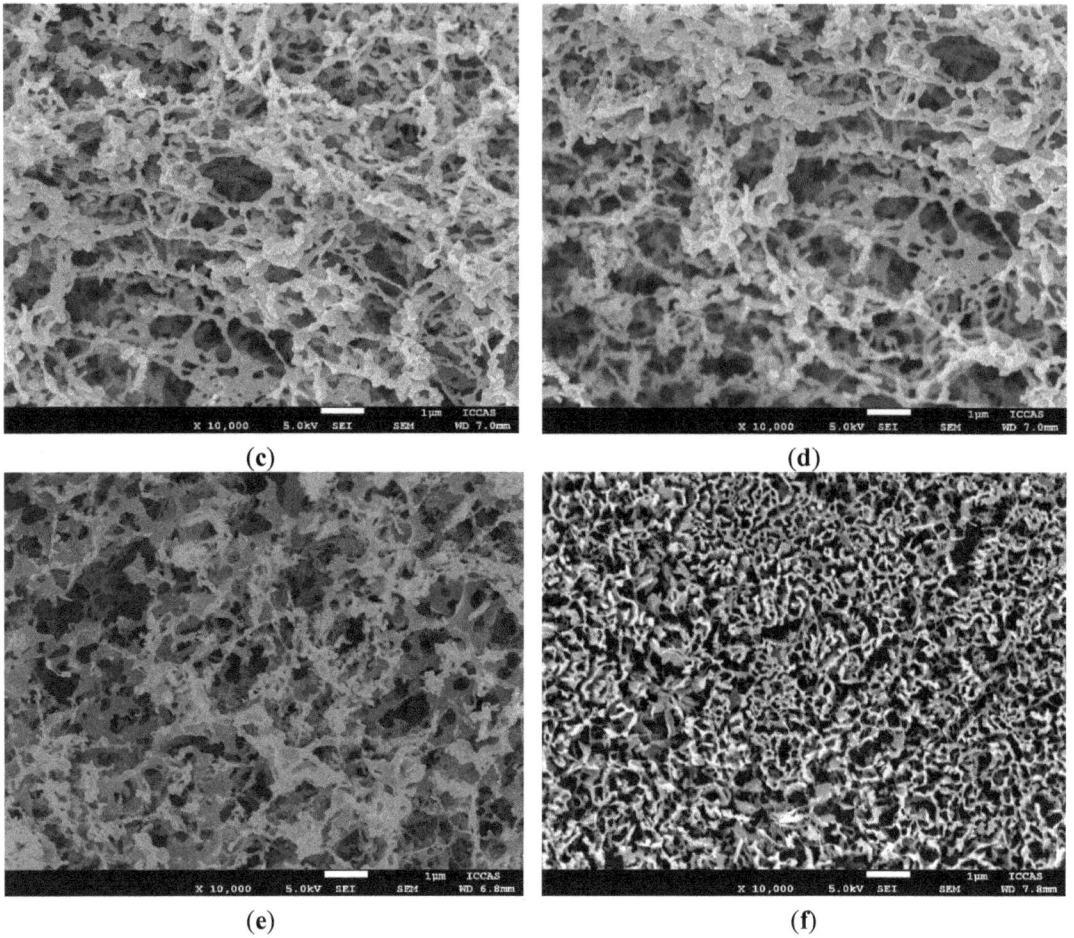

Figure 1. SEM images of the conductive polymer porous films prepared from different concentrations of polymer solutions: (**a**) 0.08%; (**b**) 0.1%; (**c**) 0.2%; (**d**) 0.4%; (**e**) 0.6%; and (**f**) 0.8%. The scale bar is 1 μm.

3.2. The Porosity and the Pore Size Distribution

Figure 2 shows the porosity and the pore size distribution prepared from 0.8% solid loading slurries of porous films. Figure 2a shows that the total porosity is about 63%; the slope of the curve slows down after 400 nm, indicating a pore size mostly within 400 nm. The pore size of the porous conductive film showed bimodal distribution characteristics (Figure 2b) from 50 nm to 1200 nm. There are two main pore diameter distributions: pores between 50 nm and 500 nm are mainly formed by freezing-drying; pores between 600 nm and 1200 nm are mainly formed by particle accumulation.

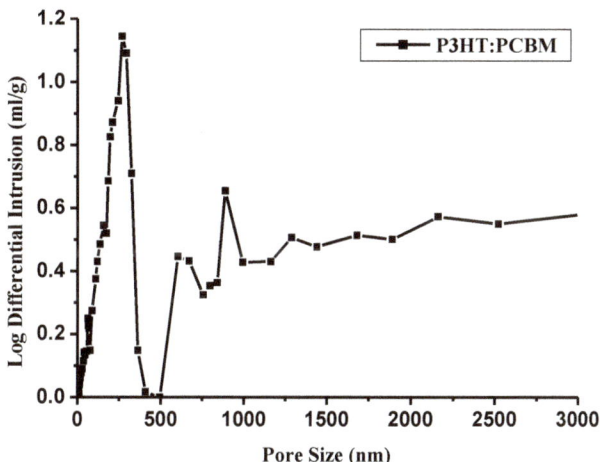

Figure 2. (**a**) Porosity and (**b**) pore size distribution of the porous conductive film.

3.3. The Electrical Conductivity

As conductive polymers, the conductivity of films was their main characteristics. The conductivity test schematic is shown in Figure 3. The conductivities of the different concentrations porous films are 8×10^{-5} S·cm^{-1}, 8.1×10^{-5} S·cm^{-1}, 7.9×10^{-5} S·cm^{-1}, 8.7×10^{-5} S·cm^{-1}, 8.6×10^{-5} S·cm^{-1} and 7.5×10^{-5} S·cm^{-1}, respectively. The conductivities for each of the porous films were similar, which belong to the range of semiconductors. Therefore, the porous films were conductive.

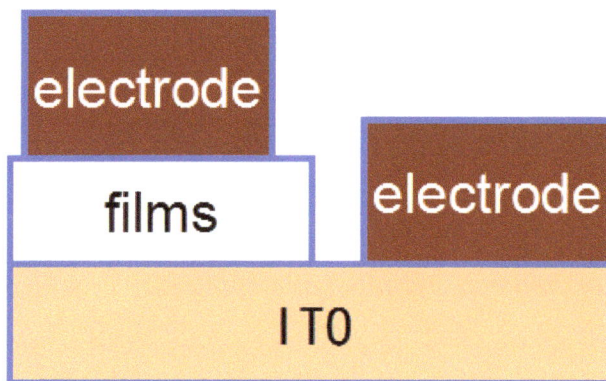

Figure 3. Structural schematic for measuring the conductivity of films.

3.4. The Electro-Responsive Wettability

Wettability is an important parameter, and porous films usually display a highly hydrophobic character due to the hydrophobic polymer matrix and the air entrapped inside the pores, which increases the surface roughness [38]. Figure 4 exhibits the static water contact angles of the as-prepared polymer porous films from P3HT:PCBM solutions with different concentrations. The CA data demonstrate that all of the porous films fabricated by the freeze drying method are hydrophobic, and the CA increases with increasing the solution concentrations, which indicates that the hydrophobicity of porous films is increased. The CAs of porous films are larger than those of the smooth P3HT:PCBM film fabricated by spin-coating, as shown in Figure 5. For example, when the solution concentration is 0.08%, the porous film is hydrophobic (CA = 105.9° ± 1.1°; Figure 4a); however, the CA of the smooth film is hydrophilic (CA = 43.6° ± 2.1°; Figure 5a). The hydrophobicity of porous film can be ascribed to the air trapped in the pores, which can prevent the intrusion of water into the pores and result in the larger contact angle. Comparing Figures 4 and 5, we can see clearly that the CA of the porous film is always larger than that of the smooth film prepared from the same solution. With increasing the solution concentration, the CA increases both for these two kinds of films due to increasing the amount of the hydrophobic polymers. The highest hydrophobicity of the porous films with a CA of 143.9° ± 2.7° (Figure 4f) can be obtained because the film prepared from a solution of 0.8% has a macroporous structure (Figure 1f). Furthermore, we also measured the water contact angle of the conductive film, which has been prepared for three months. The obtained contact angle did not change compared with that of the fresh film. The results demonstrate that the prepared conductive film has good stability.

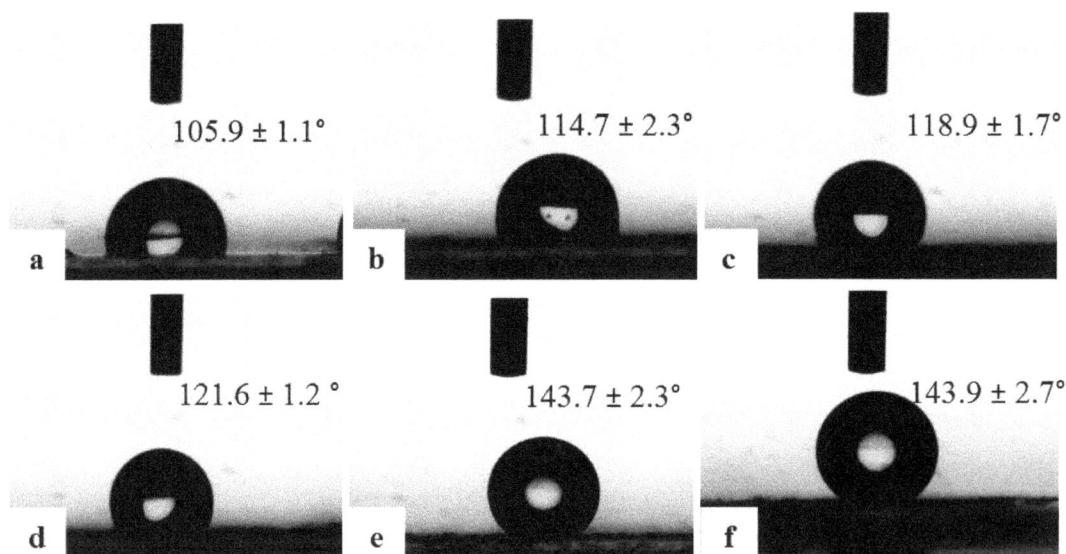

Figure 4. The water contact angle photos of the conductive polymer porous film prepared from different concentrations of the polymer solutions: (**a**) 0.08%; (**b**) 0.1%; (**c**) 0.2%; (**d**) 0.4%; (**e**) 0.6%; and (**f**) 0.8%.

Interestingly, there is an obvious wettability change for the resultant conductive polymer porous film when induced by electric fields. The electrowetting phenomenon was systematically investigated for the polymer porous film surface. Figure 6 shows the CA photos of the porous film prepared from the solution of 0.8% at different voltages, and the CA *versus* voltage curve of electrowetting is shown in Figure 7. We can see that the porous structure surface had a CA of about 144.7° at the initial stage (0 V). The CA began to decrease slowly when the voltage ranged from 2 V to 24 V; however, the CA sharply decreased to 25° when the voltage increased to 26 V. The results demonstrate that the electrowetting happened when the applied voltage ranged from 2 V to 26 V. The phenomenon indicates that the voltage can affect the CA of the porous structure significantly, and the higher voltage induced the smaller CA. Finally, a remarkable wettability transition was obtained with a CA change as large as about 120°. Therefore, the prepared polymer porous film can be transferred from a highly hydrophobic surface to a highly hydrophilic surface via applying a voltage of 26 V. In addition, we also studied the advancing and receding angles of the porous film prepared from the solution of 0.8% at different voltages (Figure 8). Obviously, the water advancing and receding angles of the polymer porous surface decreased with the increase of the voltage. The initial advancing angle and initial receding angle were 158.3° ± 1.1° and 157.1° ± 1.1°, respectively. Additionally, they decreased to 80.1° ± 1.4° and 66.3° ± 1.1° after applying a voltage of 24 V, respectively. The results testify that the water advancing and receding angles of the as-prepared porous film can be effectively controlled from relatively high to relatively low by varying the applied voltage, which is due to the increase of the surface tension of the droplets induced by the voltage.

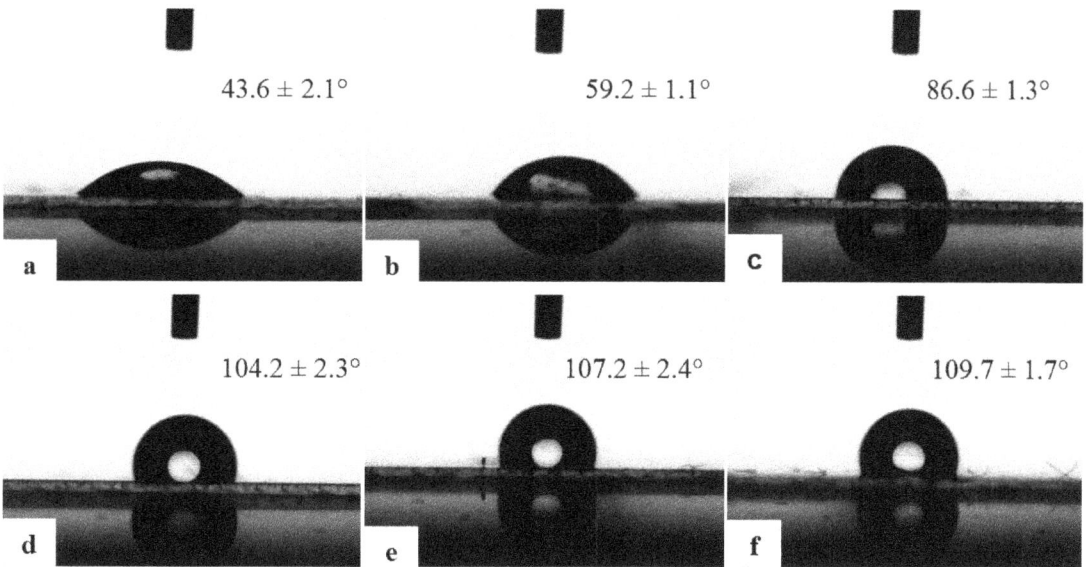

43.6 ± 2.1°   59.2 ± 1.1°   86.6 ± 1.3°

a   b   c

104.2 ± 2.3°   107.2 ± 2.4°   109.7 ± 1.7°

d   e   f

Figure 5. The water contact angle photos of the smooth polymer film spin-coated from different concentrations of the polymer solutions: (**a**) 0.08%; (**b**) 0.1%; (**c**) 0.2%; (**d**) 0.4%; (**e**) 0.6%; and (**f**) 0.8%.

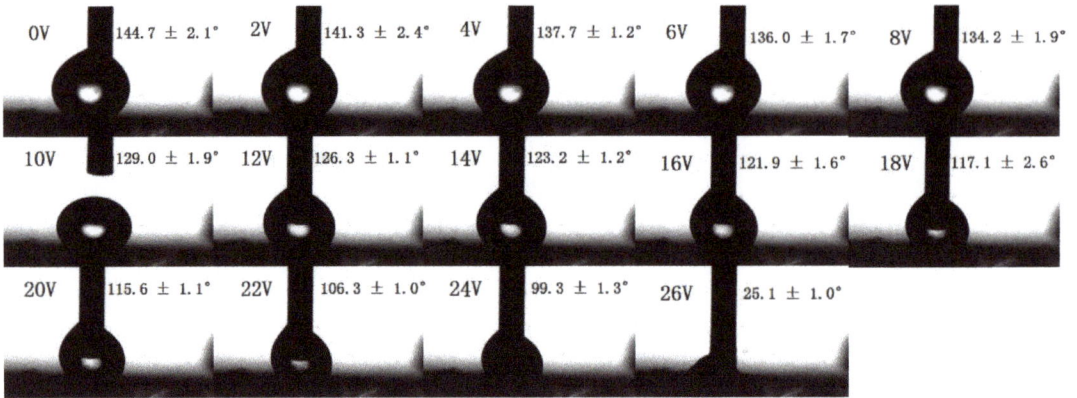

Figure 6. The water contact angle photos of the conductive polymer porous film prepared from the polymer solution of 0.8% at different voltages.

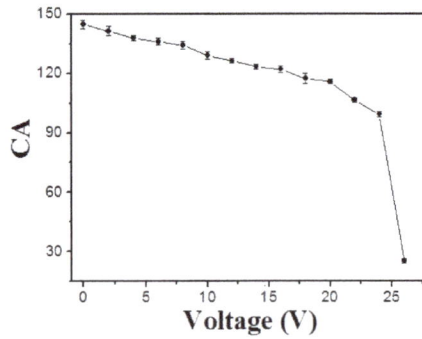

Figure 7. The contact angles (CAs) of the porous film prepared from the solution of 0.8% at different voltages.

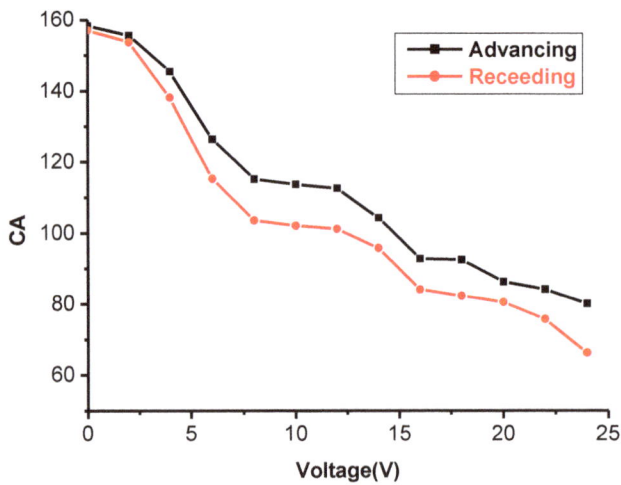

Figure 8. The advancing and receding angles of the porous film prepared from the solution of 0.8% at different voltages.

This wettability change could be explained by the Wenzel equation:

$$\cos \theta_1 = f \frac{\sigma_{sv} - \sigma_{sl}}{\sigma_{lv}} \tag{1}$$

in which θ_1 is the Wenzel equilibrium contact angle, f is the surface roughness of porous film and σ_{sv}, σ_{sl} and σ_{lv} are the interfacial energies of solid-vapor, solid-liquid and liquid-vapor, respectively. Upon applying a voltage dU, an electric double layer builds up spontaneously at the solid-liquid interface, in which positive and negative charges accumulate on the conductive polymer porous film surface and the liquid side of the interface, respectively [39]. This spontaneous accumulation process leads to a reduction of the (effective) interfacial tension σ_{sl}^{eff}, which can be obtained from the following formulas:

$$d\sigma_{sl}^{eff} = -\rho_{sl} dU \tag{2}$$

where $\rho_{sl} = \rho_{sl}$ (U) is the surface charge density of the counter-ions on the liquid side [40]. The voltage dependence of σ_{sl}^{eff} can be calculated by integrating Equation (2), which demonstrates that applying a voltage will decrease the interfacial tension σ_{sl}^{eff}. Combining with Young's Equation (1), $\cos \theta_Y$ will increase with decreasing σ_{sl}^{eff}, the contact angle thus will decrease upon the application of a voltage.

3.5. Tunable Water Adhesion Properties of the Polymer Porous Film

Adhesive force is a kind of ability of a material adhering to another material surface, which depends on not only surface structure and chemical composition, but also the external conditions, such as temperature, humidity, radiation, vibration, voltage, and so on; wherein the voltage is an important factor that affects the adhesion. In this paper, the adhesive force was defined as the force required to lift the water droplet off the substrate and can be assessed by a highly sensitive micromechanical balance system. We investigated the adhesive behaviors of the conductive polymer porous film prepared from the solution of 0.8% by applying different voltages. The adhesive force *versus* applied voltage curve is exhibited in Figure 9. Obviously, the water adhesion of the polymer porous surface increased with the increase of the voltage. The initial adhesive force was 122 μN and increased to 169 μN after applying a voltage of 27 V. The results testify that the water adhesive force of the as-prepared porous film can be effectively controlled from relatively low to relatively high adhesion by varying the applied voltage, which is due to the increase of the surface tension of the droplets induced by the voltage.

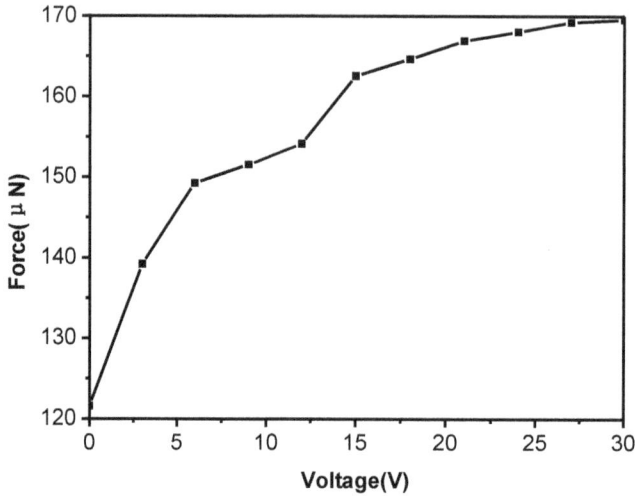

Figure 9. The adhesion force of the conductive polymer porous film at different voltages.

4. Conclusions

The conductive polymer porous film composed of P3HT and PCBM was fabricated by the freeze drying method. The morphology of the prepared porous film, electrowetting and adhesive forces induced by the applied voltage were investigated. The SEM images show that the macroporous structured film could be formed by arranging the lamellar polymer layers in different orientations, and the pore diameters ranged from 50 nm to 500 nm when the solution concentration of P3HT:PCBM was 0.8%. The electrowetting phenomenon of the prepared porous film happened when the applied voltage ranged from 2 V to 26 V, which caused the initial hydrophobic porous surface to change into a hydrophilic surface. A CA change as large as about 120° occurred, which is due to the reduction of the interfacial tension of the solid-liquid. In addition, the water adhesive force of the porous film increased clearly from the initial 122 µN to 169 µN when a voltage of 21 V was applied. The obtained conductive polymer porous film with tunable wettability and relatively high water adhesion will be very useful for manipulating liquid collection and transportation.

Acknowledgments

This work was supported by the National Research Fund for Fundamental Key Projects (Grant No. 2014CB931802) and the National Natural Science Foundation of China (Grant No. 21103146).

Author Contributions

Yuqi Teng finished the experiment and data processing; Yuqi Zhang write the article; Liping Heng and Xiangfu Meng schemed the experiment; Qiaowen Yang and Lei Jiang took part in the discussion.

Conflicts of Interest

The authors declare no conflict of interest.

References

1. Dai, Z.; Lee, C.-S.; Kim, B.-Y.; Kwak, C.-H.; Yoon, J.-W.; Jeong, H.-M.; Lee, J.-H. Honeycomb-like periodic porous $LaFeO_3$ thin film chemiresistors with enhanced gas-sensing performances. *ACS Appl. Mater. Interfaces* **2014**, *6*, 16217–16226.
2. Tuller, M.; Or, D.; Dudley, L.M. Adsorption and capillary condensation in porous media: Liquid retention and interfacial configurations in angular pores. *Water Resour. Res.* **1999**, *35*, 1949–1964.
3. Mondal, K.; Kumar, J.; Sharma, A. Self-organized macroporous thin carbon films for supported metal catalysis. *Colloids Surf. A* **2013**, *427*, 83–94.
4. Zhang, L.L.; Zhao, X.; Stoller, M.D.; Zhu, Y.; Ji, H.; Murali, S.; Wu, Y.; Perales, S.; Clevenger, B.; Ruoff, R.S. Highly conductive and porous activated reduced graphene oxide films for high-power supercapacitors. *Nano Lett.* **2012**, *12*, 1806–1812.
5. Hu, L.; Pasta, M.; Mantia, F.L.; Cui, L.; Jeong, S.; Deshazer, H.D.; Choi, J.W.; Han, S.M.; Cui, Y. Stretchable, porous, and conductive energy textiles. *Nano Lett.* **2010**, *10*, 708–714.
6. O'Brien, F.J.; Harley, B.A.; Yannas, I.V.; Gibson, L. Influence of freezing rate on pore structure in freeze-dried collagen-GAG scaffolds. *Biomaterials* **2004**, *25*, 1077–1086.
7. Mikos, A.G.; Sarakinos, G.; Leite, S.M.; Vacant, J.P.; Langer, R. Laminated three-dimensional biodegradable foams for use in tissue engineering. *Biomaterials* **1993**, *14*, 323–330.
8. Nam, Y.S.; Park, T.G. Porous biodegradable polymeric scaffolds prepared by thermally induced phase separation. *J. Biomed. Mater. Res.* **1999**, *47*, 8–17.
9. Schugens, C.; Maquet, V.; Grandfils, C.; Jérôme, R.; Teyssié, P. Biodegradable and macroporous polylactide implants for cell transplantation: 1. Preparation of macroporous polylactide supports by solid-liquid phase separation. *Polymer* **1996**, *37*, 1027–1038.
10. Schugens, C.; Maquet, V.; Grandfils, C.; Jérôme, R; Teyssié, P. Polylactide macroporous biodegradable implants for cell transplantation. II. Preparation of polylactide foams by liquid-liquid phase separation. *J. Biomed. Mater. Res.* **1996**, *30*, 449–461.
11. Imhof, A.; Pine, D.J. Ordered macroporous materials by emulsion templatin. *Nature* **1997**, *389*, 948–951
12. Cameron, N.R. High internal phase emulsion templating as a route to well-defined porous polymers. *Polymer* **2005**, *46*, 1439–1449.
13. Barg, S.; Soltmann, C.; Andrade, M.; Koch, D.; Grathwohl, G. Cellular ceramics by direct foaming of emulsified ceramic powder suspensions. *J. Am. Ceram. Soc.* **2008**, *91*, 2823–2829.
14. Fu, Q.; Rahaman, M.N.; Bal, B.S.; Brown, R.F.; Day, D.E. Mechanical and *in vitro* performance of 13-93 bioactive glass scaffolds prepared by a polymer foam replication technique. *Acta Biomater.* **2008**, *4*, 1854–1864.

15. Fu, H.; Fu, Q.; Zhou, N.L.; Huang, W.; Rahamana, M.N.; Wang, D.; Liu, X. *In vitro* evaluation of borate-based bioactive glass scaffolds prepared by a polymer foam replication method. *Mater. Sci. Eng. C* **2009**, *29*, 2275–2281.

16. Heng, L.; Wang, B.; Li, M.; Zhang, Y.; Jiang, L. Advances in fabrication materials of honeycomb structure films by the breath-figure method. *Materials* **2013**, *6*, 460–482.

17. Heng, L.; Li, J.; Li, M.; Tian, D.; Fan, L.-Z.; Jiang, L.; Tang, B.Z. Ordered honeycomb structure surface generated by breath figures for liquid reprography. *Adv. Funct. Mater.* **2014**, doi:10.1002/adfm.201401342.

18. Zhai, S.; Ye, J.-R.; Wang, N.; Jiang, L.-H.; Shen, Q. Fabrication of porous film with controlled pore size and wettability by electric breath figure method. *J. Mater. Chem. C* **2014**, *2*, 7168–7172.

19. Lee, J.T.Y.; Chow, K.L. SEM sample preparation for cells on 3D scaffolds by freeze-drying and HMDS. *Scanning* **2011**, *33*, 1–14.

20. Hou, Q.; Grijpma, D.W.; Feijen, J. Preparation of interconnected highly porous polymeric structures by a replication and freeze drying process. *J. Biomed. Mater. Res. B* **2003**, *67B*, 732–740.

21. Aranaz, I.; Gutiérrez, M.C.; Ferrer, M.L.; del Monte, F. Preparation of chitosan nanocompositeswith a macroporous structure by unidirectional freezing andsubsequent freeze-drying. *Mar. Drugs* **2014**, *12*, 5619–5642.

22. Qian, L.; Zhang, H. Controlled freezing and freeze drying: A versatile route for porous and micro-/nano-structured materials. *J. Chem. Technol. Biotechnol.* **2011**, *86*, 172–184.

23. Munch, E.; Saiz, E.; Tomsia, A.P. Architectural control of freeze-cast ceramics through additives and templating. *J. Am. Ceram. Soc.* **2009**, *92*, 1534–1539.

24. Deville, S.; Saiz, E.; Tomsia, A.P. Ice-templated porous alumina structures. *Acta Mater.* **2007**, *55*, 1965–1974.

25. Madihally, S.V.; Matthew, H.W.T. Porous chitosan scaffolds for tissue engineering. *Biomaterials* **1999**, *20*, 1133–1142.

26. Cho, C.H.; Eliason, J.F.; Matthew, H.W.T. Application of porous glycosaminoglycan-based scaffolds for expansion of human cord blood stem cells in perfusion culture. *J. Biomed. Mater. Res.* **2008**, *86A*, 98–107.

27. Zhang, Z.; Sèbe, G.; Rentsch, D.; Zimmermann, T.; Tingaut, P. Ultralightweight and flexible silylated nanocellulose sponges for the selective removal of oil from water. *Chem. Mater.* **2014**, *26*, 2659–2668.

28. Wen, L.; Hou, X.; Tian, Y.; Nie, F.-Q.; Song, Y.; Zhai, J.; Jiang, L. Bioinspired smart gating of nanochannels toward photoelectric-conversion systems. *Adv. Mater.* **2010**, *22*, 1021–1024.

29. Zhang, X.; Jin, M.; Liu, Z.; Tryk, D.A.; Nishimoto, S.; Murakami, T.; Fujishima, A. Superhydrophobic TiO_2 surfaces: Preparation, photocatalytic wettability conversion, and superhydrophobic–superhydrophilic patterning. *J. Phys. Chem. C* **2007**, *111*, 14521–14529.

30. Faustini, M.; Nicole, L.; Boissière, C.; Innocenzi, P.; Sanchez, C.; Grosso, D. Hydrophobic, antireflective, self-Cleaning, and antifogging sol–gel coatings: An example of multifunctional nanostructured materials for photovoltaic cells. *Chem. Mater.* **2010**, *22*, 4406–4413.

31. O'Brien, F.J.; Harley, B.A.; Yannas, I.V.; Gibson, L.J. The effect of pore size on cell adhesion in collagen-GAG scaffolds. *Biomaterials* **2005**, *26*, 433–441.

32. Qu, L.; Dai, L.; Stone, M.; Xia, Z.; Wang, Z.L. Carbon nanotube arrays with strong shear binding-on and easy normal lifting-off. *Science* **2008**, *322*, 238–242.

33. Cho, W.K.; Choi, I.S. Fabrication of hairy polymeric films inspired by geckos: Wetting and high adhesion properties. *Adv. Funct. Mater.* **2008**, *18*, 1089–1096.

34. Feng, L.; Zhang, Y.; Xi, J.; Zhu, Y.; Wang, N.; Xia, F.; Jiang, L. Petal effect: A superhydrophobic state with high adhesive force. *Langmuir* **2008**, *24*, 4114–4119.

35. Heng, L.; Meng, X.; Wang, B.; Jiang, L. Bioinspired design of honeycomb structure interfaces with controllable water adhesion. *Langmuir* **2013**, *29*, 9491–9498.

36. Treat, N.D.; Brady, M.A.; Smith, G.; Toney, M.F.; Kramer, E.J.; Hawker, C.J.; Chabinyc, M.L. Interdiffusion of PCBM and P3HT reveals miscibility in a photovoltaically active blend. *Adv. Energy Mater.* **2011**, *1*, 82–89.

37. Chen, D.; Nakahara, A.; Wei, D.; Nordlund, D.; Russell, T.P. P3HT/PCBM bulk heterojunction organic photovoltaics: Correlating efficiency and morphology. *Nano Lett.* **2011**, *11*, 561–567.

38. Stenzel-Rosenbaum, M.H.; Davis, T.P.; Fane A.G.; Chen, V. Porous polymer films and honeycomb structures made by the self-organization of well-defined macromolecular structures created by living radical polymerization techniques. *Angew. Chem. Int. Ed.* **2001**, *40*, 3428–3432.

39. Mugele, F.; Baret, J.-C. Electrowetting: From basics to applications. *J. Phys.: Condens. Matter* **2005**, *17*, R705–R774.

40. Wixforth, A.; Strobl, C.; Gauer, C.; Toegl, A.; Scriba, J.; von Guttenberg, Z. Acoustic manipulation of small droplets. *Anal. Bioanal. Chem.* **2004**, *379*, 982–991.

Characterization of Fibrin and Collagen Gels for Engineering Wound Healing Models

Oihana Moreno-Arotzena, Johann G. Meier, Cristina del Amo and José Manuel García-Aznar

Abstract: Hydrogels are used for 3D *in vitro* assays and tissue engineering and regeneration purposes. For a thorough interpretation of this technology, an integral biomechanical characterization of the materials is required. In this work, we characterize the mechanical and functional behavior of two specific hydrogels that play critical roles in wound healing, collagen and fibrin. A coherent and complementary characterization was performed using a generalized and standard composition of each hydrogel and a combination of techniques. Microstructural analysis was performed by scanning electron microscopy and confocal reflection imaging. Permeability was measured using a microfluidic-based experimental set-up, and mechanical responses were analyzed by rheology. We measured a pore size of 2.84 and 1.69 µm for collagen and fibrin, respectively. Correspondingly, the permeability of the gels was $1.00 \cdot 10^{-12}$ and $5.73 \cdot 10^{-13}$ m^2. The shear modulus in the linear viscoelastic regime was 15 Pa for collagen and 300 Pa for fibrin. The gels exhibited strain-hardening behavior at *ca.* 10% and 50% strain for fibrin and collagen, respectively. This consistent biomechanical characterization provides a detailed and robust starting point for different 3D *in vitro* bioapplications, such as collagen and/or fibrin gels. These features may have major implications for 3D cellular behavior by inducing divergent microenvironmental cues.

Reprinted from *Materials*. Cite as: Moreno-Arotzena, O.; Meier, J.G.; del Amo, C.; García-Aznar, J.M. Characterization of Fibrin and Collagen Gels for Engineering Wound Healing Models. *Materials* **2015**, *8*, 1636-1651.

1. Introduction

Wound healing demonstrates the capacity of skin to regenerate in an orchestrated manner. However, pathological healing processes, such as fibrosis, hypertrophic scars or ulcers, can lead to major disabilities or even death and have a high global incidence [1]. As a reference, 3–6 million people in the United States were affected by these disorders in 2010 [2].

The healing process is the result of a complex interaction of many factors that regulate the development of wounds. This complexity has been addressed by means of diverse approaches. *In vivo* [3], *in vitro* [4,5] and *in silico* [6] studies have been performed to elucidate fundamental wound healing mechanisms. *In vitro* assays have been developed to analyze reepithelialization [7,8]. However, a more complex 3D process occurs in full-thickness injury healing [9]. Novel methodologies based on microfluidic techniques are being developed that offer unique features for the rational design of physiologically relevant *in vitro* systems [10], which could be directed to mimic wound healing processes. Hydrogels have been employed to resemble the extracellular matrix (ECM) [11,12].

Although progress has been made, knowledge of the most adequate conditions to recreate the local microenvironment of wound healing remains lacking.

In a reductionist simplification of the environmental complexity of the wound healing scenario, three main actors can be distinguished: cells, environmental signaling and the ECM. Most previous studies have focused on analyzing the cell-environment signaling interaction [13]. The regulatory role of the ECM is considered critical for cellular processes, e.g., wound healing [14,15]. The ECM is a 3D fibrillar network that provides both architectural scaffolding and a heterogeneous signaling distribution in the whole cell vicinity. The biomechanical cues arising from the ECM play a fundamental role in the modulation of cellular behavior and mechanotransduction in wound healing [15]. The contributions of matrix stiffness and microstructure on cell behavior have been widely demonstrated [16,17]. Recent evidence also indicates a relevant role of interstitial flow [18–20] and ECM confinement level [21–23] on basic cellular processes, e.g., 3D cell migration. Therefore, to develop accurate biomimetic *in vitro* models, it is crucial to select the most adequate material to resemble the ECM *in vivo*.

In wound healing, the primary matrices are the fibrin clot and the granulation tissue, which is mainly formed by newly-deposited collagen [15]. To represent these local microenvironments, biomimetic hydrogels composed of fibrin or collagen I, respectively, are typically used [24,25]. These proteins are very useful because they self-assemble at a proper ionic strength [26]. However, for applications with the objective of recreating wound healing environments, a profound knowledge of the biomechanical and biophysical properties of both hydrogels is required.

In this regard, multiple studies have analyzed the microstructural features and bulk stiffness of similar hydrogels [27–30]. As elements of the microstructure, fiber arrangement and diameter have been extensively studied by scanning electron microscopy (SEM) [31] and confocal reflection imaging (CRI) [32]. Rheological [29,30,33,34], axial tensile tests [28,35] and other techniques [36,37], as well as assays at the individual fiber level [38,39] have been performed. A limited number of studies have also quantified the hydraulic resistance of gel scaffolds to fluid flow. These studies have focused on improving the nutrient diffusion in scaffolds for tissue engineering applications [40,41] or analyzing 3D cell migration [18,42].

The application of collagen and fibrin hydrogels as scaffolds in tissue engineering and *in vitro* experiments and their biomechanical characterization have increased remarkably [27,29,35]. However, these studies have employed a wide diversity of hydrogel compositions and different measurement methods [43]. Modification of the gel composition, polymerization temperature or pH alters various biophysical properties [44]. These variations hinder the application of hydrogels in the controlled representation of microenvironments for wound healing experiments, as well as the analysis of the impact of these parameters on the cell response.

Due to the infeasibility of addressing all possible combinations, in this work, we chose generalized and defined compositions for the gel scaffolds. The chosen specific collagen and fibrin gels have been widely used as physiologically-relevant matrix representations [33,43,45], in applications such as wound healing. Our main aim was to establish a quantitative evaluation of the functional behavior of both hydrogels under experimental conditions that were as comparable as possible. Therefore, we assessed not only the microstructural and rheological properties of both scaffolds, but also their hydraulic resistance to fluid flow. We obtained coherent and corresponding datasets for each hydrogel composition using four complementary experimental techniques that have not previously been used in combination for these gels. This work reports the complete characterization of relevant parameters for biomimetic matrices for 3D *in vitro* assays and primarily focuses on mimicking wound regeneration using widely-used collagen and a fibrin hydrogel compositions. The presented methodology could also be suitable for the study of other scaffolds and their variations.

2. Experimental Section

2.1. Preparation of Fibrin and Collagen Gels

2.1.1. Fibrin Gels

Plasminogen-, fibronectin- and factor XIII-depleted human fibrinogen (American Diagnostica GmbH) was diluted in buffer (50 mM Tris, 100 mM NaCl and 5 mM EDTA) as indicated by the provider. The fibrinogen was mixed with human FXIII (American Diagnostica GmbH) and allowed to polymerize in the presence of human alpha-thrombin (American Diagnostica GmbH), $CaCl_2$ (Sigma) and cell culture media FGM-2 BulletKit (Lonza). Finally, the hydrogels were hydrated and stored in an incubator for 24 h before initiating any experiment. The pH of the gels was 7.4, and the concentration of each constituent per final volume was 3.3 $mg\cdot mL^{-1}$ fibrinogen, 22 $\mu g\cdot mL^{-1}$ FXIII, 1 $U\cdot mL^{-1}$ thrombin and 5 mM $CaCl_2$.

2.1.2. Collagen Gels

The procedure for constructing collagen gels was adapted from a previous work by Shin *et al.* [46]. Collagen type I (BD Biosciences) was buffered to a final concentration of 2 $mg\cdot mL^{-1}$ with 10× DPBS (Gibco) supplemented with calcium and magnesium, cell culture media FGM-2 BulletKit (Lonza) and cell culture-grade water (Lonza). The pH of the dilution was adjusted to 7.4 with NaOH. Mixtures were allowed to polymerize inside humid chambers at 37 °C. Next, the gels were hydrated and stored in an incubator for 24 h before experimentation.

2.2. Scanning Electron Microscopy

Hydrogels were fixed with 2.5% glutaraldehyde (Sigma-Aldrich) followed by 1% electron microscopy grade osmium tetroxide (Ted Pella, Inc., Redding, CA, USA). The hydrogels were subsequently dehydrated in 30%, 50%, 70%, 80% and 95% ethanol solutions, respectively. Gels were freeze-fractured in liquid nitrogen before a final dehydration step in 100% ethanol. The gels were finally subjected to critical point drying using a Baltec CPD030. The samples were sputter-coated with gold-palladium for 4 min using an Emitech K550, resulting in a layer thickness of 15 nm. The samples were visualized by high-resolution imaging with a Merlin field emission scanning electron microscope (FESEM) from Zeiss with a beam voltage of 1 kV and a magnification of 80–120 kX.

2.3. Confocal Reflection Imaging

Confocal reflection was performed using a Leica SP2 equipped with a 63×/1.4 N.A. (numerical aperture) oil immersion lens. The samples were excited at 488 nm with an argon laser and detected at 479–498 nm.

2.4. Microstructural Analysis

To elucidate the microstructural features of the 3D networks, confocal reflection imaging (CRI) and SEM images were acquired as previously described. The void ratio, pore size and fiber radius were evaluated using the free software ImageJ [47]. For the fiber radius and pore size measurements, a straight line and measurement tools were employed. For void ratio analysis, confocal reflection images were binarized. The fiber-to-pore ratio was subsequently calculated from the areas of white (pores) and black (fibers) pixels within the binary images. Three independent sets were examined for each hydrogel, and the data are presented as the mean ± SEM.

2.5. Permeability Experiments

To measure Darcy's permeability (K) of the hydrogels, a specific microfluidic-based experimental set-up was employed. This set-up reproduced the hydraulic environment of the *in vitro* physiological studies of wound healing. A microfluidic platform was used to assess the permeability values of both hydrogels.

The gels were allowed to polymerize within the microfluidic devices, which were fabricated as described by Shin *et al.* [46]. Medium reservoir tubes were inserted into the channel outlets (shown in Figure 1) as described by Sudo *et al.* [48]. The difference in height of the media columns on both sides of the gel caused a pressure gradient of 500 or 13 mm of H_2O for fibrin and collagen, respectively.

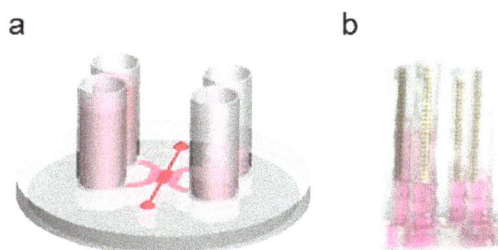

Figure 1. Microfluidic-based experimental set-up for permeability measurements. (**a**) The schematic shows the arrangement of the media columns with respect to the geometry. The design comprises a central gel cage (fuchsia) and two main media channels (pink), which are connected to the corresponding media columns. (**b**) A picture of an actual experiment demonstrates a pressure difference of approximately 100 Pa across the gel induced by the height difference between the media columns on both sides of the geometry.

From Darcy's law, the relationship between the pressure difference and the permeability is as follows:

$$\Delta P(t) = \Delta P(0) \cdot e^{-ct} \tag{1}$$

where t is time, $\Delta P(t)$ is the pressure difference at each time point and $\Delta P(0)$ is the initial pressure difference. The constant c is related to the permeability K, as shown below [48]:

$$K(m^2) = \frac{c \cdot \mu \cdot L \cdot A_r}{\rho \cdot g \cdot A} \tag{2}$$

where μ and ρ are the viscosity and density of the fluid, respectively, L is the length of the gel through which the pressure drop is established, A_r is the area of the media reservoirs, g is the acceleration due to gravity and A is the cross-sectional flow area.

Therefore, Equation (1) states that, for a given initial pressure difference, the pressure difference will tend to equilibrium with an exponential decay. Based on this interpretation, by tracing the experimental pressure difference drop over time, the measured data points were fitted using Equation (1), and the value of c was obtained for each hydrogel. Finally, to characterize the interstitial resistance to flow, Equation (2) was solved for K, and Darcy's permeability values were determined.

2.6. Rheology

For the rheological measurements, a Bohlin Gemini 200 HR Nano rheometer was used. The lower torque limit of the instrument was 3 nN·m in oscillation. All tests were performed using a cone-plate geometry with a diameter of 40 mm, a cone angle of 1° and a truncation height of 30 μm. The temperature was maintained at 37 °C ± 0.1 °C using a Peltier plate.

The samples were pipetted onto the rheometer plate by filling its gap, as demonstrated in Figure 2. To prevent evaporation, the shear gap was covered with a 0.1 Pa·soil, used for calibration.

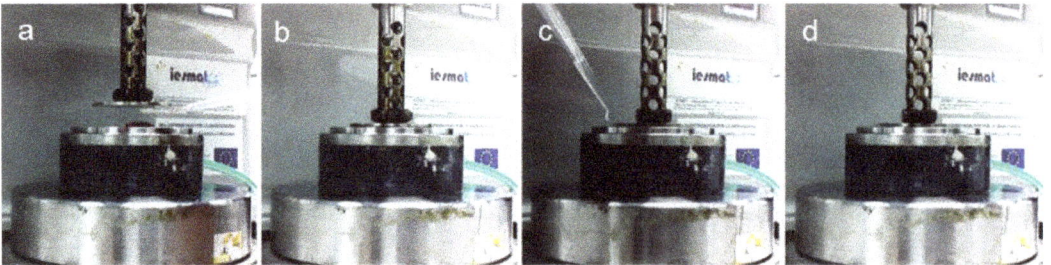

Figure 2. Image sequence of the gel filling process on the rheometer. Once the sample was mixed, it was pipetted *in situ* onto the rheometer plate (**a**); then, the gap was adjusted (**b**), and the sample was covered with oil (**c**) to prevent evaporation; the set-up was then ready to conduct the experiment (**d**).

The curing reaction was traced by measuring the evolution of the shear modulus over time at a constant temperature of 37 °C, an oscillation frequency of 1 rad·s^{-1} and an applied strain amplitude of 0.005. The dependence of the sample moduli on the oscillatory strain amplitude was measured at a constant temperature of 37 °C for excitation frequencies of 0.1 Hz and 0.01 Hz. The strain amplitude was varied in a logarithmically-equidistant interval of 10 measurement points per decade from 0.001 to 1. For each point, data from 6 periods were accumulated.

3. Results and Discussion

3.1. Microstructural Study

Confocal reflection and SEM images were acquired to visualize both the collagen and fibrin hydrogels. As shown in Figures 3 and 4, the lattices of fibrin were more entangled than those of collagen. The collagen networks exhibited twisted patterns formed by bundled fibers, consistent with that reported by Lai et al. [28,35]. The assembled fibrin fibers were straighter and appeared more individually, consistent with its role in physiological clot structures [49].

In addition to this qualitative assessment, microstructural features, such as the void ratio, pore size and fiber radius, were estimated from the images and are summarized in Table 1. As shown in the images, the fiber density of the fibrin networks was greater than that of the collagen matrices, leading to void ratios of approximately 71% and 80%, respectively. Accordingly, the pore size was 1.69 μm for fibrin and 2.84 μm for collagen. The fiber radii were approximately 79 and 66 nm for fibrin and collagen, respectively. For the collagen fibers, the variation of the measured data was quite high, probably due to the variability introduced by the characteristic bundling.

Table 1. Microstructural characteristics of the hydrogels [†].

	Collagen	Fibrin
Void ratio (%)	80.15 ± 1.82	71.46 ± 1.00
Pore size (μm)	2.84 ± 0.94	1.69 ± 0.33
Fiber radius (nm)	79.51 ± 33.16	66.53 ± 13.57

[†] Data are the mean ± SEM.

3.2. Permeability Quantification

Previous work has primarily focused on analyses of microstructure and bulk stiffness due to the key roles of these properties in basic cellular events in 2D [50]. However, wound healing primarily occurs under 3D conditions, and there is accumulating evidence that the confinement of cells has a crucial role on their behavior [21,22,51,52]. A bulk quantity that relates to the confinement feature of hydrogels is their hydraulic resistance. The hydraulic resistance of hydrogel matrices not only controls the transport of nutrients and the shear stress exerted on cells, but also regulates the directional migration of the cells [42,53]. Therefore, in this characterization, we include the assessment of the hydraulic resistance to fluid flow of the scaffolds, because it is key to the rational design and interpretation of physiologically-relevant microsystems.

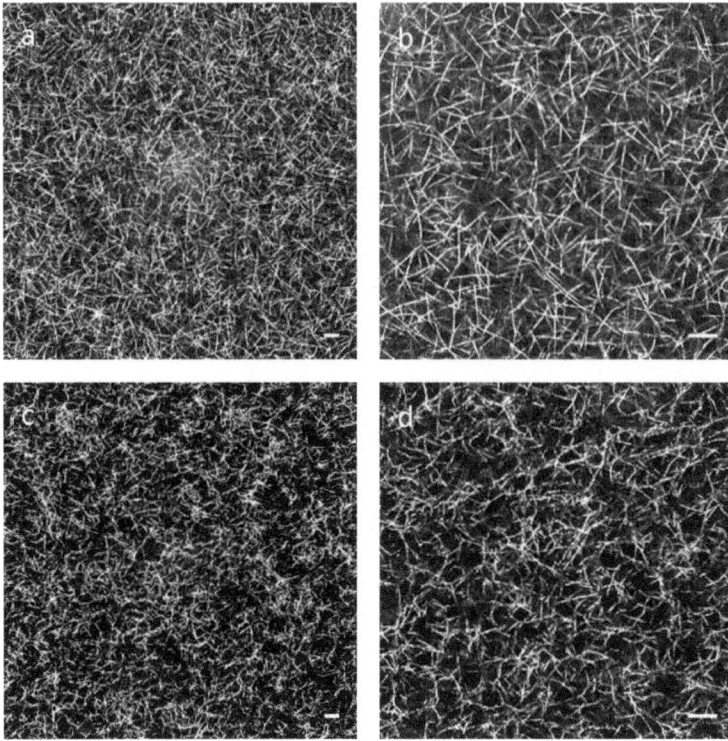

Figure 3. 3D network of the hydrogels. The confocal reflection images show the arrangement of the fibrillar networks for (**a**,**b**) fibrin and (**c**,**d**) collagen gels. (**a**) and (**c**) show a general view; (**b**) and (**d**) are zoomed images of the right-bottom corner of the previous images, respectively. Fibers composing the collagen networks are twisted, whereas the fibers in the fibrin appear straighter. Scale bars correspond to 10 μm.

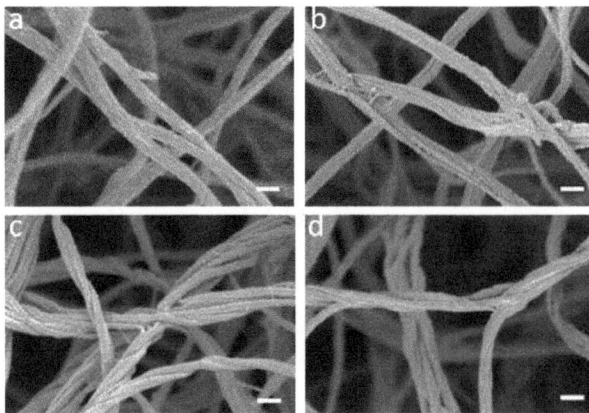

Figure 4. Fiber layout within hydrogels. SEM images of (**a**,**b**) fibrin and (**c**,**d**) collagen gels acquired at magnifications of 80–120 kX show the morphological features of the fibers. The collagen fibers exhibit characteristic bundling, whereas the fibrin fibers are formed more individually. The scale bars correspond to 200 nm.

To quantify the hydrogel permeability, we generated an initial pressure difference across both hydrogels. As shown in Figure 5, the decrease in pressure difference over time was then monitored. The obtained data points were then fitted to an exponential function formatted as in Equation (1), and R^2 values of 0.96 and 0.98 for fibrin and collagen, respectively, were obtained. The exponent coefficient values (c) from the fitting were 0.13 h^{-1} or $4.00 \cdot 10^{-05}$ s^{-1} for fibrin and 0.24 h^{-1} or $7.00 \cdot 10^{-05}$ s^{-1} for collagen. Likewise, by solving Equation (2) for K, we calculated Darcy's permeability for fibrin and collagen: $5.73 \cdot 10^{-13}$ and $1.00 \cdot 10^{-12}$ m^2, respectively. The values of c and K are provided in Table 2. These parameters establish the resistance that a particular porous matrix exerts on the convective fluid flow, *i.e.*, the velocity with which a pressure difference will tend toward equilibrium. Figure 6 illustrates, for a given initial pressure difference, the pressure difference decay for both hydrogels. Pressure equilibrium is achieved more rapidly in the collagen gels than the fibrin gels, indicating less resistance to flow and consistent with the greater void ratio and pore size of the collagen gel observed in the image analysis and other previous measurements [41,42].

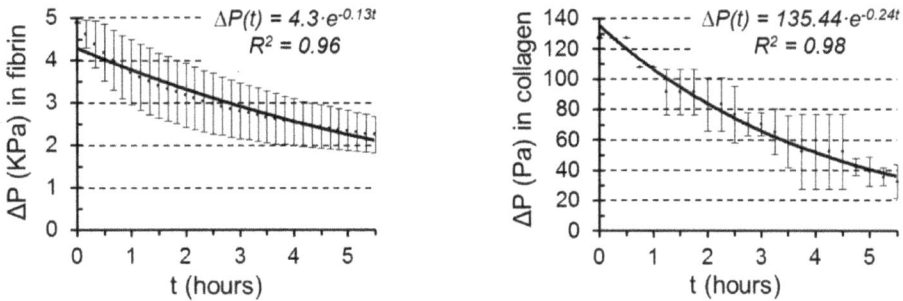

Figure 5. Pressure difference drop over time in hydrogels. The experimental data points of the pressure difference *vs.* time were plotted and fitted to an exponential function for fibrin (**left**) and collagen (**right**) gels. The resulting expressions and R^2 values are also indicated. The exponent coefficient determines the value of the permeability (K).

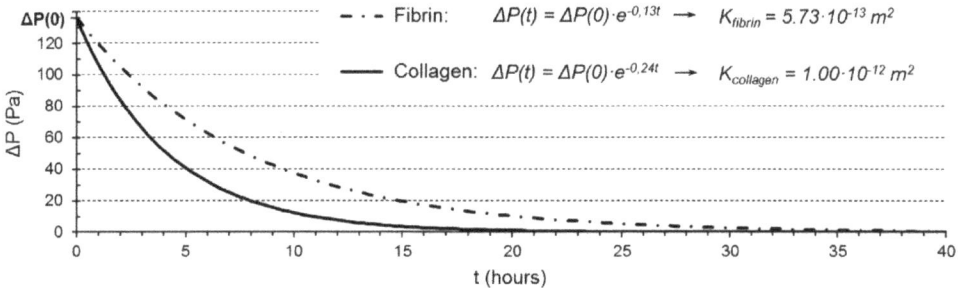

Figure 6. Comparison of the pressure difference drop for both hydrogels. The curves show the pressure difference drop over time for a given initial pressure difference for the collagen and fibrin hydrogels. Compared with collagen, the lower permeability (K) of the fibrin results in a slower function decay.

Table 2. Resistance to fluid flow.

	Collagen	Fibrin
Exponent coefficient, c (s^{-1})	$7.00 \cdot 10^{-05}$	$4.00 \cdot 10^{-05}$
Darcy's permeability, K (m^2)	$1.00 \cdot 10^{-12}$	$5.73 \cdot 10^{-13}$

3.3. Mechanical Response

The polymerization was traced for 3 h, beginning immediately after the gel solution was pipetted and set within the rheometer plate. Figure 7 shows the evolution of the shear modulus (G') over time at a constant temperature of 37 °C. The biexponential function given in Equation (3) fits the measured data points well, thus suggesting two processes with distinct reaction rates (a fast and a slow rate), t_1 and t_2, resulting in an increase in the value of the modulus:

$$G'(t) = G'_1\left(1 - e^{-\left(\frac{t-x_c}{t_1}\right)}\right) + G'_2\left(1 - e^{-\left(\frac{t-x_c}{t_2}\right)}\right) \tag{3}$$

where G'_1 and G'_2, are the associated parameters that characterize the modulus contribution of each process and x_c is the parameter that adjusts the time scale to the beginning of the reactions.

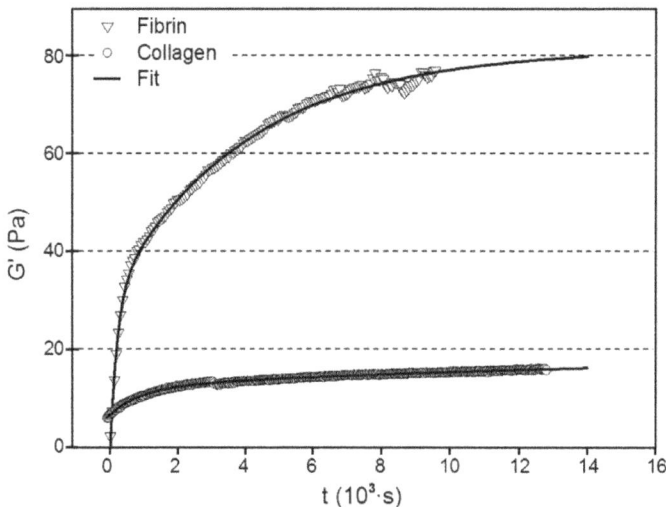

Figure 7. Time evolution of the shear modulus (G') of the hydrogels. The temperature was maintained at 37 °C; the excitation frequency was 1 rad·s^{-1}; and the strain amplitude was 0.005. The lines are fits to the biexponential Equation (3). Note that the polymerization of the gels was not yet complete, *cf.* Figure 8.

The fitted parameter values are shown in Table 3. The fast and slow processes may be associated with the fibrillogenesis of the filaments (fast) and the growth and crosslinking of these filaments (slow) to form a fiber network. Clearly, fibrin cured faster than collagen under the conditions studied and achieved significantly greater modulus values. Furthermore, the fast and the slow processes are more easily distinguishable in fibrin compared with collagen. Fibrin gels are cross-linked and stabilized by FXIII; accordingly, the complete cross-linking of a blood clot during coagulation takes longer than the formation of its fibrillar backbone [29], which could explain the differences in the polymerization kinetics of the hydrogels (see Figure 7). This fact could also explain why confocal scanning micrographs do not show much difference with or without ligation [38].

Table 3. Parameters used to fit Equation (3) to the measured data in Figure 7.

	Fibrin		Collagen	
	Value	*Error*	*Value*	*Error*
G'_1	31.12	0.48	11.70	0.10
t_1	211.87	11.09	896.75	22.16
G'_2	50.78	0.35	7.11	0.30
t_2	4,045.83	113.48	13,919.40	1,431.71
x_c	117.52	4.63	−591.82	15.62

After tracing the curing, the samples remained at rest *in situ* for 24 h before initiating any subsequent rheological experiments. Strain sweep assays were performed for excitation frequencies of 0.1 and 0.01 Hz. Figure 8 outlines the dependence of the measured elastic (G') and viscous (G'') shear moduli on the applied strain amplitude for the fibrin and collagen gels. For the measured excitation frequencies, the registered moduli were similar within the experimental error. The mechanical response of the hydrogels tested by means of oscillatory strain amplitude sweeps revealed the shear modulus in the linear viscoelastic regime and the onset of strain hardening at greater strains (non-linear regime) of the individual hydrogels.

In the linear viscoelastic regime, for the fibrin networks, we measured an elastic shear modulus of 300 Pa, which matches other published values [27,29,33,45]. For the collagen gels, a value of 15 Pa was determined. These data agree with previous measurements obtained using analogous gel preparations [30,54]. Comparing fibrin to collagen, we attributed the dramatic increase in stiffness to FXIII, consistent with its role in acute clots to prevent bleeding problems [49], because the protein concentration difference between the hydrogels and the difference in the microstructural parameters appeared too small to explain the difference in modulus alone.

Both materials were characterized by substantial strain hardening, which occurred within the strain range of 10%–100% and 50%–100% for fibrin and collagen, respectively. On a physical level, the strain hardening of biological hydrogels can be interpreted in analogy to the polymer network theory of semiflexible chains. Semiflexible chains are characterized by similar magnitudes of persistence length and contour length. In networks, the relevant contour length is the distance between network junctions. Such semiflexible chains do not form loops and knots, yet are sufficiently flexible to have significant thermal bending fluctuations [30]. Therefore, in a simplified picture, the onset of strain hardening relates to the straightening of these semiflexible filaments in the network upon straining. An earlier onset of strain hardening suggests a lower degree of freedom and lower thermal fluctuations as a result of a greater ratio of the persistence to the contour length or, in other words, a straighter filament.

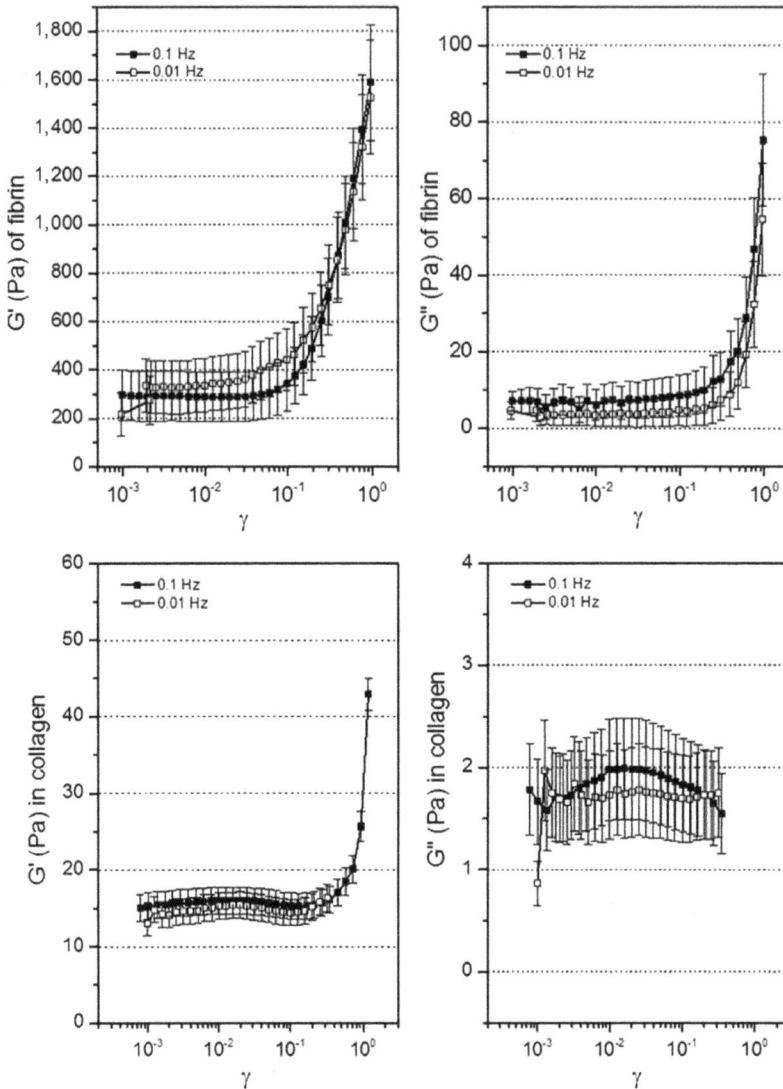

Figure 8. Strain sweeps of hydrogels. Elastic (G') and viscous (G'') shear moduli are shown as a function of the strain amplitude (γ) in parts per unit at frequencies of 0.1 Hz and 0.01 Hz. The temperature during the experiments was maintained at 37 °C. Strain hardening of fibrin and collagen occurred at 10% and 50% strain, respectively. In the elastic range, the elastic shear modulus was 300 and 15 Pa, respectively.

It is tempting to associate the straighter appearance of the fibrin filaments relative to the collagen filaments, as observed in the confocal reflection images (*cf.*, Figure 3), with the somewhat earlier onset of strain hardening in the fibrin and, consequently, to the greater ratio of the persistence to the contour length in the fibrin hydrogel.

In fact, strain hardening is generic to any network composed of semiflexible filamentous proteins [29,30,38,55]. Many soft tissues, such as blood clots, stiffen as they are strained to prevent large deformations that could threaten tissue integrity [30,56,57]. Therefore, strain hardening performs an essential physiological function. Moreover, during wound healing *in vivo*, the ECM is exposed to repeated strain due to cellular contractile machinery, cellular motility, blood flow or interstitial flow [29]. Likewise, at a cellular level, cells embedded in 3D exert forces that cause deformations of 10%–50% in the cell surroundings (20 microns far from the cell margin) and 10% away from the cell [58]. Therefore, we assume that, during wound healing, cells settle their contractile activity within the linear range. In addition, the characterization of the strain hardening and its onset is, in our opinion, an important parameter to control the creation of biomimetic hydrogels for wound healing applications.

4. Conclusions

We have presented a consistent characterization of two widely-used hydrogel compositions based on fibrin and collagen I as biomimetic environments for *in vitro* wound healing studies. The microstructural parameters obtained from microscopic techniques that probed the local and 2D environments correlate with the permeabilities and mechanical values measured using experimental techniques capable of measuring bulk properties in 3D. The studied fibrin hydrogel is characterized by a lower void ratio, lower permeability and a significantly greater shear modulus compared with the collagen hydrogel.

Distinct biomechanical properties differentially regulate migration in 2D and 3D [59]. Although 2D system mechanisms are generally established, the effects of different biomechanical properties of hydrogel materials, such as matrix stiffness [60], microarchitecture [61] or confinement [22], on 3D migratory patterns remain to be elucidated. In this sense, the presented functional characterization of the two widely-used fibrin and collagen hydrogels provides complementary and coherent experimental parameters for the biomechanical properties of the studied hydrogels as a basis for an interpretation of cellular studies in 3D for wound healing experiments. Therefore, the data presented here will open new possibilities for future models, both *in silico* and *in vitro*, of the main mechanisms that regulate wound healing. Furthermore, microfluidic systems, which are biomimetic 3D models in which hydrogels are embedded, will provide multiple possibilities for improving the recreation of the wound healing microenvironment and enable the incorporation of cell cultures, as well as co-cultures of different cell types.

Finally, our experimental approach provides a method for the measurement of the most relevant biomechanical properties of hydrogels, enabling a systematic study of the influence of the numerous combinations of compositions and preparation conditions on these scaffolds and, consequently, cell behavior.

Acknowledgments

This study was supported by the European Research Council (ERC) through Project ERC-2012-StG 306751 and the Spanish Ministry of Economy and Competitiveness (DPI2012-38090-C03-01). The authors acknowledge Roger Kamm and William Polacheck for their technical assistance with the permeability experiments.

Author Contributions

Oihana Moreno-Arotzena designed and performed the experiments and drafted the manuscript. Johann G. Meier performed the rheological measurements and revised a major portion of the manuscript. Cristina del Amo participated in performing the permeability measurements. José Manuel García-Aznar conceived of the study and participated in its design and coordination and helped draft the manuscript. All authors have read and have approved the final manuscript.

Conflicts of Interest

The authors declare no conflict of interest.

References

1. Diegelmann, R.F.; Evans, M.C. Wound Healing: An Overview of Acute, Fibrotic and Delayed Healing. *Front. Biosci.* **2004**, *9*, 283–289.
2. Guo, S.; Dipietro, L.A. Factors Affecting Wound Healing. *J. Dent. Res.* **2010**, *89*, 219–229.
3. Gahagnon, S.; Mofid, Y.; Josse, G.; Ossant, F. Skin Anisotropy *in vivo* and Initial Natural Stress Effect: A Quantitative Study Using High-Frequency Static Elastography. *J. Biomech.* **2012**, *45*, 2860–2865.
4. Kamolz, L.-P.; Keck, M.; Kasper, C. Wharton's Jelly Mesenchymal Stem Cells Promote Wound Healing and Tissue Regeneration. *Stem Cell Res. Ther.* **2014**, *5*, 62.
5. Moreno-Arotzena, O.; Mendoza, G.; Cóndor, M.; Rüberg, T.; García-Aznar, J.M. Inducing Chemotactic and Haptotactic Cues in Microfluidic Devices for Three-Dimensional *in vitro* Assays. *Biomicrofluidics* **2014**, *8*, 064122.
6. Valero, C.; Javierre, E.; García-Aznar, J.M.; Gómez-Benito, M.J. A Cell-Regulatory Mechanism Involving Feedback between Contraction and Tissue Formation Guides Wound Healing Progression. *PLoS ONE* **2014**, *9*, e92774.
7. Anon, E.; Serra-Picamal, X.; Hersen, P.; Gauthier, N.C.; Sheetz, M.P.; Trepat, X.; Ladoux, B. Cell Crawling Mediates Collective Cell Migration to Close Undamaged Epithelial Gaps. *Proc. Natl. Acad. Sci. USA* **2012**, *109*, 10891–10896.
8. Murrell, M.; Kamm, R.; Matsudaira, P. Tension, Free Space, and Cell Damage in a Microfluidic Wound Healing Assay. *PLoS ONE* **2011**, *6*, e24283.
9. Singer, A.J.; Clark, R.A.F. Cutaneous Wound Healing. *N. Engl. J. Med.* **1999**, *341*, 738–746.
10. Polacheck, W.J.; Li, R.; Uzel, S.G.M.; Kamm, R.D. Microfluidic Platforms for Mechanobiology. *Lab. Chip* **2013**, *13*, 2252–2267.
11. Griffith, L.G.; Swartz, M.A. Capturing Complex 3D Tissue Physiology *in vitro*. *Nat. Rev. Mol. Cell Biol.* **2006**, *7*, 211–224.
12. Lambrechts, D.; Roeffaers, M.; Kerckhofs, G.; Hofkens, J.; van de Putte, T.; Schrooten, J.; van Oosterwyck, H. Reporter Cell Activity within Hydrogel Constructs Quantified from Oxygen-Independent Bioluminescence. *Biomaterials* **2014**, *35*, 8065–8077.
13. Grinnell, F. Fibroblast-Collagen-Matrix Contraction: Growth-Factor Signalling and Mechanical Loading. *Trends Cell Biol.* **2000**, *10*, 362–365.

14. Martino, M.M.; Briquez, P.S.; Ranga, A.; Lutolf, M.P.; Hubbell, J.A. Heparin-Binding Domain of Fibrin(ogen) Binds Growth Factors and Promotes Tissue Repair When Incorporated within a Synthetic Matrix. *Proc. Natl. Acad. Sci. USA* **2013**, *110*, 4563–4568.

15. Tomasek, J.J.; Gabbiani, G.; Hinz, B.; Chaponnier, C.; Brown, R.A. Myofibroblasts and Mechano-Regulation of Connective Tissue Remodelling. *Nat. Rev. Mol. Cell Biol.* **2002**, *3*, 349–363.

16. Ulrich, T.A.; de Juan Pardo, E.M.; Kumar, S. The Mechanical Rigidity of the Extracellular Matrix Regulates the Structure, Motility, and Proliferation of Glioma Cells. *Cancer Res.* **2009**, *69*, 4167–4174.

17. Miron-Mendoza, M.; Seemann, J.; Grinnell, F. The Differential Regulation of Cell Motile Activity through Matrix Stiffness and Porosity in Three Dimensional Collagen Matrices. *Biomaterials* **2010**, *31*, 6425–6435.

18. Swartz, M.A.; Fleury, M.E. Interstitial Flow and Its Effects in Soft Tissues. *Annu. Rev. Biomed. Eng.* **2007**, *9*, 229–256.

19. Ng, C.P.; Hinz, B.; Swartz, M.A. Interstitial Fluid Flow Induces Myofibroblast Differentiation and Collagen Alignment *in Vitro*. *J. Cell Sci.* **2005**, *118*, 4731–4739.

20. Polacheck, W.J.; German, A.E.; Mammoto, A.; Ingber, D.E.; Kamm, R.D. Mechanotransduction of Fluid Stresses Governs 3D Cell Migration. *Proc. Natl. Acad. Sci. USA* **2014**, *111*, 2447–2452.

21. Stroka, K.M.; Gu, Z.; Sun, S.X.; Konstantopoulos, K. Bioengineering Paradigms for Cell Migration in Confined Microenvironments. *Curr. Opin. Cell Biol.* **2014**, *30C*, 41–50.

22. Pathak, A.; Kumar, S. Independent Regulation of Tumor Cell Migration by Matrix Stiffness and Confinement. *Proc. Natl. Acad. Sci. USA* **2012**, *109*, 10334–10339.

23. Haeger, A.; Krause, M.; Wolf, K.; Friedl, P. Cell Jamming: Collective Invasion of Mesenchymal Tumor Cells Imposed by Tissue Confinement. *Biochim. Biophys. Acta* **2014**, *1840*, 2386–2395.

24. Rouillard, A.D.; Holmes, J.W. Mechanical Boundary Conditions Bias Fibroblast Invasion in a Collagen-Fibrin Wound Model. *Biophys. J.* **2014**, *106*, 932–943.

25. Miron-Mendoza, M.; Seemann, J.; Grinnell, F. Collagen Fibril Flow and Tissue Translocation Coupled to Fibroblast Migration in 3D Collagen Matrices. *Mol. Biol. Cell* **2008**, *19*, 2051–2058.

26. Buehler, M.J. Nature Designs Tough Collagen: Explaining the Nanostructure of Collagen Fibrils. *Proc. Natl. Acad. Sci. USA* **2006**, *103*, 12285–12290.

27. Rowe, S.L.; Stegemann, J.P. Interpenetrating Collagen-Fibrin Composite Matrices with Varying Protein Contents and Ratios. *Biomacromolecules* **2006**, *7*, 2942–2948.

28. Lai, V.K.; Frey, C.R.; Kerandi, A.M.; Lake, S.P.; Tranquillo, R.T.; Barocas, V.H. Microstructural and Mechanical Differences between Digested Collagen–fibrin Co-Gels and Pure Collagen and Fibrin Gels. *Acta Biomater.* **2012**, *8*, 4031–4042.

29. Münster, S.; Jawerth, L.M.; Leslie, B.A.; Weitz, J.I.; Fabry, B.; Weitz, D.A. Strain History Dependence of the Nonlinear Stress Response of Fibrin and Collagen Networks. *Proc. Natl. Acad. Sci. USA* **2013**, *110*, 12197–12202.

30. Storm, C.; Pastore, J.J.; MacKintosh, F.C.; Lubensky, T.C.; Janmey, P.A. Nonlinear Elasticity in Biological Gels. *Lett. Nat.* **2005**, *435*, 191–194.

31. Gersh, K.C.; Edmondson, K.E.; Weisel, J.W. Flow Rate and Fibrin Fiber Alignment. *J. Thromb. Haemost.* **2010**, *8*, 2826–2828.

32. Hartmann, A.; Boukamp, P.; Friedl, P. Confocal Reflection Imaging of 3D Fibrin Polymers. *Blood Cells. Mol. Dis.* **2006**, *36*, 191–193.

33. Ryan, E.A.; Mockros, L.F.; Weisel, J.W.; Lorand, L. Structural Origins of Fibrin Clot Rheology. *Biophys. J.* **1999**, *77*, 2813–2826.

34. Baniasadi, M.; Minary-Jolandan, M. Alginate-Collagen Fibril Composite Hydrogel. *Materials (Basel)* **2015**, *8*, 799–814.

35. Lai, V.K.; Lake, S.P.; Frey, C.R.; Tranquillo, R.T.; Barocas, V.H. Mechanical Behavior of Collagen-Fibrin Co-Gels Reflects Transition from Series to Parallel Interactions with Increasing Collagen Content. *J. Biomech. Eng.* **2012**, *134*, 011004.

36. Yang, Y.-L.; Leone, L.M.; Kaufman, L.J. Elastic Moduli of Collagen Gels Can Be Predicted from Two-Dimensional Confocal Microscopy. *Biophys. J.* **2009**, *97*, 2051–2060.

37. Lim, B.B.C.; Lee, E.H.; Sotomayor, M.; Schulten, K. Molecular Basis of Fibrin Clot Elasticity. *Structure* **2008**, *16*, 449–459.

38. Collet, J.-P.; Shuman, H.; Ledger, R.E.; Lee, S.; Weisel, J.W. The Elasticity of an Individual Fibrin Fiber in a Clot. *Proc. Natl. Acad. Sci. USA* **2005**, *102*, 9133–9137.

39. Shen, Z.L.; Dodge, M.R.; Kahn, H.; Ballarini, R.; Eppell, S.J. Stress-Strain Experiments on Individual Collagen Fibrils. *Biophys. J.* **2008**, *95*, 3956–3963.

40. Karande, T.S.; Ong, J.L.; Agrawal, C.M. Diffusion in Musculoskeletal Tissue Engineering Scaffolds: Design Issues Related to Porosity, Permeability, Architecture, and Nutrient Mixing. *Ann. Biomed. Eng.* **2004**, *32*, 1728–1743.

41. Wufsus, A.R.; Macera, N.E.; Neeves, K.B. The Hydraulic Permeability of Blood Clots as a Function of Fibrin and Platelet Density. *Biophys. J.* **2013**, *104*, 1812–1823.

42. Polacheck, W.J.; Charest, J.L.; Kamm, R.D. Interstitial Flow Influences Direction of Tumor Cell Migration through Competing Mechanisms. *Proc. Natl. Acad. Sci. USA* **2011**, *108*, 11115–11120.

43. Antoine, E.E.; Vlachos, P.P.; Rylander, M.N. Review of Collagen I Hydrogels for Bioengineered Tissue Microenvironments: Characterization of Mechanics, Structure, and Transport. *Tissue Eng. Part B Rev.* **2014**, *20*, 683–696.

44. Lin, S.; Gu, L. Influence of Crosslink Density and Stiffness on Mechanical Properties of Type I Collagen Gel. *Materials (Basel).* **2015**, *8*, 551–560.

45. Stabenfeldt, S.E.; Gourley, M.; Krishnan, L.; Hoying, J.B.; Barker, T.H. Engineering Fibrin Polymers through Engagement of Alternative Polymerization Mechanisms. *Biomaterials* **2012**, *33*, 535–544.

46. Shin, Y.; Han, S.; Jeon, J.S.; Yamamoto, K.; Zervantonakis, I.K.; Sudo, R.; Kamm, R.D.; Chung, S. Microfluidic Assay for Simultaneous Culture of Multiple Cell Types on Surfaces or within Hydrogels. *Nat. Protoc.* **2012**, *7*, 1247–1259.

47. Rasband, W.S. ImageJ. *U. S.* National Institutes of Health, Bethesda, MD, USA. 1997. Available online: http://imagej.nih.gov/ij/ (accessed on 5 April 2015).

48. Sudo, R.; Chung, S.; Zervantonakis, I.K.; Vickerman, V.; Toshimitsu, Y.; Griffith, L.G.; Kamm, R.D. Transport-Mediated Angiogenesis in 3D Epithelial Coculture. *FASEB J.* **2009**, *23*, 2155–2164.

49. Weisel, J.W. The Mechanical Properties of Fibrin for Basic Scientists and Clinicians. *Biophys. Chem.* **2004**, *112*, 267–276.

50. Engler, A.J.; Sen, S.; Sweeney, H.L.; Discher, D.E. Matrix Elasticity Directs Stem Cell Lineage Specification. *Cell* **2006**, *126*, 677–689.

51. Prentice-Mott, H.V.; Chang, C.-H.; Mahadevan, L.; Mitchison, T.J.; Irimia, D.; Shah, J.V. Biased Migration of Confined Neutrophil-like Cells in Asymmetric Hydraulic Environments. *Proc. Natl. Acad. Sci. USA* **2013**, *110*, 21006–21011.

52. Wolf, K.; Friedl, P. Extracellular Matrix Determinants of Proteolytic and Non-Proteolytic Cell Migration. *Trends Cell Biol.* **2011**, *21*, 736–744.

53. Shi, Z.-D.; Tarbell, J.M. Fluid Flow Mechanotransduction in Vascular Smooth Muscle Cells and Fibroblasts. *Ann. Biomed. Eng.* **2011**, *39*, 1608–1619.

54. Ulrich, T.A.; Jain, A.; Tanner, K.; MacKay, J.L.; Kumar, S. Probing Cellular Mechanobiology in Three-Dimensional Culture with Collagen-Agarose Matrices. *Biomaterials* **2010**, *31*, 1875–1884.

55. Pritchard, R.H.; Huang, Y.Y.S.; Terentjev, E.M. Mechanics of Biological Networks: From the Cell Cytoskeleton to Connective Tissue. *Soft Matter* **2014**, *10*, 1864–1884.

56. Shah, J.V.; Janmey, P.A. Strain Hardening of Fibrin Gels and Plasma Clots. *Rheol. Acta* **1997**, *36*, 262–268.

57. Mitsak, A.G.; Dunn, A.M.; Hollister, S.J. Mechanical Characterization and Non-Linear Elastic Modeling of Poly(glycerol Sebacate) for Soft Tissue Engineering. *J. Mech. Behav. Biomed. Mater.* **2012**, *11*, 3–15.

58. Legant, W.R.; Miller, J.S.; Blakely, B.L.; Cohen, D.M.; Genin, G.M.; Chen, C.S. Measurement of Mechanical Tractions Exerted by Cells in Three-Dimensional Matrices. *Nat. Methods* **2010**, *7*, 969–971.

59. Doyle, A.D.; Petrie, R.J.; Kutys, M.L.; Yamada, K.M. Dimensions in Cell Migration. *Curr. Opin. Cell Biol.* **2013**, *25*, 642–649.

60. Ehrbar, M.; Sala, A.; Lienemann, P.; Ranga, A.; Mosiewicz, K.; Bittermann, A.; Rizzi, S.C.; Weber, F.E.; Lutolf, M.P. Elucidating the Role of Matrix Stiffness in 3D Cell Migration and Remodeling. *Biophys. J.* **2011**, *100*, 284–293.

61. Kubow, K.E.; Conrad, S.K.; Horwitz, A.R. Matrix Microarchitecture and Myosin II Determine Adhesion in 3D Matrices. *Curr. Biol.* **2013**, *23*, 1607–1619.

Polypropylene Biocomposites with Boron Nitride and Nanohydroxyapatite Reinforcements

Kai Wang Chan, Hoi Man Wong, Kelvin Wai Kwok Yeung and Sie Chin Tjong

Abstract: In this study, we develop binary polypropylene (PP) composites with hexagonal boron nitride (hBN) nanoplatelets and ternary hybrids reinforced with hBN and nanohydroxyapatite (nHA). Filler hybridization is a sound approach to make novel nanocomposites with useful biological and mechanical properties. Tensile test, osteoblastic cell culture and dimethyl thiazolyl diphenyl tetrazolium (MTT) assay were employed to investigate the mechanical performance, bioactivity and biocompatibility of binary PP/hBN and ternary PP/hBN-nHA composites. The purpose is to prepare biocomposite nanomaterials with good mechanical properties and biocompatibility for replacing conventional polymer composites reinforced with large hydroxyapatite microparticles at a high loading of 40 vol%. Tensile test reveals that the elastic modulus of PP composites increases, while tensile elongation decreases with increasing hBN content. Hybridization of hBN with nHA further enhances elastic modulus of PP. The cell culture and MTT assay show that osteoblastic cells attach and proliferate on binary PP/hBN and ternary PP/hBN-20%nHA nanocomposites.

Reprinted from *Materials*. Cite as: Chan, K.W.; Wong, H.M.; Yeung, K.W.K.; Tjong, S.C. Polypropylene Biocomposites with Boron Nitride and Nanohydroxyapatite Reinforcements. *Materials* **2015**, *8*, 992-1008.

1. Introduction

In recent years, the global demand for artificial human bone replacements has been ever increasing due to a surge in the number of patients suffering from aging, bone disease and injury. Most orthopedic implants are made of metallic materials, including austenitic 316L stainless steel, cobalt-chromium and titanium-based alloys, due to their high mechanical strength and good ductility. However, the Young's modulus of such metallic alloys far exceeds that of human bones. This creates a stress shielding effect of the surrounding bone tissue, causing the implant to carry a higher proportion of the applied load. Consequently, bone resorption and loosening of the metallic implant can result in the failure of the replacement. Moreover, human body fluids with about 0.9 wt% sodium chloride at 37 °C are hostile to metallic alloys. Thus, metallic implants may undergo electrochemical dissolution or corrosion upon exposure to human body fluids, releasing metallic ions that induce inflammatory response, allergy and cytotoxicity.

In general, nickel ion is the main cause of allergy, followed by cobalt and chromium [1,2]. Furthermore, Cr^{3+} and Co^{2+} ions can bind to several cellular proteins, induce oxidation and impair their biological function, thereby causing cell death and tissue damage [3]. Generally, stainless steels exhibit good wear and corrosion resistance in aqueous environments due to the formation of thin passive oxide/hydroxide films on their surfaces [4–6]. Unfortunately, chloride ions can breakdown the passive films of stainless steels, causing pitting and crevice corrosion and creating anodic dissolution in localized regions of steels. Cobalt-chromium and 316L steel are susceptible to localized corrosion in physiological saline Ringer's solution. Ti-based alloy, such as Ti-6Al-4V, is more corrosion resistant, but inferior wear behavior is its main disadvantage [7]. These potential risks and health hazards with metallic devices have motivated materials scientists to search for other materials with good biocompatibility and no cytotoxicity.

Polymers usually find useful applications in biomedical sectors, due to their being lightweight, the ease of fabrication and the relatively low cost [8–11]. Their tensile stress and modulus can be monitored by adding fillers of micrometer sizes [12–18]. Thus, the composite approach is an effective route for producing polymer biomaterials with desired mechanical properties for bone replacements. As an example, Bonfield and coworkers added 40 vol% hydroxyapatite microfillers to high-density polyethylene to form HAPEX^{TM} composite [19,20]. Hydroxyapatite resembles the mineral component of human bones and is responsible for their mechanical strength and stiffness. However, synthetic hydroxyapatite microparticles (mHA) usually debond from the polymer matrix during the tensile test [21]. Moreover, large mHA particles usually break into small fragments upon tensile loading. These effects directly cause ineffective load transfer from the matrix to mHA fillers, resulting in low tensile strength and failure of the composite [21].

Recent advances in nanoscience and nanotechnology have led to the development and creation of functional nanomaterials with unique chemical, physical mechanical and biological characteristics. In the past decade, the use of nanomaterials in healthcare and biomedical sectors has been rapidly growing due to their potential applications in antimicrobial, bioimaging, drug delivery and orthopedic sectors [22–24]. As recognized, bone tissues are composed of nano-hydroxyapatite (nHA) platelets and collagen fibers. Accordingly, synthetic nanohydroxyapatite particles have been used as reinforcing fillers for non- and degradable polymers to form nanocomposites for biomedical applications, e.g., bone plates and bone scaffolds [25–31]. The attachment and growth of osteoblasts are significantly enhanced on the surface of nHA fillers. In addition, the filler loadings in thermoplastic polymers can be drastically reduced by adding nHA particles. The incorporation of 20 wt% nHA (6.67 vol%) to polypropylene (PP) gives rise to good biocompatibility [32]. The nHA/polymer nanocomposites exhibit better mechanical properties over conventional mHA/polymer composites at the same filler loading level.

In previous studies, we simultaneously added nHA and carbon nanotubes (CNTs) or carbon nanofibers (CNFs) to PP to form biocomposites for bone replacements [32,33]. The addition of low CNT/CNF loadings to nHA/PP composites further enhances their mechanical performance due to the large aspect ratio and remarkable high stiffness of carbonaceous nanofillers. Generally, CNTs are compatible with biological cells, provided that they are firmly embedded within the matrix of polymer composites. However, standalone or individual CNT suspension is reported to be particularly toxic to biological cells, and the cytotoxicity increases with increasing nanotube doses [34–36]. This is because needle-like CNTs can penetrate through the cell membrane and finally reside in the nucleus. BN sheets can also be rolled up into nanotubes with cellular seeding and growth behaviors similar to those of CNTs [37]. In certain cases, boron nitride nanotubes are even more cytotoxic than CNTs [38,39]. Hexagonal boron nitride (hBN) with a layered structure like graphite generally exhibits excellent lubricant behavior, superior thermal and chemical stability and good biocompatibility [40,41]. hBN and titania have been used as a filler for chitosan acting as a protective coating for stainless steel substrate [42]. In this study, we attempted to use planar hBN and nanohydroxyapatite to reinforce PP to form biocomposites for bone replacements. Boron nitride with platelet morphology was selected in order to avoid the cytotoxicity of tubular boron nitride. The main interest in employing layered hBN in polymer hybrid composites was their high chemical stability, good processability and good biological activity [43]. No information is available in the literature on the biocompatibility and tensile behavior of the hBN platelet/PP nanocomposites and hBN-nHA/PP hybrids for biomedical applications. To the best of our knowledge, the present work is the first investigation of the development of PP nanocomposites reinforced with hBN platelets and nHA rods and their biocompatibility.

2. Experimental Section

2.1. Materials

Nanostructured & Amorphous Materials Inc. (Houston, TX, USA) supplied hBN powders for this study. Figure 1 shows the TEM image of hBN powders. TEM examination was performed using a Philips CM-20 TEM microscope (Philips, Amsterdam, The Netherlands) attached to an energy dispersive X-ray spectrometer. Apparently, hBN exhibited a platelet feature with sizes ranging from about 30 to 150 nm. Nanohydroxyapatite powders with a rod-like feature, *i.e.*, a length of about 100 nm and a width of 20 nm, were purchased from Nanjing Emperor Nano Materials (Nanjing, China). Figure 2 shows the TEM image of nHA with rod-like feature. Polypropylene pellets for injection molding purpose (Mophlen HP 500N) were obtained from Basell (Jubail, Saudi Arabia). In general, surface coupling agents can enhance interfacial bonding between the filler and the matrix; however, they may induce cytotoxicity [44]. Both hBN and nHA fillers were not treated with organic surface coupling agents in order to avoid cytotoxicity of the biocomposites.

2.2. Preparation of Nanocomposites

Extrusion and injection molding are versatile and effective processes for large-scale manufacturing of PP-based composites [13,15,45]. Table 1 summarizes typical compositions of binary PP/hBN and ternary PP/hBN-nHA composites. Prior to melt-mixing, PP pellets were dried in an oven at 60 °C for 24 h. Melt compounding was first performed by feeding the material mixtures into a Brabender twin-screw extruder at a screw rotation speed of 30 rpm. The six barrel zones of this extruder were heated to temperatures ranging from 190 to 230 °C. Extruded products were sliced into small pellets and loaded into the Brabender extruder again. The purpose was to obtain effective mixing and homogeneous dispersion of reinforcing fillers in the polymer matrix. The products were pelletized again, dried overnight in an oven and subsequently molded into dog-bone tensile bars and circular disks using an injection molder (Toyo TI-50H, Akashi, Japan). The disks were primarily used for cell seeding and proliferation measurements. The composites were twice extruded for filler homogenization followed by injection molding; the matrix material could be degraded slightly. However, applied stress was mainly carried by the fillers of composites, thus minute matrix degradation would not affect overall mechanical performance of the composites greatly. Furthermore, nHA fillers were found to be very effective at improving the dimensional and thermal stability of PP [32]. Such a "three-step processing" strategy was also adopted by other researchers for making composite materials with improved physical and mechanical properties [46].

Figure 1. TEM micrographs of hexagonal boron nitride (hBN) showing platelet feature with sizes ranging from about 30 to 150 nm.

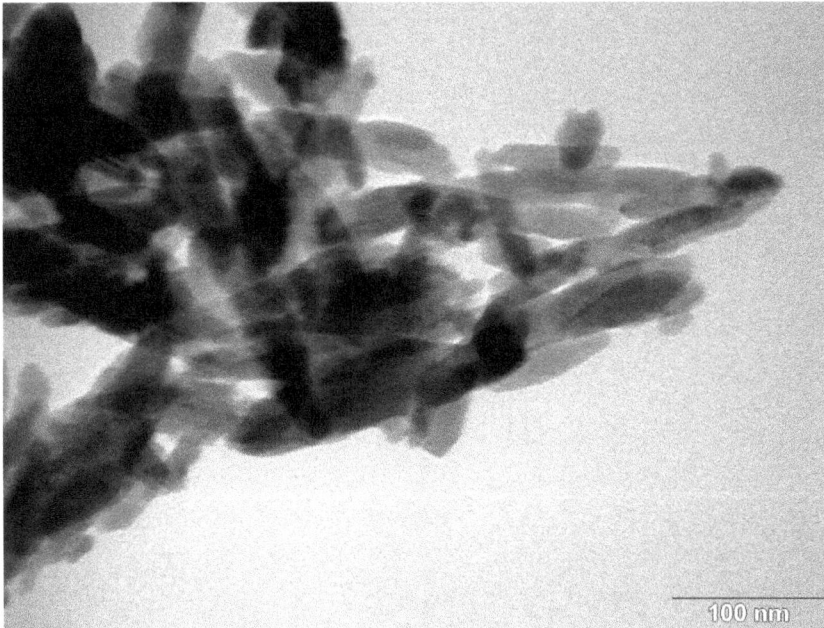

Figure 2. TEM micrograph of nHA showing rod-like feature with an average length of 90 nm.

Table 1. The compositions of the polypropylene (PP) composites studied. nHA, nanohydroxyapatite.

Specimen	PP (wt%)	hBN (wt%)	nHA (wt%)
PP/5% hBN	95	5	0
PP/10% hBN	90	10	0
PP/15% hBN	85	15	0
PP /20% hBN	80	20	0
PP/20% nHA	80	0	20
PP/5% hBN-20% nHA	75	5	20
PP/10% hBN-20% nHA	70	10	20
PP/15% hBN-20% nHA	65	15	20

2.3. Material Characterization

Scanning electron microscopy (SEM) was employed to examine the morphologies of fillers, composites and osteoblasts. Both field-emission SEM (Jeol JSM-6335F, Tokyo, Japan) and conventional SEM (Jeol JSM 820, Tokyo, Japan) were used for this purpose. The surfaces of composite samples were deposited with a thin carbon film. Tensile experiments were performed at room temperature using an Instron tester (Model 5567, Norwood, MA, USA) at a crosshead speed of 10 mm·min^{-1} in accordance with ASTM D638-08 [47]. Young's modulus of the samples was determined from the linear region of stress-strain curves. Five samples of each composition were used for testing, and the average values were evaluated.

2.4. Cell Seeding and Proliferation

Human osteoblasts (Saos-2) were seeded in Dulbecco's Modified Eagle's Medium (DMEM) supplemented with 10% fetal bovine serum, penicillin and streptomycin. The samples (4 × 4 × 1 mm) for cell cultivation and proliferation tests were sliced from injection molded disks into small rectangles. They were rinsed with 70% ethanol and phosphate-buffered saline (PBS) solutions. A suspension of Saos-2 containing 10^4 cells was seeded on these samples placed in a 96-well plate and then kept in a humidified incubator with 5% CO_2 in air at 37 °C for 4 and 7 days, respectively. The culture medium was refreshed every two days. Following the incubation, the samples were washed with PBS and fixed with 10% formaldehyde, followed by dehydration through a series of graded ethanol solutions and critical point drying. Once dry, they were deposited with a thin gold film and placed inside SEM.

The proliferation of osteoblasts on all specimens was assessed using the 3-(4,5-dimethylthiazol-2-yl)-2,5-diphenyltetrazolium bromide (MTT) assay in 96-well plates. A cell suspension with 10^4 cells was introduced to cultured plates with and without samples followed by incubation in a humidified atmosphere of 5% carbon dioxide in air at 37 °C for 4 and 7 days, respectively. The culture medium was refreshed every 2 days. At selected cultivation periods, the medium was aspirated, then 10 μL of MTT solution (5 mg MTT:1 mL DMEM) was added to each well and incubated for 4 h for at 37 °C. At this stage, the tetrazolium ring of MTT salt was cleaved by the succinic dehydrogenase in mitochondria of viable osteoblasts, forming insoluble formazan crystals. The formazan was lastly dissolved in 10% sodium dodecyl sulfate (SDS)/0.01 M hydrochloric acid (100 μL). The absorbance or optical density (OD) of dissolved formazan was quantified spectrophotometrically at a wavelength of 570 nm using a multimode detector (Beckman Coulter DTX 880, Fullerton, CA, USA), with a reference wavelength of 640 nm. Wells with culture medium, MTT, SDS and osteoblasts were used as the control, while wells without osteoblastic cells were employed as a blank background. The samples were used for each test, and the results were expressed in terms of mean ± standard deviation (SD). MTT tests were repeated at least twice.

3. Results and Discussion

3.1. Morphology and Mechanical Behavior

Figure 3a,b shows representative SEM images of binary PP/5% hBN and PP/15%hBN composites. Apparently, hBN fillers are dispersed uniformly in the PP matrix of these composites. The morphology of the typical PP/15% hBN-20%nHA hybrid at low and high magnifications is shown in Figure 4a,b, respectively. Similarly, hBN fillers are distributed homogeneously in the polymer matrix. However, aggregates of nHA fillers can be observed in the PP/15% hBN-20%nHA hybrid, due to its high nHA content, *i.e.*, 20 wt%. Large nHA content is added to the PP hybrid in order to promote the adhesion and proliferation of osteoblasts. PP homopolymer is bioinert and, thus, ineffective for anchoring osteoblasts on its surface. As mentioned before, bone tissues are composed of nHA platelets and collagen fibers. Homogeneous dispersion of nHA in the PP matrix can be achieved by adding low nHA contents. However, such PP nanocomposites are unsuitable for biomedical implant applications due to their low biocompatibility. From our previous study, a minimum nHA content of 20 wt% is required for achieving good bioactivity and biocompatibility of the PP composites [32].

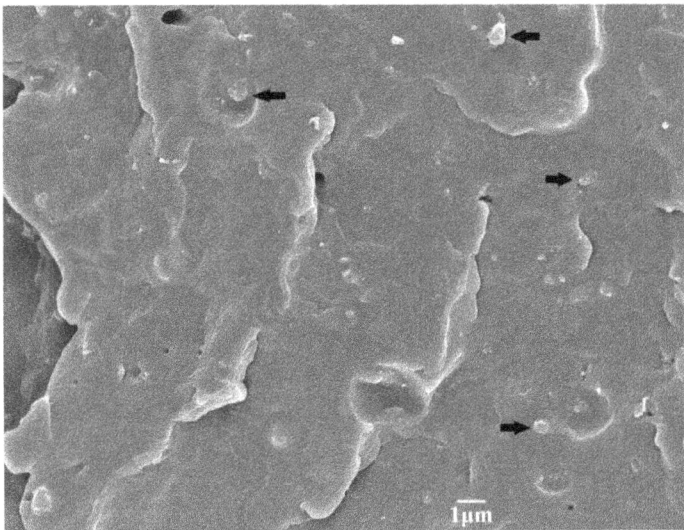

(a)

Figure 3. *Cont.*

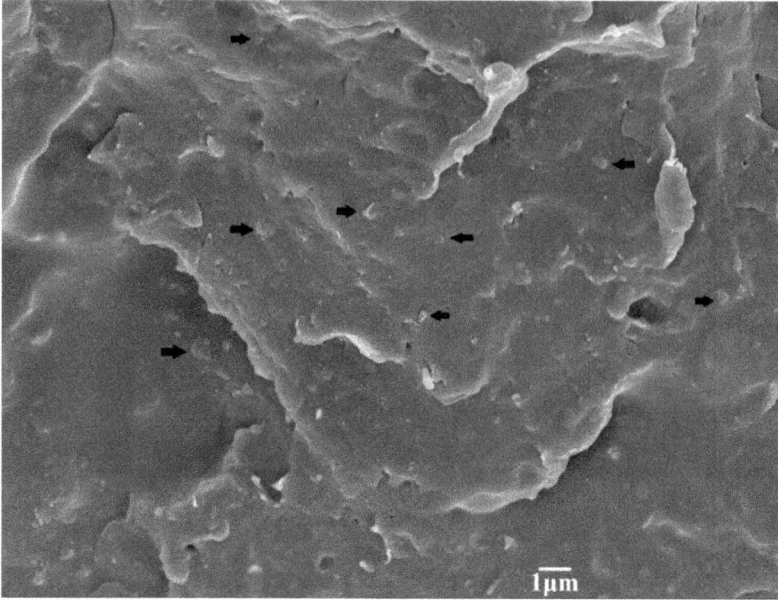

(b)

Figure 3. SEM micrographs showing fractured surfaces of (**a**) PP/5% hBN and (**b**) PP/15% hBN composites with a uniform dispersion of hBN fillers. Black arrow: hBN.

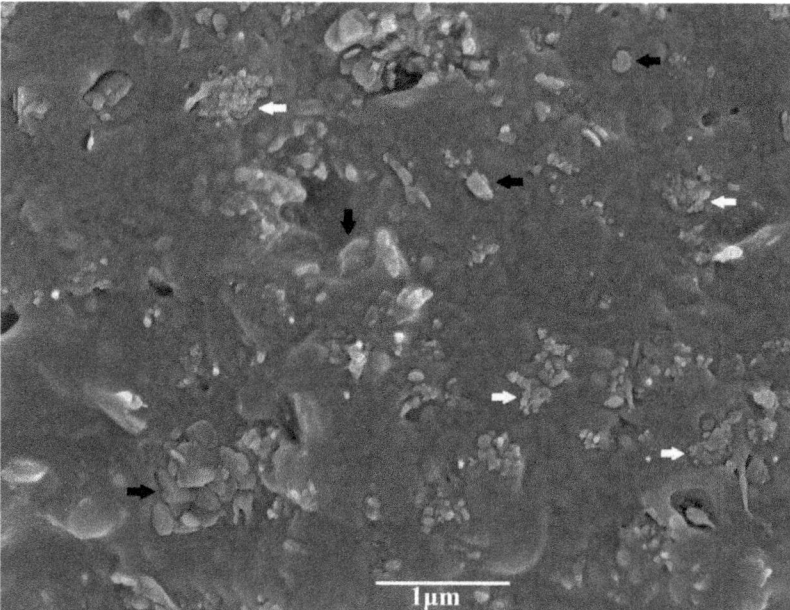

(a)

Figure 4. *Cont.*

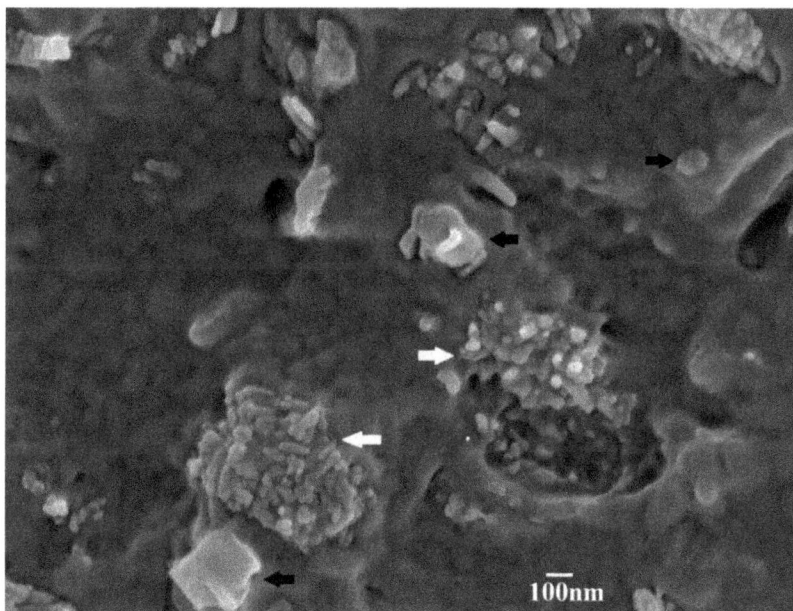

(b)

Figure 4. SEM micrographs showing fractured surface of PP/15% hBN-20% nHA hybrid at (**a**) low and (**b**) high magnifications. nHA aggregates can be readily seen. Black arrow: hBN; white arrow: nHA.

The tensile test results for all specimens studied are tabulated in Table 2. For comparison, the tensile properties of high-density polyethylene (HDPE) composites reinforced with 10, 20, 30 and 40 vol% mHA (4.14 μm), as well as human cortical bone are also listed in this Table [44,48]. It is apparent that the hBN additions up to 20 wt% are beneficial for improving the elastic modulus of PP composites. Moreover, hBN-nHA filler hybridization further increases the modulus of composites, as expected. The PP/15%hBN-20%nHA hybrid displays a maximum modulus of 2.38 GPa, being a 69% improvement over PP. The Young's modulus *vs.* hBN content plots for binary PP/hBN composites and ternary PP/hBN-nHA hybrids are summarized in Figure 5. From Table 2, PP exhibits a large elongation at break (>600%). The additions of 5% and 10% hBN to PP do not impair its tensile ductility. However, the elongation of PP drops to 300% by adding 15 wt% hBN and further reduces to 43% with the addition of 20% hBN. This is due to the interaction between the hBN fillers and the matrix restricts the movement of PP polymer chains at high filler loadings. This behavior is commonly observed in PP composites reinforced with fillers of micro- and nano-scale dimensions [45,49–51]. Hybridization of BN and nHA fillers decreases the tensile elongation of PP composites markedly, especially at high filler contents. Table 2 also reveals that conventional HDPE composites require 20 vol% mHA content to achieve a modulus of 1600 MPa and 30 vol% mHA to reach 2730 MPa. However, the modulus of PP reaches 1615 MPa by adding only 4.5 vol% hBN and further increases to 2383 MPa by adding 7.0 vol% (15 wt%) hBN and 6.67 vol% (20 wt%) nHA. The total hybrid filler content in PP/15 wt% hBN-20 wt% nHA composite is 13.67 vol%, being smaller than that of the HDPE/30 vol% mHA composite. The modulus of the PP/15 wt% hBN-20 wt% nHA

194

hybrid is close to that of the HDPE/30 vol% mHA composite, but the tensile strength of the former is 27.5% higher than that of the latter. It is noted that the biocompatibility of the HDPE/30 vol% mHA composite is unsatisfactory; thus, filler loading of 40 vol% mHA is needed to fabricate conventional HAPEX^{TM} composite. The PP/15 wt% hBN-20 wt% nHA hybrid exhibits a higher tensile strength, but lower stiffness than the HAPEX^{TM} composite. HAPEX^{TM} can only be used for non-loading maxillofacial bone-replacements, due to its stiffness being below the modulus of load-bearing cortical bone of humans [52].

Table 2. Mechanical properties of PP/hBN and PP/hBN-nHA biocomposites.

Specimen	Elastic Modulus, MPa	Tensile Stress, MPa	Elongation at Break, %
PP	1,414 ± 40	26.2 ± 0.5	>600
PP/5 wt% (2.2 vol%) hBN	1,536 ± 34	26.8 ± 0.3	>600
PP/10 wt% (4.5 vol%) hBN	1,615 ± 35	26.6 ± 0.4	>600
PP/15 wt% (7.0 vol%) hBN	1,666 ± 21	26.4 ± 0.3	300
PP/20 wt% (9.7 vol%) hBN	1,758 ± 33	26.5 ± 0.4	43.0 ± 6.3
PP/20 wt% (6.67 vol%) nHA	2,226 ± 33	30.9 ± 0.4	9.9 ± 0.6
PP/5 wt% hBN-20 wt% nHA	2,222 ± 68	27.0 ± 0.1	9.0 ± 2.0
PP/10 wt% hBN-20 wt% nHA	2,276 ± 42	28.6 ± 0.3	7.1 ± 0.6
PP/15 wt% hBN-20 wt% nHA	2,383 ± 18	26.9 ± 0.2	5.8 ± 0.3
HDPE/10 vol% mHA [44]	980 ± 20	17.3 ± 0.3	>200
HDPE/20 vol% mHA [44]	1,600 ± 20	17.8 ± 0.1	34.0 ± 9.5
HDPE/30 vol% mHA [44]	2,730 ± 10	19.5 ± 0.2	6.4 ± 0.5
HDPE/40 vol% mHA [44]	4,290 ± 17	20.7 ± 1.6	2.6 ± 0.4
Cortical bone [48]	7000–30,000	----------	1–3

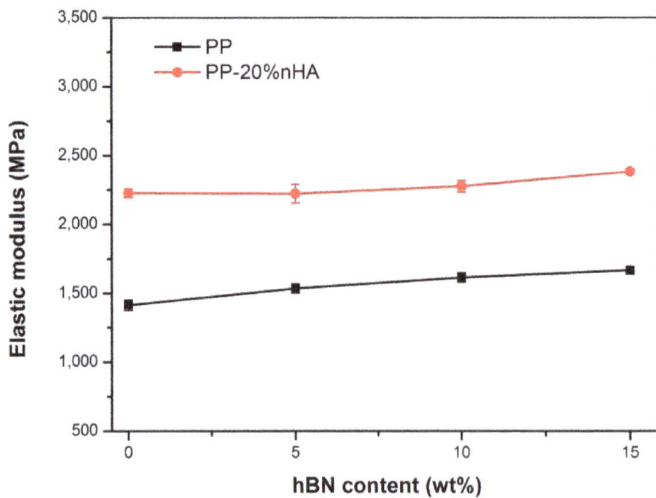

Figure 5. Elastic modulus *vs.* hBN content for PP/hBN and PP/hBN-nHA composites showing the stiffening effect of hBN.

Table 2 reveals that the elastic modulus of the composites increases slowly by adding hBN fillers. In other words, there exists no abrupt increase in the modulus of PP composites due to hBN additions, *i.e.*, no mechanical percolation. Mechanical percolation is found in the composites reinforced with fillers of very large aspect ratios, such as carbon nanotubes [53,54]. In that case, the mechanical percolation model [55,56] can be used to describe a sudden increase in the elastic or shear modulus of the composites reinforced with CNTs of very large aspect ratios. From Figure 1, the aspect ratio (width/thickness) of hBN is estimated to be ~20. Thus, hBN with a low aspect ratio cannot link with itself to form a percolative network in the PP matrix by increasing the filler content up to a critical value. Figure 3b clearly shows the absence of a percolative network in the composite with high hBN content, *i.e.*, 15 wt%.

3.2. Cell Culture and Growth

Figure 6a,b shows respective SEM images of the PP/5% hBN and PP/10% hBN composites after seeding with osteoblastic cells for four days. Apparently, the number of adhered cells on these samples increases with increasing hBN content. Similarly, hybrid composites also provide effective sites and support for the attachment of osteoblastic cells (Figure 7a,b). From these micrographs, the cells spread flatly on the surfaces of composite specimens. Osteoblasts anchor firmly on the sample surfaces via long filopodia. For the PP/15%hBN-20%nHA hybrid composite, cells are densely packed and piled up on each other, such that the entire composite surface is nearly covered with osteoblasts after seeding for four days (Figure 7b).

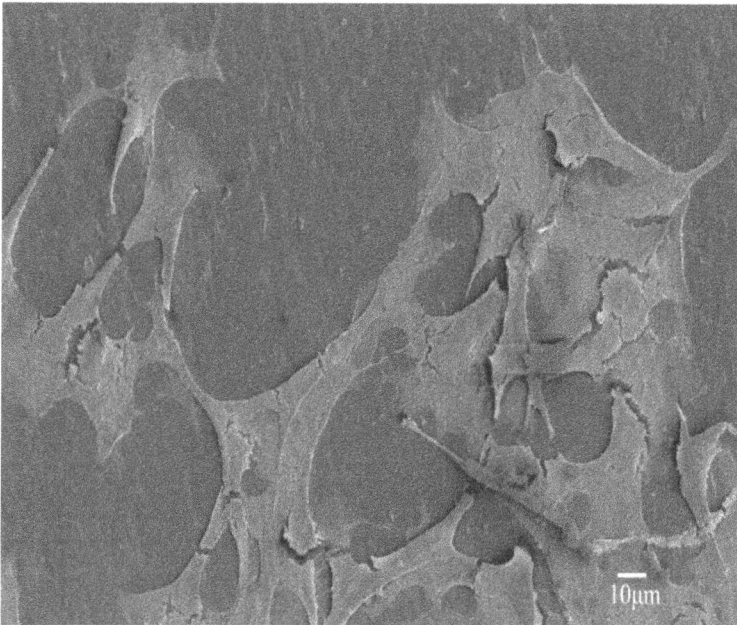

(a)

Figure 6. *Cont.*

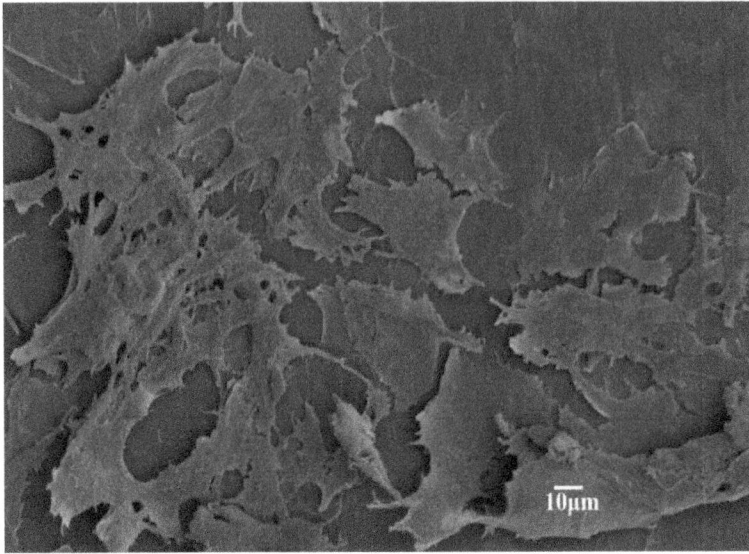

(b)

Figure 6. SEM images of (**a**) PP/5% hBN and (**b**) PP/10% hBN composites cultured with osteoblasts for four days showing spreading of bone cells on the composite surfaces.

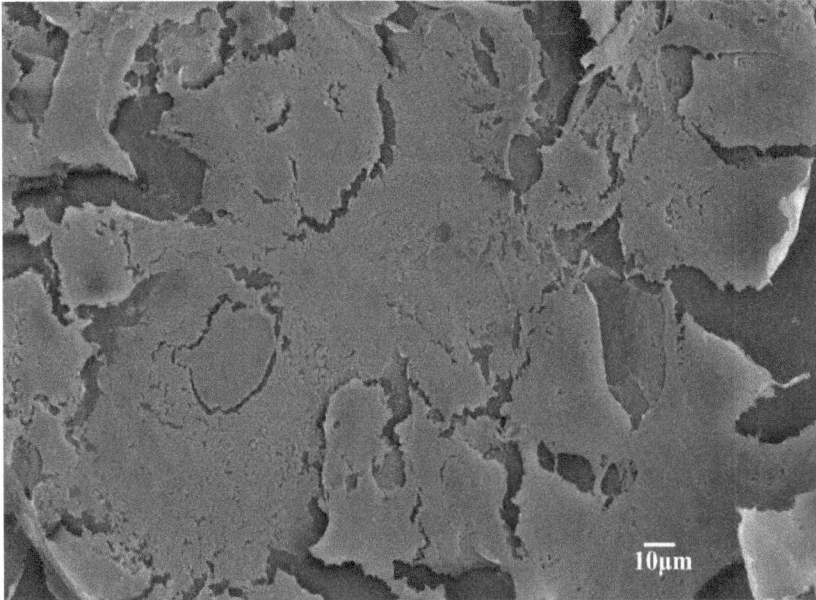

(a)

Figure 7. *Cont.*

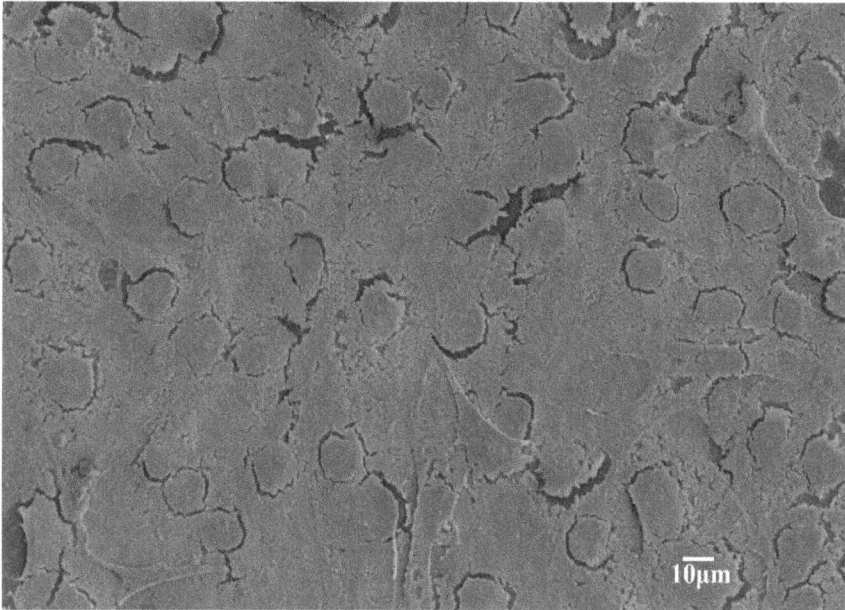

(b)

Figure 7. SEM micrographs of (**a**) PP/5% hBN-20% nHA and (**b**) PP/15% hBN-20%nHA hybrid composites after seeding with osteoblasts for four days. Osteoblasts almost cover entire surface of the specimens.

As was recognized, bone matrix is composed of nHA platelets and collagen fibers. Thus, synthetic nHA mimics the nanostructure of the inorganic phase of bone tissue and can serve as an effective seeding site for the osteoblasts. Nanohydroxyapatite with large surface areas facilitates material-bone cell interactions via protein absorption [24]. Upon adhesion to a substrate, the cell probes its environment and moves using nanometer-scale processes, such as filopodia. These interactive events lead to the formation of new bone cells. Webster *et al.* reported that synthetic nHA is biocompatible and very effective for enhancing the attachment and growth of osteoblasts. Moreover, ceramic materials, like alumina and titania, with grain sizes greater than 100 nm, have long been appreciated for their biocompatibility [57]. BN ceramic material also exhibits good biocompatibility [40], thereby promoting osteoblast adhesion.

Cell proliferation is an important health indicator for osteoblasts for ensuring the good biocompatibility of medical implants. The *in vitro* biocompatibility of binary PP/hBN composites and ternary PP/hBN-nHA hybrids was examined by the MTT assay. This assay is widely used to determine the mitochondria activity of the cells. Cytotoxicity is expressed as the percentage of cell viability by using the following relation:

$$\text{Cell viability (\%)} = 100 \left[\text{OD of sample cells/OD of control} \right] \tag{1}$$

The cellular viability of composite specimens is shown in Figure 8. It can be seen that cellular viability for the PP/hBN and PP/hBN-nHA composites increases with cell culture time from four to seven days. These specimens show low viability values. From the literature, the MTT test tends to give lower cellular viability due to formazan crystals clumping together with BN tubes and the water-insoluble nature of MTT-formazan [37,39]. However, (2-(4-iodopheneyl)-3-(4-nitophenyl)-5-(2,4-disulfophenyl)-2H-tetrazolium monosodium salt (WST-1) assay can yield higher cellular viability because of the water-soluble nature of its formazan product. It is likely that the interference of hBN with MTT-formazan also causes lower cellular viability of the PP/hBN and PP/hBN-nHA composites (28%–30%). Finally, the PP/hBN-nHA hybrid composite system exhibits a slightly higher proliferation rate than binary PP/hBN composites for all testing time intervals. This derives from the synergistic effect of the individual components of hybrid fillers. The results imply that all composite materials studied have no toxicity for the adhesion and growth of osteoblastic cells. From the tensile, cell culture and proliferation tests, it can be concluded that the PP/15%hBN-20% nHA hybrid has great potential for application in maxillofacial surgery.

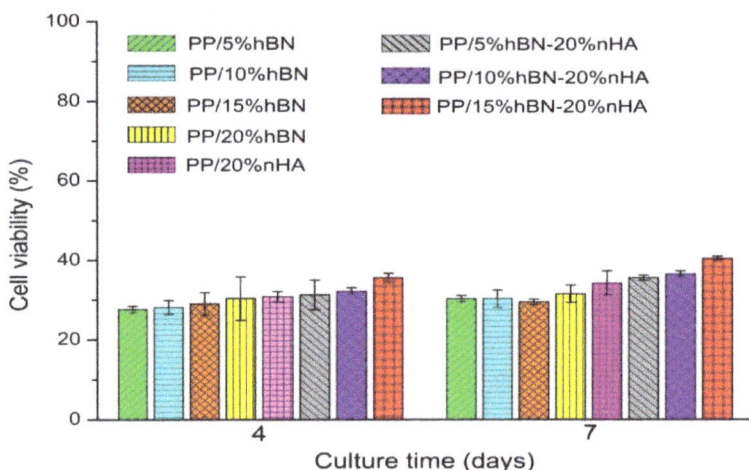

Figure 8. Cell viability of osteoblasts grown on PP/20% nHA, PP/hBN and PP/hBN-nHA composites after seeding for four and seven days. PP/15%hBN-20%nHA hybrid exhibits the highest viability.

4. Conclusions

This article presented the design and testing of binary and hybrid composites for human bone replacements. Binary PP/hBN and ternary PP/hBN-20%nHA composites were successfully prepared by melt mixing and injection molding techniques. Hybrid nanocomposites inherited the property of individual fillers by producing materials with better biocompatibility and mechanical properties. The results showed that the elastic modulus of PP composites increases with increasing the hBN content. Hybridization of hBN with nHA further enhances the elastic modulus of PP composites. The hBN and/or nHA additions reduce tensile ductility of the PP biocomposites. Finally, cell cultivation and MTT assay results revealed that the osteoblasts can attach and proliferate on binary PP/BN and ternary PP/BN-20%nHA composites.

Acknowledgments

This work was supported by the Hong Kong Research Grant Council, the General Research Fund (No. 718913) and the National Natural Science Foundation of China General Program (No. 31370957).

Author Contributions

Kai Wang Chan fabricated composite specimens, examined the specimens before and after cell cultivation tests using SEM and TEM, as well as carried out tensile and MTT assay measurements. Hoi Man Wong performed cell cultivation tests. Kelvin Wai Kwok Yeung supervised cell cultivation and MTT tests. Sie Chin Tjong designed the project and wrote the manuscript.

Conflicts of Interest

The authors declare no conflict of interest.

References

1. Jacobs, J.J.; Gilbert, J.L.; Urban, R.M. Corrosion of metal orthopedic implants. *J. Bone Joint. Surg. Am.* **1998**, *80A*, 268–282.
2. Hallab, N.; Meritt, K.; Jacobs, J.J. Metal sensitivity in patients with orthopedic implants. *J. Bone Joint. Surg. Am.* **2001**, *83A*, 428–436.
3. Scharf, B.; Clement, C.C.; Zolla, V.; Perino, G.; Yan, B.; Elci, S.G.; Purdue, E.; Goldring, S.; Macaluso, F.; Cobelli, N.; *et al.* Molecular analysis of chromium and cobalt-related toxicity. *Sci. Rep.* **2014**, *4*, 5729:1–5729:12.
4. Tjong, S.C.; Hoffman, R.W.; Yeager, E.B. Electron and ion spectroscopic studies of the passive film on iron-chromium alloys. *J. Electrochem. Soc.* **1982**, *129*, 1662–1668.
5. Tjong, S.C.; Yeager, E.B. ESCA and SIMS studies of the passive film on iron. *J. Electrochem. Soc.* **1981**, *128*, 2251–2254.
6. Tjong, S.C.; Lau, K.C. Abrasion resistance of stainless steel composites reinforced with hard TiB_2 particles. *Compos. Sci. Technol.* **2000**, *60*, 1141–1146.
7. Schmutz, P.; Quach-Vu, N.C.; Gerber, I. Metallic medical implants: Electrochemical characterization of corrosion processes. *Interface* **2008**, *17*, 35–40.
8. Tjong, S.C.; Meng, Y.Z. Effect of reactive compatibilizers on the mechanical properties of polycarbonate/poly(acrylonitrile-butadiene-styrene) blends. *Eur. Polym. J.* **2000**, *36*, 123–129.
9. Meng, Y.Z.; Hay, A.S.; Jian, X.G.; Tjong, S.C. Synthesis and properties of poly(aryl ether sulfone)s containing the phthalazinone moiety. *J. Appl. Polym. Sci.* **1998**, *68*, 137–143.
10. Meng, Y.Z.; Tjong, S.C.; Hay, A.S.; Wang, S.J. Synthesis and proton conductivities of phosphonic acid containing poly-(arylene ether)s. *J. Polym. Sci. Part A Polym. Chem.* **2001**, *39*, 3218–3226.

11. Du, L.C.; Meng, Y.Z.; Wang, S.J.; Tjong, S.C. Synthesis and degradation behavior of poly(propylene carbonate) derived from carbon dioxide and propylene oxide. *J. Appl. Polym. Sci.* **2004**, *92*, 1840–1846.

12. Orefice, R.; Clark, A.; West, J.; Brennan, A.; Hench, L. Processing, properties, and *in vitro* bioactivity of polysulfone-bioactive glass composites. *J. Biomed. Mater. Res. A.* **2007**, *80*, 565–580.

13. Gao, C.; Gao, J.; You, X.; Huo, S.; Li, X.; Zhang, Y.; Zhang, W. Fabrication of calcium sulfate/PLLA composite for bone repair. *J. Biomed. Mater. Res. A* **2005**, *73*, 244–253.

14. Fung, K.L.; Li, R.K.Y.; Tjong, S.C. Interface modification on the properties of sisal fiber-reinforced polypropylene composites. *J. Appl. Polym. Sci.* **2002**, *85*, 169–176.

15. Tjong, S.C.; Liu, S.L.; Li, R.K.Y. Mechanical properties of injection moulded blends of polypropylene with thermotropic liquid crystalline polymer. *J. Mater. Sci.* **1996**, *31*, 479–484.

16. Tjong, S.C.; Meng, Y.Z. Performance of potassium titanate whisker reinforced polyamide-6 composites. *Polymer* **1998**, *39*, 5461–5466.

17. Liang, J.Z.; Li, R.K.Y.; Tjong, S.C. Tensile properties and morphology of PP/EPDM/glass bead ternary composites. *Polym. Compos.* **1999**, *20*, 413–422.

18. Li, X.H.; Tjong, S.C.; Meng, Y.Z.; Zhu, Q. Fabrication and properties of poly(propylene carbonate)/calcium carbonate composites. *J. Polym. Sci. Part B Polym. Phys.* **2003**, *41*, 1806–1813.

19. Huang, J.; Disilvio, L.; Wang, M.; Tanner, K.E.; Bonfield, W. *In vitro* mechanical and biological assessment of hydroxyapatite-reinforced polyethylene composite. *J. Mater. Sci. Mater. Med.* **1997**, *8*, 775–779.

20. Guild, F.J.; Bonfield, W. Predictive modeling of the mechanical properties and failure processes in hydroxyapatite- polyethylene (Hapex^TM) composite. *J. Mater. Sci. Mater. Med.* **1998**, *9*, 497–502.

21. Abu Bakar, M.S.; Cheang, P.; Khor, K.A. Mechanical properties of injection molded hydroxyapatite-polyetheretherketone biocomposites. *Compos. Sci. Technol.* **2003**, *63*, 421–425.

22. Duncan, T.V. Applications of nanotechnology in food packaging and food safety: Barrier materials, antimicrobials and sensors. *J. Colloid. Interf. Sci.* **2011**, *363*, 1–24.

23. Hule, R.A.; Pochan, D.J. Polymer nanocomposites for biomedical applications. *MRS Bull.* **2007**, *32*, 354–358.

24. Zhang, J.; Webster, T.J. Nanotechnology and nanomaterials: Promises for improved tissue regeneration. *Nano Today* **2009**, *4*, 66–80.

25. Liu, H.; Webster, T.J. Mechanical properties of dispersed ceramic nanoparticles in polymer composites for orthopedic applications. *Int. J. Nanomed.* **2010**, *5*, 299–313.

26. Thein-Han, W.W.; Shah, J.; Misra, R.D.K. Superior *in vitro* biological response and mechanical properties of an implantable nanostructured biomaterial: Nanohydroxyapatite-silicone rubber composite. *Acta Biomater.* **2009**, *5*, 2668–2679.

27. Boissard, C.I.R.; Bourban, P.E.; Tami, A.E.; Alini, M.; Egli, D. Nanohydroxyapatite poly(ester urethane) scaffold for bone tissue engineering. *Acta Biomater.* **2009**, *5*, 3316–3327.

28. Li, Y.; Liu, C.; Zhai, H.; Zhu, G.; Pan, H.; Xu, X.; Tang, R. Biomimetic graphene oxide–hydroxyapatite composites *via in situ* mineralization and hierarchical assembly. *RSC Adv.* **2014**, *4*, 25398–25403.

29. Rajeswari, D.; Gopi, D.; Ramya, S.; Kavitha, L. Investigation of anticorrosive, antibacterial and *in vitro* biological properties of a sulphonated poly(etheretherketone)/strontium, cerium co-substituted hydroxyapatite composite coating developed on surface treated surgical grade stainless steel for orthopedic applications. *RSC Adv.* **2014**, *4*, 61525–61536.

30. Jiang, L.X.; Jiang, L.Y.; Xu, L.J.; Han, C.T.; Xiong, C.D. Effect of a new surface-grafting method for nano-hydroxyapatite on the dispersion and the mechanical enhancement for poly(lactide-*co*-glycolide). *Express Polym. Lett.* **2014**, *8*, 133–141.

31. Reves, B.T.; Jenning, J.A.; Burngardner, J.D.; Harrgard, W.O. Osteoinductivity assessment of BMP-2 loaded composite chitosan-nano-hydroxyapatite scaffolds in a rat muscle pouch. *Materials* **2011**, *4*, 1360–1374.

32. Liao, C.Z.; Li, K.; Wong, H.M.; Tong, W.Y.; Yeung, K.W.K.; Tjong, S.C. Novel polypropylene biocomposites reinforced with carbon nanotubes and hydroxyapatite nanorods for bone replacements. *Mater. Sci. Eng. C Mater. Biol. Appl.* **2013**, *33*, 1380–1388.

33. Liao, C.Z.; Wong, H.M.; Yeung, K.W.K.; Tjong, S.C. The development, fabrication and mechanical characterization of polypropylene composites reinforced with carbon nanofiber and hydroxyapatite nanorod hybrid fillers. *Int. J. Nanomed.* **2014**, *9*, 1299–1310.

34. Muller, J.; Huaux, F.; Lison, D. Respiratory toxicity of carbon nanotubes: How worried should we be? *Carbon* **2006**, *44*, 1048–1056.

35. Porter, A.; Gass, M.; Muller, K.; Skepper, J.N.; Midgley, P.A.; Welland, M. Direct imaging of single-walled carbon nanotubes in cells. *Nat. Nanotechnol.* **2007**, *2*, 713–717.

36. Muller, K.H.; Koziol, K.K.; Skepper, J.N.; Midgley, P.A.; Welland, M.E.; Porter, A. Toxicity and imaging of multi-walled carbon nanotubes in human macrophage cells. *Biomaterials* **2009**, *30*, 4152–4160.

37. Ciofani, G.; Danti, S.; D'Alessandro, D.; Moscato, S.; Menciassi, A. Assessing cytotoxicity of boron nitride nanotubes: Interference with the MTT assay. *Biochem. Biophy. Res. Commun.* **2010**, *394*, 405–411.

38. Horvath, L.; Magrez, A.; Golberg, D.; Zhi, C.; Bando, Y.; Smajda, R.; Horvath, E.; Forro, L.; Schwaller, B. *In vitro* investigation of the cellular toxicity of boron nitride nanotubes. *ACS Nano* **2011**, *5*, 3800–3810.

39. Ciofani, G.; Danti, S.; Genchi, G.G.; D'Alessandro, D.; Pellequer, J.; Odorico, M.; Mattoli, V.; Giorgi, M. Pilot *in vivo* toxicological investigation of boron nitride nanotubes. *Int. J. Nanomed.* **2012**, *7*, 19–24.

40. Zhang, H.; Yamazaki, T.; Zhi, C.; Hanagata, N. Identification of boron nitride nanosphere-binding peptide for intracellular delivery of CpG olideoxynucleotides. *Nanoscale* **2012**, *4*, 6343–6350.

41. Lipp, A.; Schwetz, K.A.; Hunold, K. Hexagonal boron nitride: Fabrication, properties and applications. *J. Eur. Ceram. Soc.* **1989**, *5*, 3–9.
42. Raddaha, N.S.; Cordero-Arias, L.; Cabanas-Polo, S.; Virtanen, S.; Roether, J.A.; Boccaacini, A.R. Electrophoretic deposition of chitosan/h-BN and chitosan/h-BN/TiO$_2$ composite coatings on stainless steel (316L) substrates. *Materials* **2014**, *7*, 1814–1829.
43. Lu, F.; Wang, F.; Cao, L.; Kong, C.Y.; Huang, X. Hexagonal boron nitride nanomaterials: Advanced towards bioapplications. *Nanosci. Nanotechnol. Lett.* **2012**, *4*, 949–961.
44. Wang, M.; Joseph, R.; Bonfield, W. Hydroxyapatite-polyethylene composites for bone substitution: Effects of ceramic particle size and morphology. *Biomaterials* **1998**, *19*, 2357–2366.
45. Tjong, S.C.; Meng, Y.Z. Impact-modified polypropylene/vermiculite nanocomposites. *J. Polym. Sci. Part B Polym. Phys.* **2003**, *41*, 2332–2341.
46. Coskunses, F.I.; Yilmazer, U. Preparation and characterization of low density polyethylene/ethylene methyl acrylate glycidyl methacrylate/organoclay nanocomposites. *J. Appl. Polym. Sci.* **2011**, *120*, 3087–3097.
47. *Standard Test Method for Tensile Properties of Plastics*; ASTM D638-08; ASTM: West Conshohocken, PA, USA, 2008.
48. Hench, L.L.; Wilson, J. *An Introduction to Bioceramics*; World Scientific: Singapore, 1993.
49. Tjong, S.C.; Bao, S.P. Fracture toughness of high density polyethylene/SEBS-g-MA/montmorillonite nanocomposites. *Compos. Sci. Technol.* **2007**, *67*, 314–323.
50. Tjong, S.C.; Meng, Y.Z. Morphology and mechanical characteristics of compatibilized polyamide 6-liquid crystalline polymer composites. *Polymer* **1997**, *38*, 4609–4615.
51. Xie, X.L.; Li, B.G.; Pan, Z.R.; Li, R.K.Y.; Tjong, S.C. Effect of talc/MMA *in situ* polymerization on mechanical properties of PVC-matrix composites. *J. Appl. Polym. Sci.* **2001**, *80*, 2105–2112.
52. Downes, R.N.; Vardy, S.; Tanner, K.E.; Bonfield, W. Hydroxyapatite polyethylene composite in orbital surgery. *Bioceramics* **1991**, *4*, 239–246.
53. Dalmas, F.; Cavaille, J.Y.; Gauthier, C.; Chazeau, L.; Dendievel, R. Viscoelastic behavior and electrical properties of flexible nanofiber filled polymer nanocomposites. Influence of processing conditions. *Compos. Sci. Technol.* **2007**, *67*, 829–839.
54. Capadona, J.R.; van Den Berg, O.; Capadona, L.; Schroeter, M.; Rowan, S.J.; Tyler, D.J.; Weder, C. A versatile approach for the processing of polymer nanocomposites with self-assembled nanofiber templates. *Nat. Nanotechnol.* **2007**, *2*, 765–769.
55. Quali, N.; Cavaille, J.Y.; Perez, J. Elastic, viscoelastic and plastic behavior of multiphase polymer blends. *Plast. Rubber Compos. Proces Appl.* **1991**, *16*, 55–60.
56. Kolarik, J. Simultaneous prediction of the modulus, tensile strength and gas permeability of binary polymer blends. *Eur. Polym. J.* **1997**, *34*, 585–590.
57. Webster, T.J.; Ergun, C.; Doremus, R.H.; Siegel, R.W.; Bizios, R. Enhanced functions of osteoblasts on nanophase ceramics. *Biomaterials* **2000**, *21*, 1803–1810.

Gelatin Tight-Coated Poly(lactide-*co*-glycolide) Scaffold Incorporating rhBMP-2 for Bone Tissue Engineering

Juan Wang, Dongsong Li, Tianyi Li, Jianxun Ding, Jianguo Liu, Baosheng Li and Xuesi Chen

Abstract: Surface coating is the simplest surface modification. However, bioactive molecules can not spread well on the commonly used polylactone-type skeletons; thus, the surface coatings of biomolecules are typically unstable due to the weak interaction between the polymer and the bioactive molecules. In this study, a special type of poly(lactide-*co*-glycolide) (PLGA)-based scaffold with a loosened skeleton was fabricated by phase separation, which allowed gelatin molecules to more readily diffuse throughout the structure. In this application, gelatin modified both the internal substrate and external surface. After cross-linking with glutaraldehyde, the surface layer gelatin was tightly bound to the diffused gelatin, thereby preventing the surface layer gelatin coating from falling off within 14 days. After gelatin modification, PLGA scaffold demonstrated enhanced hydrophilicity and improved mechanical properties (*i.e.*, increased compression strength and elastic modulus) in dry and wet states. Furthermore, a sustained release profile of recombinant human bone morphogenetic protein-2 (rhBMP-2) was achieved in the coated scaffold. The coated scaffold also supported the *in vitro* attachment, proliferation, and osteogenesis of rabbit bone mesenchymal stem cells (BMSCs), indicating the bioactivity of rhBMP-2. These results collectively demonstrate that the cross-linked-gelatin-coated porous PLGA scaffold incorporating bioactive molecules is a promising candidate for bone tissue regeneration.

Reprinted from *Materials*. Cite as: Wang, J.; Li, D.; Li, T.; Ding, J.; Liu, J.; Li, B.; Chen, X. Gelatin Tight-Coated Poly(lactide-*co*-glycolide) Scaffold Incorporating rhBMP-2 for Bone Tissue Engineering. *Materials* **2015**, *8*, 1009-1026.

1. Introduction

Clinical studies have shown that the adequate implantation of filling materials for treating critical bone defects resulting from traumatic injuries or tumor resections can facilitate bone regeneration [1]. Autografting is the current conventional treatment for bone defects. However, the process involves a number of drawbacks, including an additional surgical harvesting procedure, and the risk of infection or donor-site morbidity [2]. Allograft materials, such as demineralized bone matrix, also present a risk of immunologic rejection or disease transmission. These factors impede the wide application of the autografting process [3].

The biodegradable polymeric scaffolds for bone reconstruction have received significant attention because of their ability to provide a spatially and temporally appropriate environment for new bone tissue growth [4,5]. Polylactone-type biodegradable polymers, such as poly(L-lactide) (PLA), polyglycolide (PGA), and their copolymer, poly(lactide-*co*-glycolide) (PLGA), are the Food and Drug Administration (FDA)-approved matrices of scaffolds due to their low immunogenicity, non-toxicity, and adjustable degradation rate [6]. However, these polymeric scaffolds impede cell attachment and penetration due to the lack of cell anchoring sites, and they have poor hydrophilicity and low surface energy [7].

Many approaches have been developed to circumvent these drawbacks, including bulk modification and surface modification [8,9]. Bulk modification introduces functional groups to the polymer chain while it may also alter the mechanical and biodegradable properties of the material, which may lead to the undesirable results [10,11]. Surface modification allows the binding of functional groups or bioactive molecules, such as collagen, gelatin, RGD, and hydroxyapatite, onto the surfaces of scaffolds. A few of more complicated surface modifications require chemical reactions [12,13]. However, due to the lack of functional groups on the polymer backbone, it is difficult to modify the surface properties by conventional chemical methods [13]. Surface coating is a simple and general approach for surface modification. However, the bioactive molecules typically do not spread well on polylactone-type skeletons. Additionally, the interactions between the polymer backbone and the bioactive molecules are weak, which leads to an unstable surface coating [11]. To increase the coating efficiency of a polymeric scaffold, a loose polymer backbone structure can be designed, so that the bioactive molecules can easily diffuse across the skeleton and simultaneously increase the adhesiveness for other bioactive molecules to prevent coating loss.

Dunn et al. [14] first introduced an in situ-formed implant based on phase separation triggered by a solvent/non-solvent exchange, a process that has been applied to tissue engineering and drug delivery for many years [15–17]. A water-insoluble biodegradable polymer is first dissolved in an organic solvent that is miscible or partially miscible with water. Following immersion into an aqueous medium, phase separation occurs as the solvent diffuses toward the surrounding aqueous environment, while water penetrates into the organic phase. This process results in polymer precipitation and the formation of implants with a loosened skeleton. For example, Ellis et al. [16] produced PLGA flat sheet membranes with a finger-like structure using 1,4-dioxane and 1-methyl-2-pyrrolidinone (NMP) as solvents and water as the non-solvent. Porous structures are expected to form in the high mutual affinity-NMP-water medium. Oh et al. [17] fabricated the hydrophilic porous PLGA tubes using a modified immersion precipitation method and showed that the tubes were highly effective for the permeation of bovine serum albumin (BSA).

In this study, an immersion separation method was used to design and fabricate a loosened scaffold with skeletal structure and subsequently carried out the surface modifications by immersing the scaffold in a gelatin solution. Gelatin is derived from high molecular weight collagen by breaking the natural triple-helix structure of collagen into single-stranded molecules; it has been used in many aspects of tissue engineering because of its biocompatibility and ease of gelation [18]. Due to the loosened structure of the biopolymer skeleton, gelatin can easily spread across and over the scaffold surface. After a simple cross-linking procedure, gelatin binds tightly throughout the structure, thereby preventing the surface-coating gelatin from easily falling off. Moreover, gelatin is also an ideal carrier for protein delivery [19,20]. In a previous study, the unique release profile of recombinant human bone morphogenetic protein-2 (rhBMP-2) was assessed in gelatin-coated 3D scaffolds, showing first a transient burst and then sustained release profile [20]. Along similar lines, in this study, rhBMP-2 was incorporated by physically entrapment in a gelatin gel. This multifunctional scaffold composed of a PLGA skeleton, gelatin coating, and rhBMP-2 was further evaluated for cell adhesion, proliferation, and differentiation properties.

2. Results and Discussion

2.1. Scaffold Characterizations

2.1.1. Microstructure Detections of 3D Porous Scaffolds

The biocompatibility with cells and tissues of a material surface is determined by the interaction between the cells and the surface of material [10]. Due to their hydrophobicity, PLGA scaffolds are not able to well support cell adhesion and growth. When coated with gelatin, the scaffolds gain the hydrophilic property and cell-recognizable moiety [21–23]. However, the gelatin solution is not able to infiltrate deeply enough into the macropores of the polymeric substrate to form a stable composite; additionally, the gelatin layer on the exterior surface is unstable due to the insufficient adhesion force between the gelatin and the polymer material [23].

In this study, as depicted in Figure 1, a PLGA-based scaffold with a loosened skeleton was fabricated by phase separation triggered by a solvent/non-solvent exchange (Figure $1a_1,b_1$). The gelatin solution was able to easily penetrate into the loosened skeleton, and the surface gelatin coating was stabilized due to the cross-linking bonds with the glutaraldehyde-modified gelatin secured within the PLGA skeleton (Figure $1a_2,b_2$). No functional group of PLGA was involved in the surface modification. In contrast with the methods that use modifying groups in copolymerization, this technique maintains the bulk properties of the materials. Furthermore, growth factors like rhBMP-2 could be easily sealed in the scaffold for controlled release by immersing the PLGA scaffold in gelatin solution supplemented with rhBMP-2 (Figure $1a_3,b_3$).

Phase separation triggered by solvent/non-solvent exchange has previously been applied to fabricate porous nerve guide conduits and bone graft substitutes [15–17,24–27]. The asymmetrical porous structure is formed during the preparation of the biomaterials. Smaller pores are formed at the solvent/non-solvent contact side, when the polymer precipitates due to a higher initial polymer concentration as the non-solvent slowly diffused into the PLGA substrate. The larger pores are formed from the precipitation of the polymer at a lower polymer concentration relative to the initial contact side [17].

After soaking in gelatin and gelatin/rhBMP-2 solutions, the physicochemical properties of the scaffolds were determined and are summarized in Table 1. The PLGA/Gel and PLGA/Gel/rhBMP-2 scaffolds had gelatin contents of 13.8 ± 3.7 and 14.5 ± 4.1 wt%, respectively. After coating PLGA scaffold with gelatin or gelatin/rhBMP-2, the porosity of scaffold slightly decreased from $89.1\% \pm 8.3\%$ to $74.7\% \pm 10.1\%$ or $75.5\% \pm 7.9\%$, corresponding to a pore diameter decrease from 243.6 ± 72.8 to 219.8 ± 97.5 or 214.4 ± 106.3 μm, respectively. These results indicated that gelatin was successfully incorporated into the PLGA scaffold. More importantly, after gelatin-coating, the PLGA skeleton still retained properties compatible with bone regeneration, i.e., a porosity of 30%–90% and a pore size of 100–1000 μm, ranges that were considered ideal for the growth of bone tissue inside an implant.

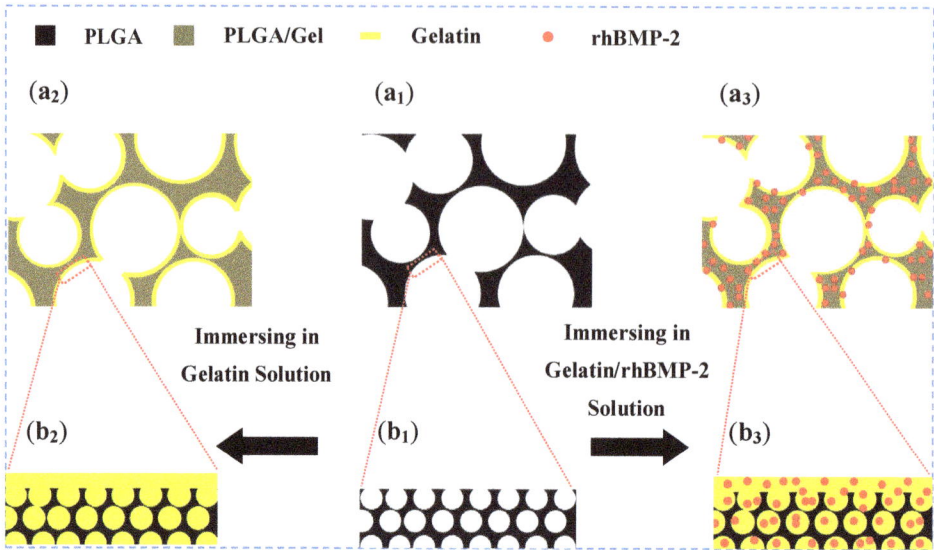

Figure 1. Schematic diagram of surface coating on PLGA scaffold (a_1,b_1), and the PLGA scaffolds coated with gelatin (PLGA/Gel; a_2,b_2) and gelatin/rhBMP-2 mixture (PLGA/Gel/rhBMP-2; a_3,b_3).

Table 1. Physical properties of scaffolds.

Scaffold	Gelatin Content (wt%)	Porosity (%)	Pore Diameter (μm)
PLGA	0	89.1 ± 8.3	243.6 ± 72.8
PLGA/Gel	13.8 ± 3.7	74.7 ± 10.1	219.8 ± 97.5
PLGA/Gel/rhBMP-2	14.5 ± 4.1	75.5 ± 7.9	214.4 ± 106.3

The microstructures of the three-dimensional (3D) porous PLGA, PLGA/Gel, and PLGA/Gel/rhBMP-2 scaffolds fabricated *via* the phase separation/particulate leaching method were observed by SEM and microscopy (Figure 2). The PLGA microstructure had well-interconnected macropores (Figure 2a), which were ideally suited for cell infiltration. As shown in Figure 2b,c, the skeleton had a honeycomb-like structure composed of microvoids with diameters of 2–4 μm, and the PLGA surface contained microscale channels, to which the internal macropores and microvoids in the skeleton were connected. The observed architecture was very favorable for the movement of proteins. Oh *et al.* [17] demonstrated that macromolecules can easily flow into the microvoids.

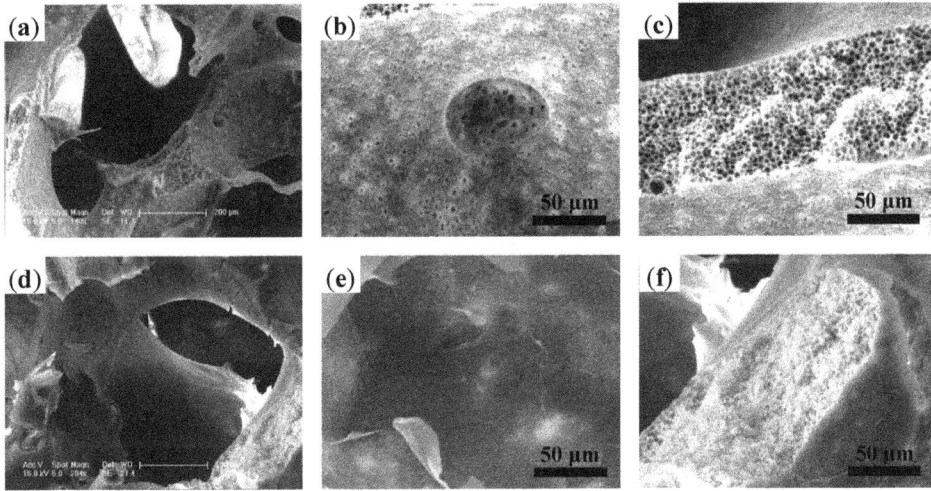

Figure 2. SEM microimages of internal structures for (**a–c**) porous PLGA scaffold and (**d–f**) PLGA scaffold with gelatin coating.

Surface modifications were carried out by submerging the PLGA scaffolds in 1% gelatin solutions. Due to the hydrophobicity of the PLGA-based material, a negative pressure was applied to facilitate the infiltration of gelatin solution into the scaffold. After cross-linking with glutaraldehyde, the gelatin was coated on the surface and stored in the microvoids, as illustrated in Figure 1a$_2$,b$_2$. Figure 2d shows the microstructure of the PLGA/Gel scaffold. The addition of gelatin did not affect the interconnections between the macropores. This observation was consistent with those of previous studies, wherein poly(ε-caprolactone) scaffolds still possessed satisfactory interconnectivity after coating with a 5% gelatin solution [20]. The surface topography of the gelatin-coated PLGA/Gel scaffold is shown in Figure 2e. A thin gelatin layer was tightly adhered to the wall, and the channels almost disappeared. The PLGA substrate after gelatin modification became more compacted (Figure 2f), indicating that the honeycomb-like structure in the substrate was partially filled with gelatin.

To further confirm the gelatin distribution in the substrate, the frozen sections of the PLGA, PLGA/FITC-Gel, and PLGA/FITC-Gel/rhBMP-2 scaffolds were observed under optical and fluorescence microscope. The inner structures of the scaffolds are shown in Figure 3, from which the gelatin coatings are clearly observed in both the PLGA/FITC-Gel and PLGA/FITC-Gel/rhBMP-2 scaffolds, as indicated by white arrows in Figure 3b,c. As shown in Figure 3e–f, the fluorescence images of the PLGA/FITC-Gel and PLGA/FITC-Gel/rhBMP-2 scaffolds showed that the gelatin surface and skeleton both emitted at high intensities, suggesting that the FITC-gelatin had penetrated into the substrate. Due to the cross-linking of gelatin within the substrate, the gelatin coating was firmly secured on the surface.

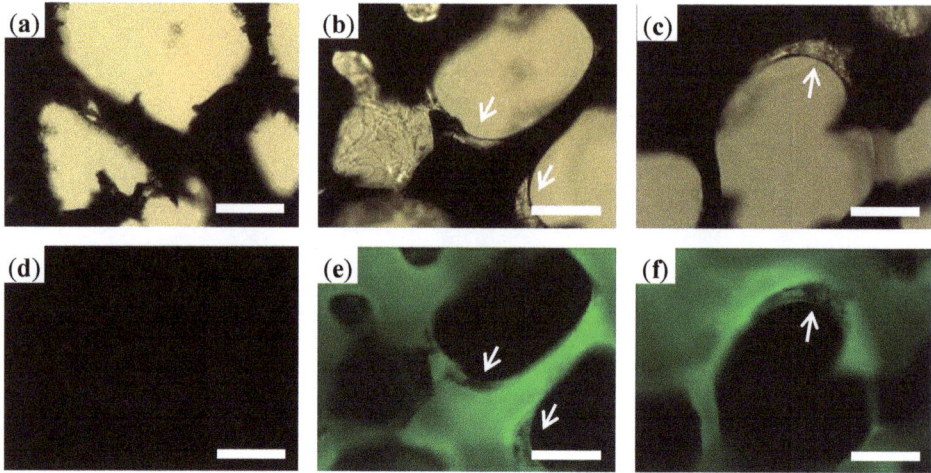

Figure 3. Fluorescence microscopic images of internal structures of porous PLGA scaffold (**a,d**), and PLGA scaffolds with gelatin (**b,e**) and gelatin/rhBMP-2 coating (**c,f**). Scale bar = 100 μm.

The stability of the gelatin coating was further evaluated by calculating the gelatin loss with time. As shown in Figure 4a, gelatin was slowly released with a cumulative release of <20% within two weeks, which was much lower than the reported cumulative loss of 30% for the gelatin-coated traditional scaffolds [20]. The SEM microimages of the PLGA and PLGA/Gel scaffolds after incubation for two weeks at 37 °C are shown in Figure 4b,c. The microstructure of the PLGA scaffold did not change after the two-week incubation and exhibited many channels on the surface. The gelatin layer in the PLGA/Gel scaffold remained tightly coated on the PLGA substrate, which corresponded to the small amounts of gelatin loss. These results indicated that the gelatin coating was stable on the surface of scaffold due to the loosened skeletal scaffold structure.

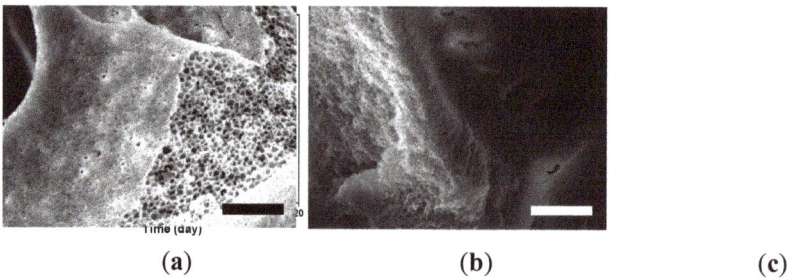

Figure 4. Gelatin leaching kinetic of coated scaffold (**a**); SEM microimages of PLGA scaffold (**b**); and PLGA scaffold with gelatin coating (**c**). Scale bar = 30 μm.

2.1.2. Mechanical Property Analyses

The porous scaffolds are designed to provide mechanical support until the regenerative tissue or organ is structurally stabilized [28]. Therefore, the appropriate mechanical properties are crucial for such porous scaffolds. Ideally, the mechanical properties are similar to that of the supported tissue. The scaffold should maintain its structural stability and integrity in an *in vivo* biomechanical environment and provide appropriate micro-stress stimulations for the implanted cells [29]. However, the compression strength of a PLGA scaffold was found to be lower than that of a human trabecular bone (*i.e.*, 9.3 ± 4.5 MPa) [30]. A previous study demonstrated that the gelatin entrapment in the scaffold improved the compressive strength of pure polymeric scaffolds [20]. Therefore, a higher compressive strength was expected after the gelatin modifications in our study. As depicted in Figure 5, the compressive strengths and elastic moduli (Ec) of the scaffolds were indeed increased. For dry samples, the compressive strength and Ec of the PLGA/Gel scaffold reached 6.15 ± 0.91 and 9.78 ± 1.62 MPa, respectively, compared with those of 3.83 ± 0.77 and 6.47 ± 0.91 MPa for the PLGA scaffold (Figure 5a,b). The mechanical properties measured in the dry state were different than those under physiological conditions, *i.e.*, in tissue fluid at 37 °C, because of the different media. As shown in Figure 5c,d, for wet-state samples, the compressive strength and Ec of the PLGA/Gel scaffold were significantly greater than those of the PLGA scaffold ($p < 0.05$ for both). Overall, the gelatin modifications improved the mechanical properties of the PLGA scaffold for clinical use.

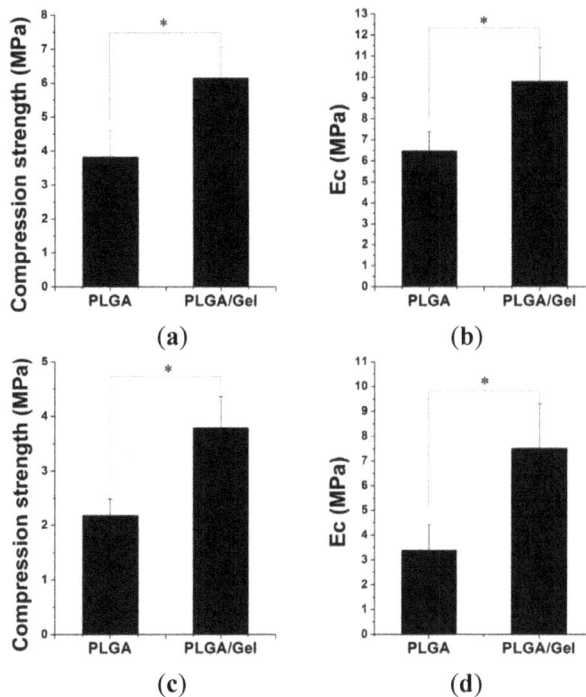

Figure 5. Compressive strength and Ec of PLGA and PLGA/Gel scaffolds under dry (**a,b**) or wet states (**c,d**). The data were represented as mean ± standard deviation (SD; $n = 3$; * $p < 0.05$).

2.1.3. Hydrophilicity Assessments

To evaluate whether the hydrophilicity of the PLGA scaffold was improved by the addition of gelatin, the water contact angles on the outer and inner surfaces were measured for the PLGA and PLGA/Gel scaffolds (Figure 6). The water contact angle of the control PLGA scaffold outer surface was significantly greater at $(93.5 \pm 5.7)°$ compared with that of the PLGA/Gel scaffold, which decreased to $(51.3 \pm 3.2)°$ ($p < 0.05$). The contact angle of the inner surface of the PLGA/Gel scaffold (i.e., $(60.1 \pm 4.2)°$) was also significantly smaller than that of the PLGA scaffold ($p < 0.05$). These observations were expected because gelatin is a more hydrophilic molecule than PLGA; thus, the decreased outer surface contact angle indicated that the gelatin molecule was successfully incorporated onto the surface of PLGA scaffold. The decreased inner surface contact angle suggested that gelatin also diffused into the substrate microvoids.

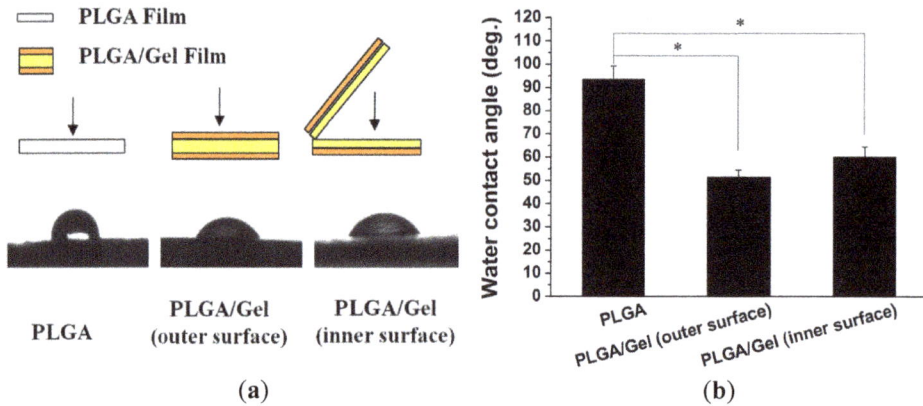

Figure 6. Test methods (**a**) and results (**a,b**) of water contact angles on outer and inner surfaces of PLGA and PLGA/Gel scaffolds. The data were represented as mean ± standard deviation (SD; $n = 3$; * $p < 0.05$).

2.2. Release Kinetics of rhBMP-2

As shown in Figure 7, the *in vitro* cumulative release behaviors of rhBMP-2 from the composites were characterized by the percentage release of rhBMP-2 as a function of time. Without gelatin modification, i.e., PLGA/rhBMP-2, a high burst release was noted; more than 80% of the loaded rhBMP-2 was released within the first 24 h. Gelatin has been extensively used in protein delivery applications. As depicted in Figure 7, rhBMP-2 showed a sustained release behavior in the PLGA/Gel/rhBMP-2 scaffold. In detail, the gelatin-coated scaffold loaded with rhBMP-2 showed a slight initial burst release within 24 h, but it was followed by a slower release over seven days.

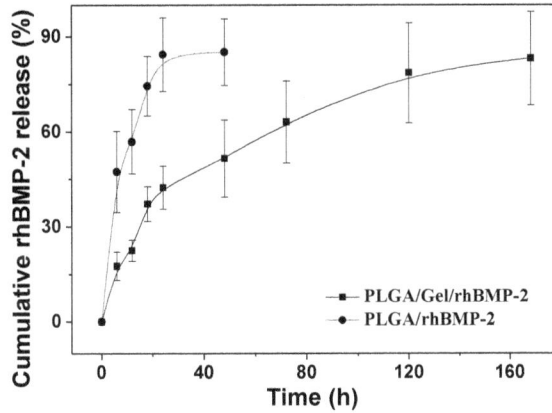

Figure 7. Release kinetics of rhBMP-2 from PLGA/rhBMP-2 and PLGA/Gel/rhBMP-2 scaffolds.

BMPs have been shown to induce bone formation by inducing chondroblastic and osteoblastic differentiations of BMSCs [31]. BMP-2 has been shown to be the most effective agent, by which to induce complete bone morphogenesis [29], and has been approved by FDA for clinical applications [32]. It is evident that the efficacy of BMP-2 in bone tissue formation is dependent on the administered dose and the delivery mode. The sustained release has been demonstrated to be sufficient for inducing bone regeneration [33]. Yamamoto *et al.* [34] demonstrated that BMP-2 showed a very serious initial burst release within 40 min of administration from glutaraldehyde-cross-linked gelatin microspheres, followed by a sustained release. In our study, the sustained release profile of the PLGA/Gel/rhBMP-2 scaffold was different, likely due to the PLGA substrate exerting some influence on the release mechanism.

2.3. Cell Adhesion, Proliferation, and Differentiation in Scaffolds

The attachment efficiencies of BMSCs cultured on various scaffolds for 3, 6, and 12 h are summarized in Figure 8a. After culture for 3 h, the cell attachment efficiencies on the PLGA/Gel and PLGA/Gel/rhBMP-2 scaffolds reached more than 30%, which was significantly greater than the percentage on the PLGA scaffold (*i.e.*, 11.8%). When cultured for 6 h, the cell attachment efficiencies on the PLGA/Gel and PLGA/Gel/rhBMP-2 scaffolds increased to approximately 80%, greater than the 45.9% reported for the PLGA scaffold. It was believed that the modification of a PLGA scaffold with gelatin increased direct cell-material binding by increasing the surface hydrophilicity, thus, facilitating an increase in early cell adhesion. The BMSC proliferation was quantitatively monitored using the MTT assay to measure the metabolic activity of the total population of cells for one, three, and seven days. As shown in Figure 8b,c, each scaffold supported the proliferation of BMSCs within seven days. The PLGA/Gel/rhBMP-2 scaffold exhibited the highest cellular activity, whereas the PLGA scaffold showed the lowest at three and seven days, suggesting that the gelatin and rhBMP-2 modifications promoted the cell attachment and proliferation. It has been reported that the expression of integrin $\beta 1$, which is promoted by rhBMP-2 [35], is required for cell spreading, adhesion, and proliferation [36,37].

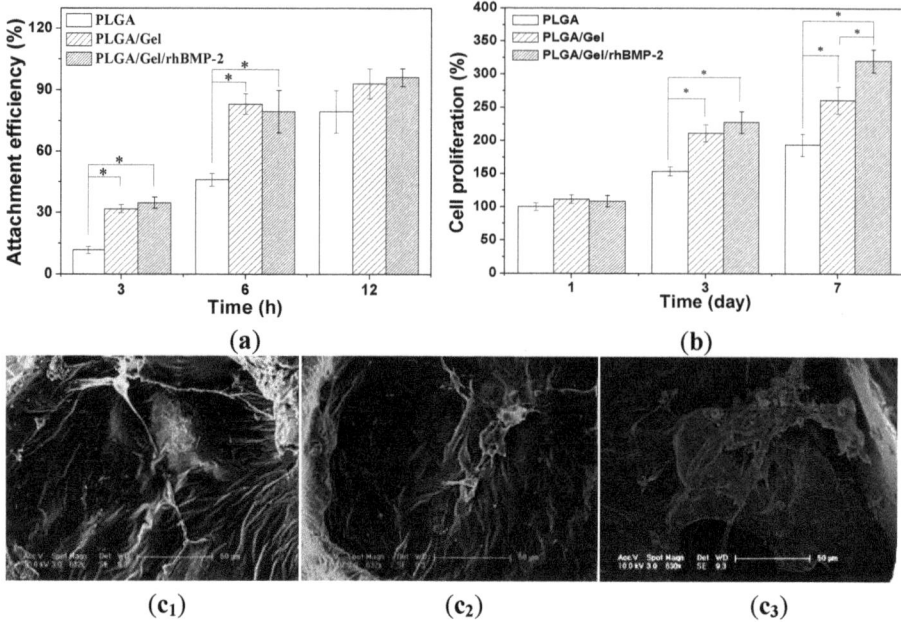

Figure 8. *In vitro* growth of BMSCs in scaffolds. Cell attachment efficiency on PLGA, PLGA/Gel, and PLGA/Gel/rhBMP-2 scaffolds at 3 and 6 h (**a**); Cell proliferation on PLGA, PLGA/Gel, and PLGA/Gel/rhBMP-2 on Day 1, 3, and 7 (**b**); and SEM microimages of cell attachment on PLGA (c_1), PLGA/Gel (c_2), and PLGA/Gel/rhBMP-2 (c_3) on Day 7 (**c**). The data were represented as mean ± standard deviation (SD; $n = 3$; * $p < 0.05$).

The cell-biomaterial interactions have been demonstrated to exert a considerable influence on the differentiation and function of BMSCs [38]. To investigate the osteogenic differentiation of BMSCs on different scaffolds, alkaline phosphatase (ALP) activity and calcium deposition were measured. ALP is a membrane enzyme commonly recognized as a marker of osteoblastic differentiation. Figure 9a shows the ALP activities of the BMSCs cultured on the different scaffolds after seven and 14 days. The significantly higher ALP activity was detected in cells cultured on the PLGA/Gel/rhBMP-2 scaffold than those on the PLGA and PLGA/Gel scaffolds ($p < 0.05$ for both). Calcium deposition was measured by alizarin red staining. As shown in Figure 9b, the quantification of ARS indicated that the deposition of calcium minerals in the PLGA/Gel/rhBMP-2 scaffold was significant higher than in the other scaffolds ($p < 0.05$). Moreover, after culturing for 21 days, the deposition of calcium mineral in the PLGA/Gel scaffold was greater than that in the PLGA scaffold ($p < 0.05$). These results demonstrated that the PLGA/Gel/rhBMP-2 and PLGA/Gel scaffolds promoted the osteogenic differentiation of BMSCs. The quantitative analyses of the expressions of the osteogenesis-related genes (*i.e.*, collagen-I (COL-I) and osteopontin (OPN)) were performed by quantitative reverse transcription-polymerase chain reaction (qRT-PCR) (Figure 10), which supported the results observed from the calcium deposition measurements. The expression levels of COL-I and OPN were higher in the

PLGA/Gel/rhBMP-2 scaffold than those in the PLGA/Gel and PLGA scaffolds. This result was attributed to the gelatin coating and the gelatin-immobilized rhBMP-2. All above data indicated that the bioactivity of rhBMP-2 was well retained during the surface-coating process, and the sustained release of rhBMP-2 showed improved BMSC differentiation.

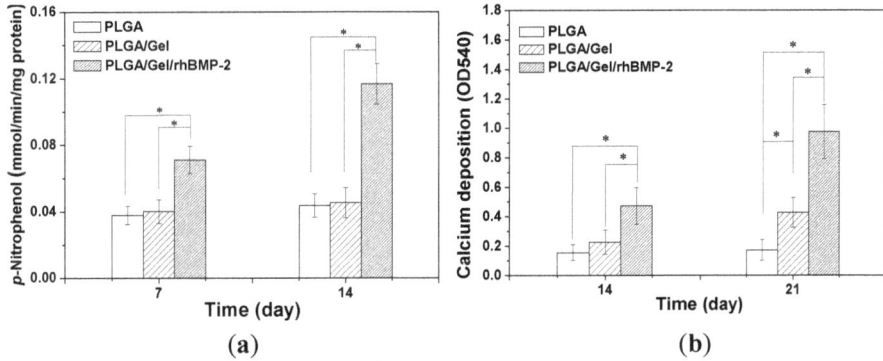

Figure 9. ALP activities of BMSCs in PLGA, PLGA/Gel, and PLGA/Gel/rhBMP-2 scaffolds during 14-day *in vitro* culture (**a**); Calcium deposition after culturing in PLGA, PLGA/Gel, and PLGA/Gel/rhBMP-2 scaffolds for 14 and 21 days (**b**). The data were represented as mean ± standard deviation (SD; $n = 3$; * $p < 0.05$).

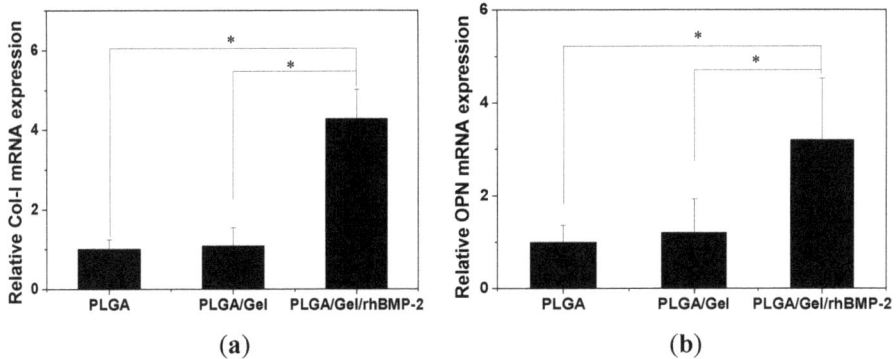

Figure 10. Quantitative analyses of osteogenesis-related gene expressions (*i.e.*, Col-I (**a**) and OPN (**b**)). The data were represented as mean ± standard deviation (SD; $n = 3$; * $p < 0.05$).

3. Experimental Section

3.1. PLGA Scaffold Fabrication

A PLGA scaffold was fabricated *via* combining phase separation and particulate leaching. Two grams of PLGA (LA:GA = 50:50, 14,000 kDa, Jinan Daigang Biomaterial Co., Ltd., Jinan, China) was added into 10.0 mL of NMP (Sigma-Aldrich Co., Shanghai, China). The mixture was continuously stirred until the polymer was thoroughly dissolved. Next, the sieved sodium chloride particulates of 100–300 µm in diameter were added into the PLGA solution. The weight ratio of the salt particulates to the polymer composites was 6:1. The mixture was cast in a homemade glass cylinder with a removable bottom. To remove NMP and salt particulates, the bottom was removed, and the mixture was immersed in distilled water for 5 days with the water exchanged every 6 h. Subsequently, any water remaining in the scaffold was exchanged by ethanol. Finally, the PLGA scaffold was obtained after 3 days of lyophilization.

3.2. PLGA/Gelatin (PLGA/Gel) and PLGA/Gel/rhBMP-2 Hybrid Scaffold Fabrication

First, 0.01 g/mL of gelatin solution was prepared by dissolving gelatin (Sigma-Aldrich Co., Shanghai, China) in distilled water at 37 °C. rhBMP-2 was supplemented to a final concentration of 5.0 µg/mL. Subsequently, the PLGA scaffold was immersed in the gelatin solution or the gelatin/rhBMP-2 solution at 37 °C under vacuum for 30 min before fumigation by glutaraldehyde at 4 °C for 24 h to complete the cross-linking reaction. To clear any residual glutaraldehyde in the scaffold, the hybrid composite was vacuum-drawn and washed with ethanol several times. Finally, the PLGA/Gel and PLGA/Gel/rhBMP-2 scaffolds were lyophilized for 3 days and stored at 4 °C in a desiccator until use.

3.3. Characterization of Scaffolds

The porosities of the scaffolds were determined using the ethanol replacement method [21]. The microstructures of the scaffolds were examined by scanning electron microscopy (SEM, Philips XL30, Eindhoven, The Netherlands). The scaffolds were fractured after snap-freezing, sputter-coated with gold, and observed at an accelerating voltage of 15 kV. To assess the gelatin distributions in the scaffolds, gelatin labeled with fluorescein isothiocyanate (FITC, Aladdin Reagent Co., Ltd., Shanghai, China) was used. The horizontal sections (100 µm thick) of the scaffolds were observed under a fluorescence microscope (TE200-U, Nikon Instruments Inc., Tokyo, Japan). The gelatin contents in the scaffolds were determined based on the change in dry weight of the scaffolds before and after modification. Cylindrical samples of 5 mm in diameter and 10 mm in height were chosen for mechanical strength tests using a universal testing machine (Instron 1121, High Wycombe, UK). The compressive strengths were measured at a crosshead speed of 2.0 mm/min. Three replicates were tested for each condition ($n = 3$).

The weight loss of the gelatin in the scaffolds was determined at 37 °C under agitation at 60 rpm over a period of 2 weeks. At predetermined time intervals, the scaffolds were freeze-dried and weighed. The percentage of gelatin loss was calculated as Gelatin loss = $(W_1 - W_2)/(W_1 - W_0) \times 100\%$, wherein W_0 is the dry weight of the original surface-coated scaffolds, W_1 is the dry weight of the modified scaffolds at baseline, and W_2 is the dry weight of the surface-coated scaffolds at time t. In addition, the scaffolds were examined by SEM after 2 weeks of incubation.

The PLGA and PLGA/Gel films were prepared for static air-water contact angle measurements using the sessile drop method on a contact angle system (VCA 2000, AST, Bellerica, MA, USA). A solution of 20% PLGA was poured in a glass plate and then immersed in double distilled water (ddH₂O) for 24 h. After exchanging NMP by water, the PLGA film was obtained. The PLGA/Gel film was prepared by exposing PLGA film to a 1% gelatin solution for 24 h. Furthermore, the PLGA/Gel film was sliced in the middle, and the hydrophilicities of the PLGA and PLGA/Gel films (outer surface and inner surface) were detected after lyophilization.

3.4. In Vitro Release Study

The PLGA scaffold was immersed in 5.0 μg/mL rhBMP-2 solution to obtain PLGA/rhBMP-2. The PLGA/Gel/rhBMP-2 and PLGA/rhBMP-2 scaffolds were used for the *in vitro* release study. Each scaffold ($5 \times 5 \times 5$ mm^3) incorporating rhBMP-2 was incubated in 5.0 mL of phosphate-buffered saline (PBS) at 37 °C under stirring at 60 rpm. At specified time intervals, 0.2 mL of the supernatant was collected, and an equal volume of fresh PBS was added. The content of rhBMP-2 was measured using an enzyme-linked immunosorbent assay (ELISA) kit (R&D System) according to the manufacturer's instruction. The release profiles were obtained by plotting the percentage of cumulatively content of released rhBMP-2 against time. The experiments were performed in triplicate.

3.5. Cell Adhesion, Proliferation, and Differentiation Assays

3.5.1. Bone Mesenchymal Stem Cell (BMSC) Isolation

Two-month-old New Zealand white rabbits were sacrificed for BMSC isolation. Bone marrow aspirate (5.0 mL) were obtained from the rabbit tibias and subsequently cultured. Briefly, the isolated cell pellets were resuspended in 5.0 mL of the complete Dulbecco's Modified Eagle's Medium (DMEM; Gibco BRL, Grand Island, NJ, USA) supplemented with 10% (v/v) fetal calf serum (FCS; Gibco BRL, Grand Island, NJ, USA) and 100 IU/mL penicillin-streptomycin (Sigma, Shanghai, China). The cells were seeded in culture dishes (Corning Costar Co., Cambridge, MA, USA) and cultured at 37 °C in an incubator with 5% carbon dioxide (CO_2). Non-adherent cells were discarded when the medium was changed after 24 h. Subsequently, the medium was replaced every other day until the cells reached 80% confluence. Then, the cells were washed with PBS, digested by 0.25% trypsin/ethylenediamine tetraacetic acid (trypsin/EDTA; Sigma, Shanghai, China), and subcultured at a 1:3 dilution under the same condition until the third passage.

3.5.2. Cell Adhesion and Proliferation Assays

For the adhesion and proliferation studies, 1×10^5 BMSCs were seeded on the scaffolds ($5 \times 5 \times 3$ mm^3). The adhesion and proliferation of the cells for each sample were determined by a standard 3-(4,5-dimethylthiazoyl-2-yl)-2,5-diphenyltetrazolium bromide (MTT) assay. At 6 and 12 h post-seeding, the culture medium was discarded, and the unattached cells were washed away with PBS. The attached cells on the scaffolds were detached by trypsin/EDTA, and the number of cells was carefully counted. Finally, the cell attachment efficiency was calculated according to the following equation: Cell attachment efficiency = N_1/N_0, where N_1 and N_0 were the numbers of attached and seeded cells, respectively.

Cell proliferation was determined on Day 1, 3, and 7. The scaffolds were incubated in an MTT solution (5.0 mg/mL in PBS) for 4 h. After the removal of the MTT solution, the acidified isopropanol (0.2 mL of 0.04 M HCl in 10 mL of isopropanol) was added to solubilize the resultant formazan product. The absorbance of the extractant at 492 nm was recorded on a Thermo Electron MK3 spectrophotometer (Thermo Scientific, Hudson, NH, USA). The relative cell number (%) was determined by comparing the absorbance to that for PLGA on Day 1. The mean value of nine readings for each sample was used as the final result. The cell proliferation on Day 7 was also observed by SEM.

3.5.3. Cell Differentiation Assays

Alkaline phosphatase (ALP) activity was determined after culturing the cells in DMEM/F12, FBS (10%, V/V) for 7 and 14 days. Briefly, the medium of each well was carefully removed. Then, the cells were washed three times with PBS, lysed in radioimmunoprecipitation assay (RIPA) buffer, frozen at -80 °C for 30 min, and thawed at 37 °C. Then, p-nitrophenol phosphate substrate (pNPP) solution was added, and the samples were incubated in the dark for 30 min at 37 °C. The reaction was terminated with 3.0 M NaOH, and the ALP activity was read on a multifunction microplate scanner (Tecan Infinite M200) at 405 nm. Measurements were compared with p-nitrophenol standards and normalized by the total protein content, which was determined with a bicinchoninic acid (BCA) kit (Pierce Biotechnologies, Rockford, IL, USA).

Calcium deposition was determined by alizarin red S (ARS) staining of the BMSCs after culture in DMEM/F12, FBS (10%) for 14 and 21 days. After three 5 min rinses in water, the scaffolds were incubated in ARS stain solution (0.1% ARS in Tris-HCl buffer, pH 8.0, Sigma, Shanghai, China) for 30 min at 37 °C. The scaffolds were then washed in distilled water three times for 5 min each. The stained samples were treated with 10% (w/v) cetylpyridinium chloride in 10.0 mM sodium phosphate for 15 min at room temperature. The absorbance of ARS at 540 nm was recorded on a Thermo Electron MK3 spectrophotometer.

The osteogenesis-related gene expression levels were quantitatively assessed using RT-qPCR for BMSCs cultured on various scaffolds incubated for 14 days. Total RNA was extracted using TRIzol Reagent (Invitrogen) according to the manufacturer's protocol. The total RNA concentration and purity were estimated using Nanodrop Plates (Tecan Infinite M200, Tecan Group Ltd., Maennedorf, Switzerland), and the RNA was reverse transcribed as described in the M-MLV manual (Promega). RNA was added to a 20.0 μL reverse transcription reaction mixture containing $5 \times$ M-MLV buffer, dNTP mixture, RNase inhibitor, RTase M-MLV, RNase free dH$_2$O, and oligo (dT) primer. The expression levels of osteogenic markers were quantified using a qPCR SYBR Green Mix Kit (Stratagene). The primer sequences specific for the target gene and the internal control gene (glyceraldehyde-3-phosphate dehydro-genase (GAPDH)) used for qRT-PCR are listed in Table 2. The specificities of the listed oligonucleotides were checked by Basic Local Alignment Search Tool (BLAST) against the rabbit RefSeq RNA database at NCBI. The qPCR amplification was performed as follows: initial heating at 95 °C for 10 min, followed by 40 cycles at 95 °C for 30 s, 58 °C for 60 s, and 72 °C for 60 s. The expression levels were determined using threshold cycles (Ct) that were determined by the iCycler iQ Detection System software. The relative transcript quantities were calculated using the ΔΔCt method. The GAPDH gene was used as a reference gene and was amplified along with the target genes from the same cDNA samples. The difference in the Ct of the sample mRNA relative to the GAPDH mRNA was defined as the ΔCt. The difference between the ΔCt of the control cells and the ΔCt of the cells grown on the substrates was defined as the ΔΔCt. The fold change in mRNA expression was expressed as $2^{-\Delta\Delta Ct}$.

Table 2. Sequences of primers for quantitative reverse transcription-polymerase chain reaction (qRT-PCR).

Gene	Forward Primer Sequence	Reverse Primer Sequence
COL-I	5′-CTCGCTCACCACCTTCTC-3′	5′-TAACCACTGCTCCACTCTG-3′
OPN	5′-CGTGGATGATATTGATGAGGATG-3′	5′-TCGTCGGAGTGGTGAGAG-3′
GAPDH	5′-GATGGTGAAGGTCGGAGTG-3′	5′-TGTAGTGGAGGTCAATGAATGG-3′

3.6. Statistical Analysis

The data were presented as mean ± standard deviation (SD). The independent and replicated experiments were used to analyze the statistical variability of the data analyzed using Student's *t*-test, and $p < 0.05$ was considered to be significant.

4. Conclusions

In this study, a gelatin tight-coated rhBMP-2-incorporated PLGA-based scaffold with a loosened skeleton was fabricated. Because of the special structure, the gelatin molecule easily diffused throughout the scaffold. By cross-linking with glutaraldehyde, the gelatin coating was tightly bound with both the internal and external surfaces of microscale channel. The modification with gelatin also significantly improved the mechanical strength and hydrophilicity of these surfaces. For gelatin-coated scaffolds with rhBMP-2, a sustained release behavior was observed *in vitro*, which enhanced the attachment, proliferation, and differentiation of BMSCs. The obtained data collectively demonstrate that the gelatin-coated PLGA scaffolds can effectively deliver bioactive factors and hold great promise for bone tissue engineering.

Acknowledgments

This work was financially supported by the National Natural Science Foundation of China (Nos. 51273081 and 51303174) and the Science and Technology Planning Project of Changchun City (No. 14KG045).

Author Contributions

We confirm that all the listed authors have participated actively in the study. Juan Wang and Dongsong Li participated in experimental operation, data collection, and manuscript writing. Tianyi Li contributed to data collection and statistical analysis. Jianxun Ding, Baosheng Li, Jianguo Liu, and Xuesi Chen designed the proposal of this work, and revised the manuscript.

Conflicts of Interest

The authors declare no conflict of interest.

References

1. Katthagen, B.D.; Pruss, A. Bone allografting. *Orthopade* **2008**, *37*, 764–771.
2. Arrington, E.D.; Smith, W.J.; Chambers, H.G.; Bucknell, A.L.; Davino, N.A. Complications of iliac crest bone graft harvesting. *Clin. Orthop. Rel. Res.* **1996**, *329*, 300–309.
3. Benichou, G. Direct and indirect antigen recognition: The pathways to allograft immune rejection. *Front. Biosci.* **1999**, *4*, D476–D480.
4. Shuang, F.; Hou, S.X.; Zhao, Y.T.; Zhong, H.B.; Xue, C.; Zhu, J.L.; Bu, G.Y.; Cao, Z. Characterization of an injectable chitosan-demineralized bone matrix hybrid for healing critical-size long-bone defects in a rabbit model. *Eur. Rev. Med. Pharmacol. Sci.* **2014**, *18*, 740–752.
5. Zhao, X.X.; Lui, Y.S.; Toh, P.W.J.; Loo, S.C.J. Sustained release of hydrophilic L-ascorbic acid 2-phosphate magnesium from electrospun polycaprolactone scaffold-A study across blend, coaxial, and emulsion electrospinning techniques. *Materials* **2014**, *7*, 7398–7408.
6. Liu, Y.; Cui, H.; Zhuang, X.; Wei, Y.; Chen, X. Electrospinning of aniline pentamer-graft-gelatin/PLLA nanofibers for bone tissue engineering. *Acta Biomater.* **2014**, *10*, 5074–5080.
7. Liu, X.H.; Holzwarth, J.M.; Ma, P.X. Functionalized synthetic biodegradable polymer scaffolds for tissue engineering. *Macromol. Biosci.* **2012**, *12*, 911–919.
8. Asadinezhad, A.; Lehocky, M.; Saha, P.; Mozetic, M. Recent progress in surface modification of polyvinyl chloride. *Materials* **2012**, *5*, 2937–2959.
9. Sengel-Turk, C.T.; Hascicek, C.; Dogan, A.L.; Esendagli, G.; Guc, D.; Gonul, N. Surface modification and evaluation of PLGA nanoparticles: The effects on cellular uptake and cell proliferation on the HT-29 cell line. *J. Drug. Deliv. Sci. Tec.* **2014**, *24*, 166–172.
10. Wang, S.G.; Cui, W.J.; Bei, J.Z. Bulk and surface modifications of polylactide. *Anal. Bioanal. Chem.* **2005**, *381*, 547–556.
11. Deng, C.; Tian, H.Y.; Zhang, P.B.; Sun, J.; Chen, X.S.; Jing, X.B. Synthesis and characterization of RGD peptide grafted poly(ethylene glycol)-*b*-poly(L-lactide)-*b*-poly(L-glutamic acid) triblock copolymer. *Biomacromolecules* **2006**, *7*, 590–596.
12. Karde, V.; Ghoroi, C. Influence of surface modification on wettability and surface energy characteristics of pharmaceutical excipient powders. *Int. J. Pharm.* **2014**, *475*, 351–363.
13. Shin, Y.M.; Jo, S.Y.; Park, J.S.; Gwon, H.J.; Jeong, S.I.; Lim, Y.M. Synergistic effect of dual-functionalized fibrous scaffold with BCP and RGD containing peptide for improved osteogenic differentiation. *Macromol. Biosci.* **2014**, *14*, 1190–1198.
14. Dunn, R.L.; Cowsar, D.R.; Vanderbilt, D.P. Biodegradable *in situ* forming implants and methods of producing the same. U.S. Patent No. 4938763, 3 July 1990.

15. Parent, M.; Nouvel, C.; Koerber, M.; Sapin, A.; Maincent, P.; Boudier, A. PLGA *in situ* implants formed by phase inversion: Critical physicochemical parameters to modulate drug release. *J. Control. Release* **2013**, *172*, 292–304.

16. Ellis, M.J.; Chaudhuri, J.B. Poly(lactic-*co*-glycolic acid) hollow fibre membranes for use as a tissue engineering scaffold. *Biotechnol. Bioeng.* **2007**, *96*, 177–187.

17. Oh, S.H.; Lee, J.H. Fabrication and characterization of hydrophilized porous PLGA nerve guide conduits by a modified immersion precipitation method. *J. Biomed. Mater. Res. A* **2007**, *80A*, 530–538.

18. Amirian, J.; Linh, N.T.B.; Min, Y.K.; Lee, B.T. The effect of BMP-2 and VEGF loading of gelatin-pectin-BCP scaffolds to enhance osteoblast proliferation. *J. Appl. Polym. Sci.* **2015**, *132*, 41241:1–41241:9.

19. Tan, S.; Fang, J.Y.; Yang, Z.; Nimni, M.E.; Han, B. The synergetic effect of hydrogel stiffness and growth factor on osteogenic differentiation. *Biomaterials* **2014**, *35*, 5294–5306.

20. Zhang, Q.; Tan, K.; Zhang, Y.; Ye, Z.; Tan, W.S.; Lang, M. *In situ* controlled release of rhBMP-2 in gelatin-coated 3D porous poly(ε-caprolactone) scaffolds for homogeneous bone tissue formation. *Biomacromolecules* **2014**, *15*, 84–94.

21. Fan, H.B.; Hu, Y.Y.; Zhang, C.L.; Li, X.S.; Lv, R.; Qin, L.; Zhu, R. Cartilage regeneration using mesenchymal stem cells and a PLGA-gelatin/chondroitin/hyaluronate hybrid scaffold. *Biomaterials* **2006**, *27*, 4573–4580.

22. Shen, H.; Hu, X.X.; Yang, F.; Bel, J.Z.; Wang, S.G. Combining oxygen plasma treatment with anchorage of cationized gelatin for enhancing cell affinity of poly(lactide-*co*-glycolide). *Biomaterials* **2007**, *28*, 4219–4230.

23. Chen, C.H.; Lee, M.Y.; Shyu, V.B.H.; Chen, Y.C.; Chen, C.T.; Chen, J.P. Surface modification of polycaprolactone scaffolds fabricated *via* selective laser sintering for cartilage tissue engineering. *Mater. Sci. Eng. C Mater. Biol. Appl.* **2014**, *40*, 389–397.

24. Schloegl, W.; Marschall, V.; Witting, M.Y.; Volkmer, E.; Drosse, I.; Leicht, U.; Schieker, M.; Wiggenhorn, M.; Schaubhut, F.; Zahler, S.; *et al.* Porosity and mechanically optimized PLGA based *in situ* hardening systems. *Eur. J. Pharm. Biopharm.* **2012**, *82*, 554–562.

25. Krebs, M.D.; Sutter, K.A.; Lin, A.S.P.; Guldberg, R.E.; Alsberg, E. Injectable poly(lactic-*co*-glycolic) acid scaffolds with *in situ* pore formation for tissue engineering. *Acta Biomater.* **2009**, *5*, 2847–2859.

26. Hakimimehr, D.; Liu, D.M.; Troczynski, T. *In-situ* preparation of poly(propylene fumarate)-hydroxyapatite composite. *Biomaterials* **2005**, *26*, 7297–7303.

27. Kempe, S.; Mader, K. *In situ* forming implants–An attractive formulation principle for parenteral depot formulations. *J. Control. Release* **2012**, *161*, 668–679.

28. Wu, L.; Zhang, J.; Jing, D.; Ding, J. "Wet-state" mechanical properties of three-dimensional polyester porous scaffolds. *J. Biomed. Mater. Res. A* **2006**, *76*, 264–271.

29. Agrawal, C.M.; Ray, R.B. Biodegradable polymeric scaffolds for musculoskeletal tissue engineering. *J. Biomed. Mater. Res.* **2001**, *55*, 141–150.

30. Liebschner, M.A.K. Biomechanical considerations of animal models used in tissue engineering of bone. *Biomaterials* **2004**, *25*, 1697–1714.

31. Urist, M.R. Bone: Formation by autoinduction. *Science* **1965**, *150*, 893–899.
32. Mckay, W.F.; Peckham, S.M.; Badura, J.M. A comprehensive clinical review of recombinant human bone morphogenetic protein-2 (INFUSE (R) Bone Graft). *Int. Orthop.* **2007**, *31*, 729–734.
33. Haidar, Z.S.; Hamdy, R.C.; Tabrizian, M. Delivery of recombinant bone morphogenetic proteins for bone regeneration and repair. Part B: Delivery systems for BMPs in orthopaedic and craniofacial tissue engineering. *Biotechnol. Lett.* **2009**, *31*, 1825–1835.
34. Yamamoto, M.; Ikada, Y.; Tabata, Y. Controlled release of growth factors based on biodegradation of gelatin hydrogel. *J. Biomater. Sci. Polym. Ed.* **2001**, *12*, 77–88.
35. Song, Y.; Ju, Y.; Morita, Y.; Xu, B.; Song, G. Surface functionalization of nanoporous alumina with bone morphogenetic protein 2 for inducing osteogenic differentiation of mesenchymal stem cells. *Mater. Sci. Eng. C Mater. Biol. Appl.* **2014**, *37*, 120–126.
36. Rowland, T.J.; Miller, L.M.; Blaschke, A.J.; Doss, E.L.; Bonham, A.J.; Hikita, S.T.; Johnson, L.V.; Clegg, D.O. Roles of integrins in human induced pluripotent stem cell growth on Matrigel and vitronectin. *Stem Cells Dev.* **2010**, *19*, 1231–1240.
37. Abraham, S.; Kogata, N.; Fassler, R.; Adams, R.H. Integrin beta1 subunit controls mural cell adhesion, spreading, and blood vessel wall stability. *Circ. Res.* **2008**, *102*, 562–570.
38. Hanson, S.; D'Souza, R.N.; Hematti, P. Biomaterial-mesenchymal stem cell constructs for immunomodulation in composite tissue engineering. *Tissue Eng. Part A* **2014**, *20*, 2162–2168.

Alginate-Collagen Fibril Composite Hydrogel

Mahmoud Baniasadi and Majid Minary-Jolandan

Abstract: We report on the synthesis and the mechanical characterization of an alginate-collagen fibril composite hydrogel. Native type I collagen fibrils were used to synthesize the fibrous composite hydrogel. We characterized the mechanical properties of the fabricated fibrous hydrogel using tensile testing; rheometry and atomic force microscope (AFM)-based nanoindentation experiments. The results show that addition of type I collagen fibrils improves the rheological and indentation properties of the hydrogel.

Reprinted from *Materials*. Cite as: Baniasadi, M.; Minary-Jolandan, M. Alginate-Collagen Fibril Composite Hydrogel. *Materials* **2015**, *8*, 799-814.

1. Introduction

Hydrogels are being extensively used for tissue engineering applications. There has been a remarkable effort in producing properties of hydrogels as close to the native tissue microenvironment as possible [1,2]. The properties of interest include physical properties, biochemical properties, and biological properties. For example, most tissues in the human body are hierarchically structured, involving length-scales from nanoscale to macroscale [3,4]. Accordingly, one of the major routes in making more biomimetic hydrogels involves adding micro/nanostructures to the host hydrogel polymer [5]. These nanostructures include polymeric, inorganic/ceramic and metallic nanoparticles, nanotubes, nanowires, graphene, nanodiamonds, *etc.* [6–17]. The components often add extra functionalities to the based hydrogel polymer, such as electrical, physical, chemical, and biological functionalities. It is desirable for the added components to impart these functionalities with minimal or no compromise to the other original properties of the host hydrogel. The obtained hydrogel is often termed a nanocomposite or hybrid hydrogel [15,18–23]. A recent review summarized the latest progress in nanocomposite hydrogels [24].

Nanofiber-reinforced hydrogels are a class of nanocomposite hydrogels, in which often electrospun polymeric nanofibers are added to the hydrogel matrix [25–33]. Recent work reported on transparent electrospun gelatin nanofibers infiltrated with alginate hydrogel for cornea tissue engineering [25]. Addition of nanofibers enhanced the elastic modulus of the hydrogel by several folds. In another study, 3D rapid prototyping technique was used to form a crossed log-pile of elastic fibers that were subsequently impregnated with an epoxy-based hydrogel [32].

In this article, we report on the synthesis and characterization of a composite fibrous hydrogel by incorporation of native type I collagen fibrils into the alginate hydrogel. The process for the synthesis of the collagen-alginate composite hydrogel is schematically shown in Figure 1. Briefly, alginate was added to solutions of collagen fibrils of various concentrations. The resulting mixture was subsequently cross-linked by calcium ions using calcium carbonate ($CaCO_3$).

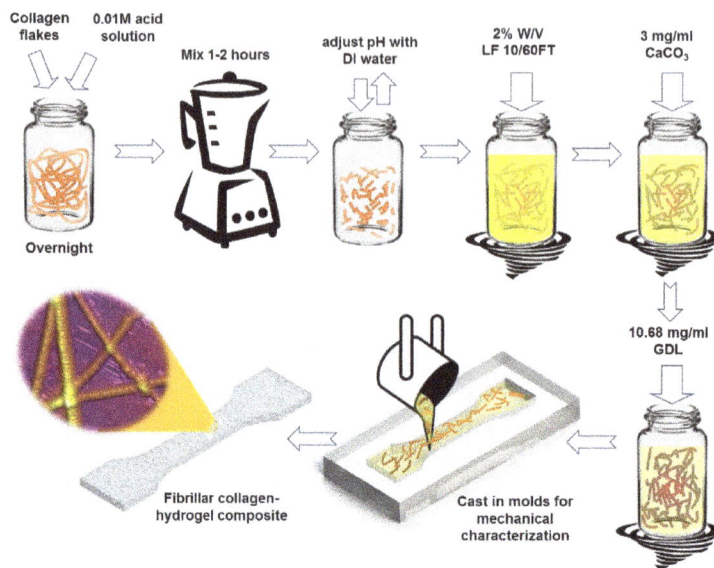

Figure 1. Schematic of the step-by-step preparation process of hybrid fibrous composite hydrogel with incorporation of native type I collagen fibrils. LF 10/60FT is a sodium alginate from FMC Biopolymer (Philadelphia, PA, USA), GDL is D-(+)-Gloconic acid δ-lactone. See Experimental Section.

Alginate hydrogel is being widely used for cell encapsulation, cell transplantation, drug delivery, and tissue engineering applications [34–36]. Although alginate possesses many favorable properties for tissue engineering applications, it lacks specific interaction with mammalian cells. Therefore, it is often functionalized with RGD-containing cell adhesion ligands [35]. Tripeptide Arg-Gly-Asp (RGD) is a common cell-recognition ligand in extra cellular matrix that binds integrins to the membrane proteins of different cell types. This cell adhesion ligand is present in collagen fibrils in tissues. Accordingly, composite hydrogel of alginate-collagen may provide the adhesion sites for cell adhesion. In addition, alginate hydrogel lacks the hierarchical fibrous structure of native tissues. Nanofibers such as electrospun nanofibers can be added to alginate hydrogel to provide the hierarchical structure of the native tissues. Native collagen has several advantages in this regard. It adds fibrous structure to the alginate hydrogel. In addition, it provides the RGD-binding sites for the cell adhesion. Finally, collagen possesses the characteristic nano-topography feature, in contrast to the smooth surface of synthetic nanofibers.

Collagen fibrils are major components of extracellular matrix (ECM) and connectives tissues such as bone and tendon, as well as in tissues such as cornea. Similar to alginate, collagen hydrogels have been used for various applications including investigation of adherence of bone marrow stromal cells, as scaffold for cartilage tissue engineering, in vascular grants, and for applications in wound healing and as a pro-angiogenic site for islet transplantation [37–41]. Native type I collagen fibrils have characteristic periodic patterns of 60–70 nm [42] (See Figure 2). In addition to the RGD binding sites, these highly periodic nano-topographical features are believed to be important for cell adhesion and growth [39,43–47]. The majority of synthetic nanofibers including electrospun nanofibers that are being used in scaffold for tissue engineering lack this important nano-topographical feature.

Figure 2 shows scanning electron microscope (SEM) and atomic force microscope (AFM) images of the native collagen fibrils. In both images the characteristic periodic banding of collagen is apparent. The subset in Figure 2D shows a line-profile taken along the dashed line that more clearly shows this periodic banding pattern with a periodicity of 60–70 nm. This periodic pattern arises from the special microstructural arrangement of collagen molecules in a "quarter stagger" arrangement [42,48]. The periodicity is 60–70 nm, which results in the so-called "gap" and "overlap" regions. The diameter of the individual collagen fibrils varies from 50 to 200 nm based on our previous study [42].

Collagen hydrogels are often prepared from collagen molecules (or triple helix structure). However, type I collagen in native tissue is fibrillar with a characteristic periodic pattern. The *in vitro* fibrillogenesis of collagen triple helix to fibril is performed by adjusting the pH from the original acidic solution to pH ~7.4, a process that is still not fully understood. Although collagen in this neutralization process self-assembles to a filamentous structure, however, the *in vitro* assembled collagen fibril may lack the characteristic banding pattern (~67 nm) of native collagen [49]. It has been shown than some of the assembled collagen in the fibrillogenesis process includes both fibrils that display a periodic banding pattern and filamentous structures that do not have this characteristic collagen striation [49].

The elastic moduli of alginate hydrogels vary considerably depending on gelling conditions and cross-linking [36]. The most common cross-linking is ionic cross-linking by Ca^{2+} ions [50], although photocross-linked alginate hydrogels have been also reported [51]. Elastic moduli of ionically cross-linked ($CaCO_3$) alginate hydrogels measured using compression experiments were reported to be from 5 to 120 kPa, depending on the concentration of the Ca^{2+} ions, which was varied from 0.5% to 5% [50]. The compression modulus of photocross-linked hydrogel was reported to be ~170 kPa [51]. Mechanical properties of collagen hydrogel covalently functionalized with three different monomers, *i.e.*, 4-vinylbenzyl chloride, glycidyl methacrylate and methacrylic anhydride were characterized using atomic force microscope (AFM) [52]. By adjusting the degree of functionalization, an elastic modulus in the range of 16–387 kPa was obtained. Similar to other reports, no collagen fibril formation was observed in these specimens [52]. A biomimetic fibrillar collagen scaffold was recently introduced. By altering the freeze drying conditions through introduction of multiple temperature gradients, collagen scaffolds with complex pore orientations, and anisotropy in pore size and alignment were produced. However, no mechanical properties were reported [53]. The mechanical properties of collagen-chitosan hydrogel were characterized using compression experiment [41]. For this hydrogel, soluble collagen molecules were used after pH adjustment, which resulted in fibril formation. The elastic modulus of the pure collagen hydrogel was measured to be 0.4 kPa as compared to 0.7 kPa for collagen-chitosan hydrogel [41].

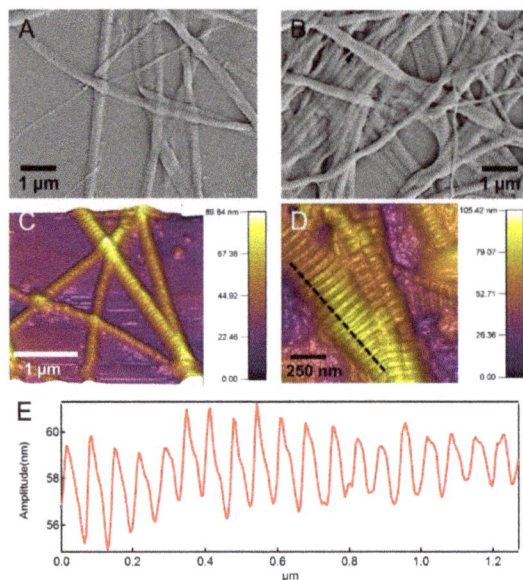

Figure 2. (A,B) Scanning electron microscope (SEM) morphology and **(C,D)** atomic force microscope (AFM) topography images of the native type I collagen fibrils used for synthesis of the composite hydrogel. The characteristic periodic pattern of 60–70 nm is apparent in images; **(E)** Shows a line-profile taken along the dashed line in **(D)** and shows this periodic pattern.

Although mechanical properties of alginate hydrogel and collagen hydrogel have been reported in the literature, the properties of the composite hydrogel are not available. In addition, for composite hydrogel of this type, mechanical properties depend on the type of loading and deformation, given their fibrous and heterogeneous microstructure. Therefore, the behavior of the composite hydrogel in shear, tension, and indentation could be different. We used rheometry (shear deformation), tensile test and atomic force microscopy (AFM)-based nanoindentation to characterize the mechanical properties of the fabricated hydrogel samples.

2. Results and Discussion

Figure 3 shows representative SEM images of the surface of the hydrogel samples. The samples were freeze-dried for observation in SEM. Figure 3A is the hydrogel sample without collagen fibrils. Figure 3B–D are SEM images of samples with collagen fibrils. Collagen fibrils are apparent on the surface of the specimens, pointed to by arrows. Collagen fibrils are several microns long and several hundred in diameter, similar to the isolated collagen fibrils shown in Figure 2. Several additional SEM images of the hydrogels are presented in Figure S1.

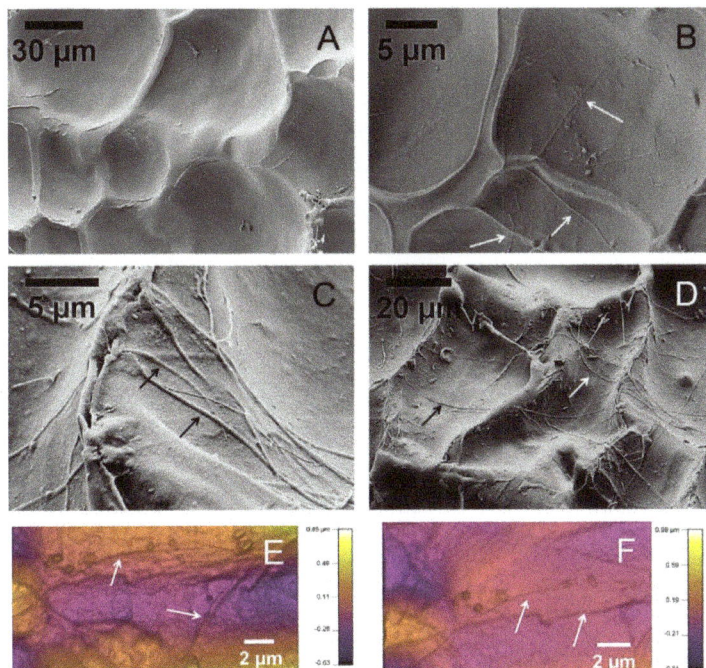

Figure 3. SEM and AFM images of the surface of the hydrogel samples. (**A**) A hydrogel sample with no collagen fibrils; (**B–D**) Composite hydrogel samples with collagen fibrils, (**B**) 1X, (**C,D**) 2X; (**E,F**) AFM topography images of collagen-alginate hydrogel samples. Arrows point to collagen fibrils.

Typical experimental results for rheological properties of the hydrogel samples with different concentration of collagen fibrils are shown in Figure 4. These display typical behavior for the five samples tested for each concentration. Overall, the behavior was consistent with less than 9% error between different samples. Rheometry measures mechanical properties of the samples in terms of storage (E') and loss moduli (E'') under combined compressive and shear deformations. Storage modulus represents the elastic energy, while the loss modulus measures the dissipative energy. The storage modulus of hydrogel with no collagen fibril (alginate hydrogel) content increases by increasing the frequency (deformation rate), as shown in Figure 4B. Addition of collagen preserves this trend. The loss modulus of hydrogel with no collagen content is initially constant and then decreases by increasing the frequency. This means that energy dissipation, which is related to the toughness of the material decreases with frequency. However, addition of collagen appears to reverse this trend. Data in Figure 4C suggest that this increase in loss modulus is more apparent for higher frequencies. In Figure 4B, the data for 0X and 1X specimens show overlap for low frequency up to 1 Hz. From 1 to 10 Hz, the 1X specimen shows larger storage modulus. The loss modulus, however, shows larger values for 1X in all frequencies. There could be a possible explanations for overlap between 0X and 1X in Figure 4B as follows: It seems that contribution of collagen to the storage modulus starts at high frequencies >1 Hz, while for 0.01 Hz to ~1 HZ, the contribution of collagen to the storage modulus is negligible. However, the contribution of collagen to the loss modulus occurs for all frequencies from 0.01 to 10 Hz. It is possible that for low concentration of 1X and in small frequencies, collagen fibrils slide within the alginate matrix. As such they will not contribute to the elastic properties; however, the sliding deformation will cause energy dissipation appearing in the loss modulus. As the frequency increases, the sliding of collagen samples may reduce, which results in enhancement of the storage modulus and decrease in the loss modulus, as shown in Figure 4C. In addition, the data for 2X and 3X concentrations appear to overlap for all frequencies. Collagen fibrils in our study have a very low concentration. Based on SEM images shown in Figures 2 and S1, the distribution of collagen in samples is not fully homogenous at different points of the sample. We believe that the overlaps in data could be an error introduced by random orientation and dispersion of collagen fibrils in the specimens. Statistical analysis (Two way ANOVA, p <0.05) followed with a Tukey test clearly confirmed significant changes in storage and loss moduli—and as a result, complex modulus increased. Overall, addition of small quantities of collagen fibrils results in a several times increase in elastic and loss moduli of the hydrogel samples, which is apparent in the results of the complex modulus ($E' + iE''$) in Figure 4D.

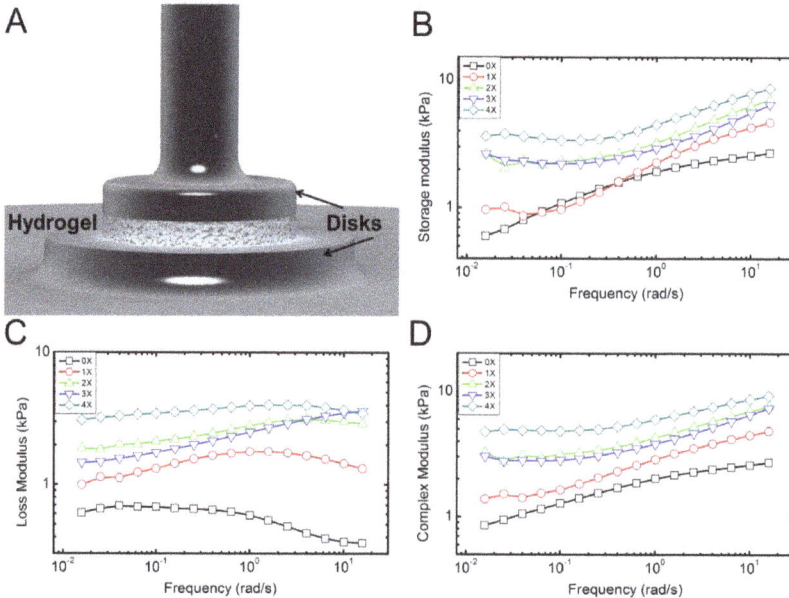

Figure 4. Characterization of the rheological properties of the hydrogel samples. (**A**) Schematic shows the disk-shaped hydrogel sample between the two circular disks of the rheometer; (**B–D**) Typical plots of the storage, loss, and complex moduli, respectively, extracted from the rheological measurements. Five samples for each concentration were tested.

Figure 5A shows a schematic of the tensile specimens in a costume-designed gripper adapter. A typical stress-strain response of the hydrogel samples is shown in Figure 5B. The response is presented in terms of engineering stress *vs.* engineering strain. Stress was obtained by dividing the force by the initial cross-sectional area of the sample. The cross-section was measured using a caliper at three different points in the gauge length of the sample and the average value was used. Strain was calculated by dividing the cross-head displacement by the original length of the sample. Stress–strain response of the samples shows a typical J-response (nonlinear) initially followed by a linear response. We used the slope of the linear section for each of the samples to obtain the elastic (Young's) modulus [34,54]. In addition, for each experiment we obtained the failure strain, the failure stress (strength), and the toughness. The toughness was calculated from the area under the stress–strain response, as shown with the shaded area in Figure 5B. From statistical analysis (Two way ANOVA, $p < 0.05$) followed with a Tukey test of tensile results, it can be concluded that, in contrast to rheological properties, addition of collagen fibril has minimal effect on the tensile properties of the composite hydrogel. There is a minimal increasing trend for elastic modulus, failure stress, and toughness. However, the failure strain does not change by addition of collagen fibrils. Table S1 gives detailed tensile properties of the tested samples. It should be noted that collagen fibrils are randomly distributed in the hydrogel samples, and are not necessarily aligned with the tension direction. With future improvement for aligning the fibrils, the tensile properties may show larger improvement.

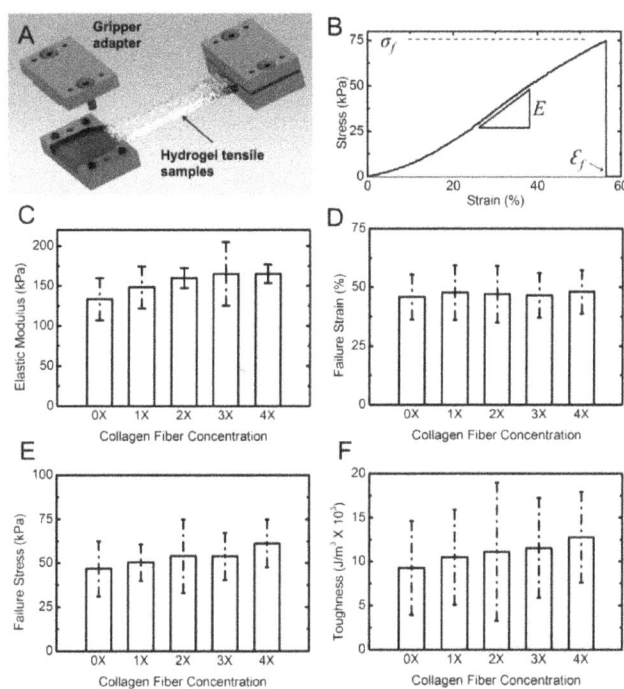

Figure 5. Tensile properties of hydrogel samples. (**A**) Schematic shows the dog-bone shaped hydrogel sample gripped using costume-designed gripper adaptor; (**B**) A typical stress-strain response from a hydrogel sample. The slope of the linear section of the response was used to extract the elastic (Young's) modulus of the hydrogel specimen; (**C**) The Young's modulus; (**D**) failure strain; (**E**) failure stress; and (**F**) toughness of hydrogel *vs.* collagen concentration. Error bars represent standard deviation (SD) for five samples for each concentration.

Figure 6 shows the results of the nanoindentation experiment. Hydrogel samples were indented using an AFM probe with a spherical tip inside a liquid medium as schematically shown in Figure 6A. Figure 6B is a typical extension-retraction response from an indentation experiment. For each sample, a map of 100 indentation points was obtained. A typical map is shown in Figure 6C. Several additional indentation maps are given in Figure S2. The color code represents the indentation modulus with the bright colors showing larger indentation modulus values.

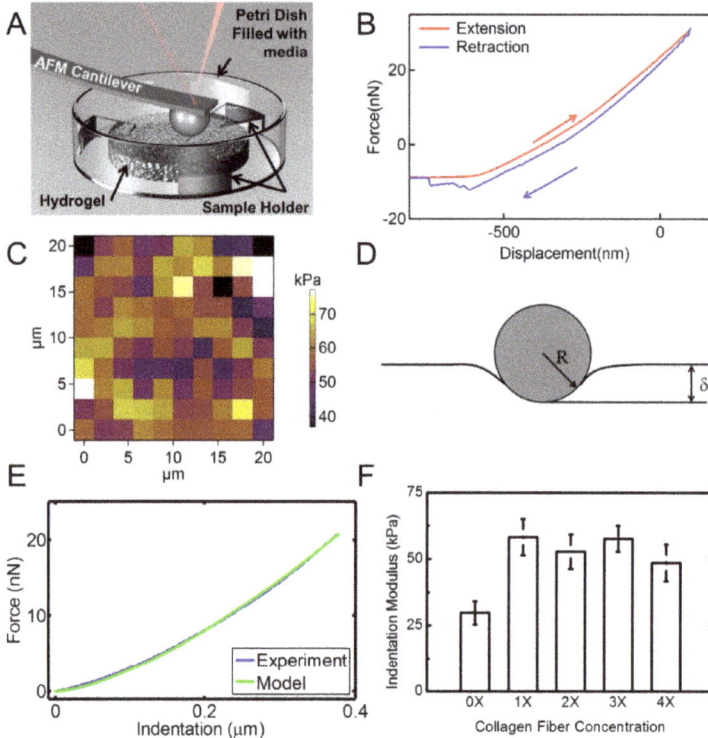

Figure 6. AFM nanoindentation experiments on hydrogel specimens. (**A**) Schematic shows the setup for indentation of hydrogel samples in a liquid media using a spherical probe; (**B**) A typical force-displacement (indentation depth) response from hydrogel; (**C**) Indentation map on a 20 μm × 20 μm area for a total of 100 points of a sample with 4X collagen concentration. The color bar represents the indentation modulus; (**D**) Schematic of the Hertzian contact mechanics model for spherical indentation into the hydrogel surface; (**E**) Comparison of the force-indentation data from an experiment modeled with Hertzian contact mechanics; (**F**) Indentation moduli extracted from nanoindentation experiments for hydrogel samples with different concentrations of collagen fibrils. Error bars in (**F**) are SD for five measurements at each concentration.

We modeled the nanoindentation of the spherical tip into the hydrogel using the Hertzian contact mechanics model to derive the indentation modulus of the sample, Figure 6D [55]. Based on the

Hertzian model, for indentation of an elastic half-space using a spherical indenter, the indentation load-displacement relation is given as [56]:

$$F = \frac{4}{3} E_r R^{1/2} \delta^{3/2} \tag{1}$$

In this equation, δ is the deformation of the samples in contact, shown in Figure 6D. F is the indentation force, E_r is the reduced elastic modulus of the tip and the sample, and R is the tip radius [57,58]. The reduced elastic modulus is given as:

$$\frac{1}{E_r} = \frac{(1 - \upsilon_t^2)}{E_t} + \frac{(1 - \upsilon_s^2)}{E_s} \tag{2}$$

where "t" and "s" represent tip and sample, respectively. We note that since the hydrogel is much softer than the SiO_2 (silicone dioxide) spherical probe, the deformation of the probe tip is negligible. In this case, the reduced elastic modulus can be replaced with the elastic modulus of the hydrogel. This model was introduced into a MATLAB code. Each force-indentation depth curve was modeled using this model. A typical fitted model is shown in Figure 6E. The Poisson's ratio of the sample was assumed to be ~0.49, which is a common assumption for hydrated materials, given the incompressible nature of water. Hence, the indentation modulus was the only fitting parameter. We conducted indentation experiments with displacement rates of 0.5–3 µm/s to ensure that the indentation rate does not have any effect on the results in this displacement range. The results for different rates in this range showed similar behavior (Figure S3). Indentation modulus *vs.* fibril concentration is shown in Figure 6F. These data are the average of different indentation rates. As observed from statistical analysis (Two way ANOVA, $p < 0.05$) followed with a Scheffe test, the addition of collagen results in a significant increase of the indentation modulus of the hydrogel samples. Data for various collagen concentrations do not show significant variations. This may be explained by considering the random orientation and dispersion of collagen fibrils in the samples.

Rheometry and tensile experiments characterize the bulk properties of the hydrogel specimens, in shear and tension, respectively. Nanoindentation measures the local properties of the samples under local compressive indentation. The nanoindentation footprint in our experiment is on the order of 2 µm. This is in the range of the size of individual cells. Therefore, these local properties could be relevant to the microenvironment of cells, when they are cultured in this composite hydrogel.

It is widely accepted that the nanofibrous structure of collagen and its nano-topographic periodic pattern is important for cell attachment and growth [43,59,60]. As such, collagen has been used in tissue engineering applications [39,44,45,61,62]. However, fibrillar collagen in this form had not been previously reported for composite hydrogels. Collagen fibrils extracted from animal tissues are still fairly expensive. The high cost and concerns with immunogenicity associated with collagen need to be overcome before they are widely used for tissue engineering applications. Recent developments on Ovine collagen may resolve both issues, but this remains to be seen. Electrospun collagen nanofibers have been shown to be denatured collagen or simply gelatin. In general, the majority of electrospun nanofibers have a smooth topography and lack the characteristic nano-topography on the surface of the native collagen fibrils.

Although rheometry, tensile test [34], and AFM-nanoindentation [63,64] have been used for characterization of hydrogels, they are often used as a single tool for characterization. However, behavior of hydrogels in shear, tension and indentation could be different, as shown in our experimental data. Therefore, we have used all of these three techniques for characterization of the fabricated composite hydrogels. Native collagen fibrils have an elastic modulus on the order of several hundreds of kP to several GPa depending on their hydrated state [42]. Our results show that addition of an even small percentage of collagen to alginate hydrogel can result in improvement of the rheological and indentation properties of the hydrogel. Further effort is required to investigate the effect of adding higher concentration of collagen fibrils as well as to attempt to align the collagen fibrils in the hydrogel matrix to produce even stronger mechanical properties.

3. Experimental Section

Protonal® LF10/60FT, Pharm-grade sodium alginate powder from FMC BioPolymer (Philadelphia, PA, USA) was used for preparation of the alginate hydrogel. Calcium Carbonate ($CaCO_3$) was purchased from RICCA Chemical Co (Arlington, TX, USA). Calcium Sulfate Dihydrate ($CaSO_4 \cdot 2H_2O$), Collagen from bovine Achilles tendon and GDL (D-(+)-Gloconic acid δ-lactone) were purchased from Sigma-Aldrich (St. Louis, MO, USA). Calcium Chloride Dihydrate ($CaCl_2 \cdot 2H_2O$) was purchased from Fisher Scientific (Fair Lawn, NJ, USA). Sulfuric acid (H_2SO_4) was purchased from VWR (Radnor, PA, USA). Polydimethylsiloxane (PDMS) was purchased from Dow Corning Co. (Midland, MI, USA).

Preparation of casting molds: Molds with different geometries were designed to prepare samples for the tensile test, AFM nanoindentation, and rheometry experiments. For the tensile test, according to ASTM standard F2900-11 [65], dumbbell shaped geometry was used (Figure S4). For AFM nanoindentation and rheometry, disk shape mold was used (Figure S5). Several master molds with identical dimensions were machined from an Acrylic sheet. PDMS mix including the base materials and curing agent was poured into master molds with negative profile and allowed to cure in an oven below 80 °C for 2 h. After curing, the PDMS was peeled off the master mold. These PDMS molds were used as molds for casting hydrogel specimens for mechanical characterization (Figure S4B).

Preparation of native collagen fibrils: A stock solution of collagen fibrils was prepared by soaking ~500 mg collagen flakes in ~200 mL of 0.01 M sulfuric acid overnight. Subsequently, the solution was mixed using a blender for one hour to break down the collagen flakes to individual fibrils. The pH of the solution was adjusted to 6.5 by substituting the original solvent with DI water. The prepared collagen solution was left in the fridge so that the larger collagen bundles from the mixing process settled down. The supernatant solution, that contained mostly individual collagen fibrils, was separated and used for the preparation of the hydrogel samples. This was confirmed by placing several droplets of the solution on a glass slide and observing under an optical microscope. Collagen concentration in the final stock solution was measured to be ~1 mg/mL, which was considered as 4X and represents the maximum concentration. To measure the collagen concentration in the final solution, four batches of 5 mL from final solution were air dried and the dried films were weighed. The average of these measurements is reported as collagen concentration of 4X stock solution. Lower collagen concentration solutions 3X, 2X and 1X (1X ≅ 0.25 mg/mL) were prepared by diluting the 4X solution with the proper amount of DI water.

Preparation of alginate hydrogel: Alginate solution was prepared by dissolving 2% *w/v* of LF 10/60FT in DI water. Then 3 mg/mL (30 mM) calcium carbonate ($CaCO_3$) powder was added to the solution and the solution was stirred for two min. 60 mM (10.68 mg/mL) GDL was added to this solution while the solution was vortexed for two min. To maintain the pH of the solution near neutral value, the ratio of $CaCO_3$ to GDL was kept 1:2 [59,60,66]. The crosslinking process was controlled by introducing GDL for the activation of calcium ions from the calcium carbonate. GDL gradually reduces the pH of the solution, which results in the release of calcium ions and crosslinking of the alginate monomers. After this step, the prepared solution was immediately poured into the PDMS molds. The molds were kept in a humidity box for 24 h for the hydrogel to completely crosslink. To ensure the data are repeatable, at least five samples were prepared and tested for each experiment. The composite hydrogel samples were prepared by adding alginate to collagen solutions with different concentrations.

Freeze-drying and SEM imaging: Several of the hydrogel specimens were flash-frozen with liquid nitrogen, and then freeze-dried using a FreeZone freeze dryer system (LABCONCO, Kansas City, MO, USA), Figure S6. A ~15 nm gold film was sputter-coated onto the surface of the freeze dried samples prior to imaging with SEM. SEM images were acquired using a Zeiss-LEO Model 1530 variable pressure SEM (Zeiss, Oberkochen, Germany).

AFM imaging: AFM images of air-dried collagen solution and freeze-dried samples were obtained in air using MFP-3D-Bio (Asylum Research, CA, USA) with a cantilever "HQ:NSC15/Al-BS" (μMesch) with 40 nN/nm stiffness in AC mode (tapping mode) with frequency of 0.5 Hz.

Tensile test experiment: Tensile test experiments were performed using an Instron 5969 machine (Norwood, MA, USA) equipped with a pneumatic gripper and a 500 N load cell. To reduce the punching effect of the pneumatic gripper on the hydrogel specimen, a gripper adaptor was designed and fabricated that provides the possibility of adjusting the gripping pressure (Figure S7). Inner surfaces of the gripper adapter were covered with cardboard to prevent sample sliding. Prior to the tensile test, the cross-section of each sample was measured with a digital caliper. The sample was loaded onto the gripper adapter with an adjusted gap. To avoid pretension on the hydrogel samples a cardboard was placed between two gripper adapter to keep them at a constant relative distance and function as a frame for the specimen prior to the test. The samples were kept inside a humidity box prior to experiments to avoid dehydration of the specimens. Each sample was loaded onto the pneumatic gripper and the frame cardboard was cut prior to initiation of the experiment. All tensile tests were performed with a quasi-static strain rate of 1%/s.

AFM Nanoindentation experiments: Nanoindentation experiments were performed on fully hydrated hydrogel samples submerged in DI water. To prevent movement and floatation of hydrogel samples under AFM, the samples were secured from the bottom side to the petri dish and were firmly held from the top side using a costume fixture (Figure S5B). For nanoindentation experiments, a soft triangular AFM cantilever with a spring constant of k ~0.32 N/m was used. The AFM probe tip had an integrated 2 μm silicon oxide (SiO_2) spherical probe tip (sQube®). Nanoindentation experiments were performed using a MFP 3D Bio-AFM (Asylum research, Santa Barbara, CA, USA). Before the nanoindentation experiment, the deflection sensitivity of the AFM cantilever was calibrated on a stiff substrate (Si). Nanoindentation experiments were performed on different areas from 10 μm × 10 μm to 90 μm × 90 μm for a total of 100 indentation points. We examined different displacement rates from 500 nm/s to 3 μm/s, which showed similar results.

Rheometry experiments: A Discovery Hybrid Rheometer (DHR-3) (TA Instruments, New Castle, DE, USA) was used to perform parallel-plates rheological experiments. The samples were disk-shaped with a diameter of 25 mm and a height of 3 mm. The storage and loss moduli of the samples were obtained in the frequency range of 0.1–100 Hz, under 0.5 N (1 kPa) compressive force.

Statistical Analysis: Statistical analysis was performed using Origin (V8.0988; OriginLab Corp, MA, USA) to determine the statistical differences. For tensile test data and nanoindentation data, statistical comparisons were performed with one-way analysis of variance (One Way ANOVA). For rheological data, since we had collagen concentration and frequency changes, statistical comparison was performed with two-way analysis of variance (Two Way ANOVA). Statistical significance for all tests was set to be at a p value <0.05.

4. Conclusions

In summary, we fabricated composite alginate-type I collagen fibril hydrogels and characterized their mechanical properties using rheometry, tensile experiment, and AFM-spherical probe nanoindentation. The results show that addition of collagen has a pronounced effect on the rheological and indentation properties of the hydrogel, while tensile properties showed minimal changes. Nanoindentation properties improve by more than 100%. Rheological properties for 4X collagen concentration showed several times improvement over alginate hydrogel with no collagen fibrils.

Supplementary Materials

Supplementary materials can be accessed at: http://www.mdpi.com/1996-1944/8/2/0799/s1.

Acknowledgments

We thank Danieli Rodrigues for access to the rheometer for rheological measurements. The work is supported by startup fund from the University of Texas at Dallas.

Author Contributions

Both authors contributed in the design of the experiments and writing the paper.

Conflicts of Interest

The authors declare no conflict of interest.

References

1. Moutos, F.T.; Freed, L.E.; Guilak, F. A biomimetic three-dimensional woven composite scaffold for functional tissue engineering of cartilage. *Nat. Mater.* **2007**, *6*, 162–167.

2. Shapiro, J.; Oyen, M. Hydrogel composite materials for tissue engineering scaffolds. *JOM* **2013**, *65*, 505–516.

3. Dvir, T.; Timko, B.P.; Kohane, D.S.; Langer, R. Nanotechnological strategies for engineering complex tissues. *Nat. Nanotechnol.* **2011**, *6*, 13–22.

4. Ratner, B.D.; Hoffman, A.S.; Schoen, F.J.; Lemons, J.E. *Biomaterials Science: An Introduction to Materials in Medicine*; Academic Press: Waltham, MA, USA, 2013.

5. Gaharwar, A.K.; Dammu, S.A.; Canter, J.M.; Wu, C.J.; Schmidt, G. Highly extensible, tough, and elastomeric nanocomposite hydrogels from poly(ethylene glycol) and hydroxyapatite nanoparticles. *Biomacromolecules* **2011**, *12*, 1641–1650.

6. Wu, C.J.; Gaharwar, A.K.; Schexnailder, P.J.; Schmidt, G. Development of biomedical polymer-silicate nanocomposites: A materials science perspective. *Materials* **2010**, *3*, 2986–3005.

7. Bordes, P.; Pollet, E.; Averous, L. Nano-biocomposites: Biodegradable polyester/nanoclay systems. *Prog. Polym. Sci.* **2009**, *34*, 125–155.

8. Dvir, T.; Timko, B.P.; Brigham, M.D.; Naik, S.R.; Karajanagi, S.S.; Levy, O.; Jin, H.W.; Parker, K.K.; Langer, R.; Kohane, D.S. Nanowired three-dimensional cardiac patches. *Nat. Nanotechnol.* **2011**, *6*, 720–725.

9. Balazs, A.C.; Emrick, T.; Russell, T.P. Nanoparticle polymer composites: Where two small worlds meet. *Science* **2006**, *314*, 1107–1110.

10. Cha, C.; Shin, S.R.; Annabi, N.; Dokmeci, M.R.; Khademhosseini, A. Carbon-based nanomaterials: Multifunctional materials for biomedical engineering. *ACS Nano* **2013**, *7*, 2891–2897.

11. Huang, G.Y.; Wang, L.; Wang, S.Q.; Han, Y.L.; Wu, J.H.; Zhang, Q.C.; Xu, F.; Lu, T.J. Engineering three-dimensional cell mechanical microenvironment with hydrogels. *Biofabrication* **2012**, *4*, doi:10.1088/1758-5082/4/4/042001.

12. Malda, J.; Visser, J.; Melchels, F.P.; Jungst, T.; Hennink, W.E.; Dhert, W.J.A.; Groll, J.; Hutmacher, D.W. 25th anniversary article: Engineering hydrogels for biofabrication. *Adv. Mater.* **2013**, *25*, 5011–5028.

13. Annabi, N.; Tamayol, A.; Uquillas, J.A.; Akbari, M.; Bertassoni, L.E.; Cha, C.; Camci-Unal, G.; Dokmeci, M.R.; Peppas, N.A.; Khademhosseini, A. 25th anniversary article: Rational design and applications of hydrogels in regenerative medicine. *Adv. Mater.* **2014**, *26*, 85–124.

14. Place, E.S.; Evans, N.D.; Stevens, M.M. Complexity in biomaterials for tissue engineering. *Nat. Mater.* **2009**, *8*, 457–470.

15. Gaharwar, A.K.; Peppas, N.A.; Khademhosseini, A. Nanocomposite hydrogels for biomedical applications. *Biotechnol. Bioeng.* **2014**, *111*, 441–453.

16. Drury, J.L.; Mooney, D.J. Hydrogels for tissue engineering: Scaffold design variables and applications. *Biomaterials* **2003**, *24*, 4337–4351.

17. Peppas, N.A.; Hilt, J.Z.; Khademhosseini, A.; Langer, R. Hydrogels in biology and medicine: From molecular principles to bionanotechnology. *Adv. Mater.* **2006**, *18*, 1345–1360.

18. Reddy, P.R.S.; Rao, K.M.; Rao, K.S.V.K.; Shchipunov, Y.; Ha, C.S. Synthesis of alginate based silver nanocomposite hydrogels for biomedical applications. *Macromol. Res.* **2014**, *22*, 832–842.

19. Spanoudaki, A.; Fragiadakis, D.; Vartzeli-Nikaki, K.; Pissis, P.; Hernandez, J.C.R.; Pradas, M.M. Nanostructured and nanocomposite hydrogels for biomedical applications. In *Surface Chemistry in Biomedical and Environmental Science (Nato Science Series II)*; Springer: Berlin, Germany, 2006; Volume 228, pp. 229–240.

20. Shin, S.R.; Bae, H.; Cha, J.M.; Mun, J.Y.; Chen, Y.C.; Tekin, H.; Shin, H.; Farshchi, S.; Dokmeci, M.R.; Tang, S.; *et al.* Carbon nanotube reinforced hybrid microgels as scaffold materials for cell encapsulation. *ACS Nano* **2012**, *6*, 362–372.

21. Shin, S.R.; Jung, S.M.; Zalabany, M.; Kim, K.; Zorlutuna, P.; Kim, S.B.; Nikkhah, M.; Khabiry, M.; Azize, M.; Kong, J.; *et al.* Carbon-nanotube-embedded hydrogel sheets for engineering cardiac constructs and bioactuators. *ACS Nano* **2013**, *7*, 2369–2380.

22. Wang, E.; Desai, M.S.; Lee, S.W. Light-controlled graphene-elastin composite hydrogel actuators. *Nano Lett.* **2013**, *13*, 2826–2830.

23. Liu, J.Q.; Chen, C.F.; He, C.C.; Zhao, L.; Yang, X.J.; Wang, H.L. Synthesis of graphene peroxide and its application in fabricating super extensible and highly resilient nanocomposite hydrogels. *ACS Nano* **2012**, *6*, 8194–8202.

24. Gaharwar, A.K.; Kishore, V.; Rivera, C.; Bullock, W.; Wu, C.J.; Akkus, O.; Schmidt, G. Physically crosslinked nanocomposites from silicate-crosslinked PEO: Mechanical properties and osteogenic differentiation of human mesenchymal stem cells. *Macromol. Biosci.* **2012**, *12*, 779–793.

25. Tonsomboon, K.; Oyen, M.L. Composite electrospun gelatin fiber-alginate gel scaffolds for mechanically robust tissue engineered cornea. *J. Mech. Behav. Biomed. Mater.* **2013**, *21*, 185–194.

26. Wilson, S.L.; Wimpenny, I.; Ahearne, M.; Rauz, S.; El Haj, A.J.; Yang, Y. Chemical and topographical effects on cell differentiation and matrix elasticity in a corneal stromal layer model. *Adv. Funct. Mater.* **2012**, *22*, 3641–3649.

27. Yang, Y.; Wimpenny, I.; Ahearne, M. Portable nanofiber meshes dictate cell orientation throughout three-dimensional hydrogels. *Nanomed. Nanotechnol. Biol. Med.* **2011**, *7*, 131–136.

28. Stephens-Altus, J.S.; Sundelacruz, P.; Rowland, M.L.; West, J.L. Development of bioactive photocrosslinkable fibrous hydrogels. *J. Biomed. Mater. Res. A* **2011**, *98A*, 167–176.

29. Xia, Y.; Zhu, H. Polyaniline nanofiber-reinforced conducting hydrogel with unique pH-sensitivity. *Soft Matter* **2011**, *7*, 9388–9393.

30. Jang, J.; Oh, H.; Lee, J.; Song, T.-H.; Jeong, Y.H.; Cho, D.-W. A cell-laden nanofiber/hydrogel composite structure with tough-soft mechanical property. *Appl. Phys. Lett.* **2013**, *102*, doi:10.1063/1.4808082.

31. Hong, Y.; Huber, A.; Takanari, K.; Amoroso, N.J.; Hashizume, R.; Badylak, S.F.; Wagner, W.R. Mechanical properties and *in vivo* behavior of a biodegradable synthetic polymer microfiber-extracellular matrix hydrogel biohybrid scaffold. *Biomaterials* **2011**, *32*, 3387–3394.

32. Agrawal, A.; Rahbar, N.; Calvert, P.D. Strong fiber-reinforced hydrogel. *Acta Biomater.* **2013**, *9*, 5313–5318.

33. Qin, Y. The preparation and characterization of fiber reinforced alginate hydrogel. *J. Appl. Polym. Sci.* **2008**, *108*, 2756–2761.

34. Drury, J.L.; Dennis, R.G.; Mooney, D.J. The tensile properties of alginate hydrogels. *Biomaterials* **2004**, *25*, 3187–3199.

35. Rowley, J.A.; Madlambayan, G.; Mooney, D.J. Alginate hydrogels as synthetic extracellular matrix materials. *Biomaterials* **1999**, *20*, 45–53.

36. Augst, A.D.; Kong, H.J.; Mooney, D.J. Alginate hydrogels as biomaterials. *Macromol. Biosci.* **2006**, *6*, 623–633.

37. Hesse, E.; Hefferan, T.E.; Tarara, J.E.; Haasper, C.; Meller, R.; Krettek, C.; Lu, L.C.; Yaszemski, M.J. Collagen type I hydrogel allows migration, proliferation, and osteogenic differentiation of rat bone marrow stromal cells. *J. Biomed. Mater. Res. A* **2010**, *94A*, 442–449.

38. Yuan, T.; Zhang, L.; Li, K.F.; Fan, H.S.; Fan, Y.J.; Liang, J.; Zhang, X.D. Collagen hydrogel as an immunomodulatory scaffold in cartilage tissue engineering. *J. Biomed. Mater. Res. B* **2014**, *102*, 337–344.

39. Huynh, T.; Abraham, G.; Murray, J.; Brockbank, K.; Hagen, P.-O.; Sullivan, S. Remodeling of an acellular collagen graft into a physiologically responsive neovessel. *Nat. Biotechnol.* **1999**, *17*, 1083–1086.

40. Lin, J.; Li, C.; Zhao, Y.; Hu, J.; Zhang, L.-M. Co-electrospun nanofibrous membranes of collagen and zein for wound healing. *ACS Appl. Mater. Interfaces* **2012**, *4*, 1050–1057.

41. McBane, J.E.; Vulesevic, B.; Padavan, D.T.; McEwan, K.A.; Korbutt, G.S.; Suuronen, E.J. Evaluation of a collagen-chitosan hydrogel for potential use as a pro-angiogenic site for islet transplantation. *PLoS ONE* **2013**, *8*, doi:10.1371/journal.pone.0077538.

42. Minary-Jolandan, M.; Yu, M.-F. Nanomechanical heterogeneity in the gap and overlap regions of type I collagen fibrils with implications for bone heterogeneity. *Biomacromolecules* **2009**, *10*, 2565–2570.

43. Hay, E.D. Extracellular-matrix. *J. Cell Biol.* **1981**, *91*, 205–223.

44. Du, C.; Cui, F.Z.; Zhu, X.D.; de Groot, K. Three-dimensional nano-hap/collagen matrix loading with osteogenic cells in organ culture. *J. Biomed. Mater. Res.* **1999**, *44*, 407–415.

45. Fujisato, T.; Sajiki, T.; Liu, Q.; Ikada, Y. Effect of basic fibroblast growth factor on cartilage regeneration in chondrocyte-seeded collagen sponge scaffold. *Biomaterials* **1996**, *17*, 155–162.

46. Liu, X.; Ma, P.X. Phase separation, pore structure, and properties of nanofibrous gelatin scaffolds. *Biomaterials* **2009**, *30*, 4094–4103.

47. Yang, C.H.; Wang, M.X.; Haider, H.; Yang, J.H.; Sun, J.-Y.; Chen, Y.M.; Zhou, J.; Suo, Z. Strengthening alginate/polyacrylamide hydrogels using various multivalent cations. *ACS Appl. Mater. Interfaces* **2013**, *5*, 10418–10422.

48. Minary-Jolandan, M.; Yu, M.-F. Uncovering nanoscale electromechanical heterogeneity in the subfibrillar structure of collagen fibrils responsible for the piezoelectricity of bone. *ACS Nano* **2009**, *3*, 1859–1863.

49. Hwang, Y.J.; Lyubovitsky, J.G. Collagen hydrogel characterization: Multi-scale and multi-modality approach. *Anal. Methods* **2011**, *3*, 529–536.

50. Kuo, C.K.; Ma, P.X. Ionically crosslinked alginate hydrogels as scaffolds for tissue engineering: Part 1. Structure, gelation rate and mechanical properties. *Biomaterials* **2001**, *22*, 511–521.

51. Jeon, O.; Bouhadir, K.H.; Mansour, J.M.; Alsberg, E. Photocrosslinked alginate hydrogels with tunable biodegradation rates and mechanical properties. *Biomaterials* **2009**, *30*, 2724–2734.

52. Tronci, G.; Grant, C.A.; Thomson, N.H.; Russell, S.J.; Wood, D.J. Multi-scale mechanical characterization of highly swollen photo-activated collagen hydrogels. *J. R. Soc. Interface* **2014**, *12*, doi:10.1098/rsif.2014.1079.

53. Davidenko, N.; Gibb, T.; Schuster, C.; Best, S.M.; Campbell, J.J.; Watson, C.J.; Cameron, R.E. Biomimetic collagen scaffolds with anisotropic pore architecture. *Acta Biomater.* **2012**, *8*, 667–676.

54. Sharabi, M.; Mandelberg, Y.; Benayahu, D.; Benayahu, Y.; Azem, A.; Haj-Ali, R. A new class of bio-composite materials of unique collagen fibers. *J. Mech. Behav. Biomed. Mater.* **2014**, *36*, 71–81.

55. Hertz, H. Berührung fester elastischer körper. *J. Reine Angew. Math.* **1881**, *92*, 156–171. (In German)

56. Israelachvili, J.N. *Intermolecular and Surface Forces*; Academic Press: Waltham, MA, USA, 2011.

57. Huang, G.; Lu, H. Measurement of young's relaxation modulus using nanoindentation. *Mech. Time-Depend. Mater.* **2006**, *10*, 229–243.

58. Fischer-Cripps, A.C. *Nanoindentation*, 3rd ed.; Springer: Berlin, Germany, 2011.

59. Elsdale, T.; Bard, J. Collagen substrata for studies on cell behavior. *J. Cell Biol.* **1972**, *54*, 626–637.

60. Eyre, D.R. Collagen: Molecular diversity in the body's protein scaffold. *Science* **1980**, *207*, 1315–1322.

61. Glowacki, J.; Mizuno, S. Collagen scaffolds for tissue engineering. *Biopolymers* **2008**, *89*, 338–344.

62. Alberti, K.A.; Xu, Q. Slicing, stacking and rolling: Fabrication of nanostructured collagen constructs from tendon sections. *Adv. Healthc. Mater.* **2013**, *2*, 817–821.

63. Markert, C.D.; Guo, X.; Skardal, A.; Wang, Z.; Bharadwaj, S.; Zhang, Y.; Bonin, K.; Guthold, M. Characterizing the micro-scale elastic modulus of hydrogels for use in regenerative medicine. *J. Mech. Behav. Biomed. Mater.* **2013**, *27*, 115–127.

64. Kohn, J.C.; Ebenstein, D.M. Eliminating adhesion errors in nanoindentation of compliant polymers and hydrogels. *J. Mech. Behav. Biomed. Mater.* **2013**, *20*, 316–326.

65. *Standard Guide for Characterization of Hydrogels used in Regenerative Medicine*; ASTM F2900-11; ASTM International: West Conshohocken, PA, USA, 2011.

66. Slaughter, B.V.; Khurshid, S.S.; Fisher, O.Z.; Khademhosseini, A.; Peppas, N.A. Hydrogels in regenerative medicine. *Adv. Mater.* **2009**, *21*, 3307–3329.

Influence of Crosslink Density and Stiffness on Mechanical Properties of Type I Collagen Gel

Shengmao Lin and Linxia Gu

Abstract: The mechanical properties of type I collagen gel vary due to different polymerization parameters. In this work, the role of crosslinks in terms of density and stiffness on the macroscopic behavior of collagen gel were investigated through computational modeling. The collagen fiber network was developed in a representative volume element, which used the inter-fiber spacing to regulate the crosslink density. The obtained tensile behavior of collagen gel was validated against published experimental data. Results suggest that the cross-linked fiber alignment dominated the strain stiffening effect of the collagen gel. In addition, the gel stiffness was enhanced approximately 40 times as the crosslink density doubled. The non-affine deformation was reduced with the increased crosslink density. A positive bilinear correlation between the crosslink density and gel stiffness was obtained. On the other hand, the crosslink stiffness had much less impact on the gel stiffness. This work could enhance our understanding of collagen gel mechanics and shed lights on designing future clinical relevant biomaterials with better control of polymerization parameters.

Reprinted from *Materials*. Cite as: Lin, S.; Gu, L. Influence of Crosslink Density and Stiffness on Mechanical Properties of Type I Collagen Gel. *Materials* **2015**, *8*, 551-560.

1. Introduction

Type I collagen network, a major component of the extracellular matrix (ECM) of connective tissues, has a profound impact on cellular and tissue behaviors. Type I collagen gels are widely used as a three-dimensional (3D) scaffold for culturing cells and engineering various tissues capable of providing optimal microenvironments in the form of physical and chemical cues [1]. The structural properties of collagen gel provide the basis of cell-scaffold interactions and were considered in many scaffold designs [2].

It was well acknowledged that microstructure configurations modulated the macroscopic properties of cross-linked fiber networks [3]. The relationship between mechanical properties of collagen gel and the quality of cross-linked fiber structure, (including fiber dimensions, fiber strength, and various polymerization reaction conditions including collagen concentration, pH, *etc.*) was documented in the literature [2,4]. Experimental studies [5–7] have shown that type I collagen gel stiffness and failure stress increased with collagen concentration, pH, or temperature during polymerization. These polymerization conditions also led to an increased fiber density, fiber length or a reduced cross-section. In addition, Zeugolis *et al.* showed that the chemical crosslinking potently altered the gel stiffness and failure stress more than physical or biological crosslinking approaches [8]. Sheu *et al.* observed that the concentration of glutaraldehyde was positively correlated with the degree of cross-linking, (*i.e.*, crosslink density) [9]. Charulatha *et al.* demonstrated that five cross-linking agents led to different crosslink density and chemical structure, as well as mechanical responses of formulated collagen membrane [10]. The various chemical structures of crosslinks are speculated to correspond to different tensile strength.

Computational models of random distributed fibers were also utilized to further inspect the mechanism of fiber network behaviors for fine-tuning the microenvironment of cell culture [11,12]. Crosslinks in two-dimensional models was simply represented as intersection points, which were constrained either as freely rotating pin joints [13–15] or welded joints [16]. The 3D crosslinks were treated as either regular fibers [12] or torsional springs where their rotational stiffness was obtained by fitting to experimental data [11]. However, the role of crosslinks on the gel mechanics was not elucidated yet in the existing models.

In this work, the role of crosslinks on the collagen gel properties was investigated through computational modeling. The collagen fiber network modulated gel behavior was validated against the experiment by Roeder *et al.* [6]. The mechanism of strain stiffening of collagen gel was elucidated. The crosslinks with varied density and stiffness corresponding to different polymerization conditions [9,10] were formulated in a 3D collagen network. These microscopic crosslink properties were then correlated with the macroscopic gel mechanics. These results could be used to guide the design of scaffold with tunable material properties.

2. Materials and Methods

A representative volume element (RVE) with the side length of 40 μm was used to represent commonly used type I collagen gel with a concentration of 1 mg/mL [17], which is equivalent to the fiber volume fraction of 0.073% (Figure 1a). 3D collagen fibers (1934 in total) were randomly distributed using the random seed algorithm [18]. The fiber is 8 μm long and could be truncated to 4 μm at the boundaries. The fiber diameter of 62 nm was based on the measurement for the collagen gel polymerized at 37 °C and pH 7.4 [5]. Each fiber was meshed with 2 μm beam elements corresponding to the mean crosslink spacing of collagen fiber networks [19]. The crosslinks were generated between nodes when their distance is less than or equal to a certain value, referred to as crosslink threshold. Figure 1b demonstrated 2360 crosslinks between fibers in Figure 1a with a threshold of 800 nm. The uncross-linked fibers were then removed, shown in Figure 1c, due to lack of contribution to the mechanics of collagen networking. The crosslink density was calculated as the number of crosslinks per collagen fiber, (e.g., 2.09 in Figure 1c).

The Young's modulus of collagen fiber was adopted as 50 MPa [11]. Crosslinks were assumed to have the same material property as collagen fibers in the baseline model. Uniaxial tension was applied along the x-direction of the RVE. No sliding motion existed between crosslink and collagen. Nonlinear finite element models were solved using ABAQUS 6.12 (Simulia, Providence, RI, USA). Various crosslink thresholds and crosslink stiffness were tested to unravel the role of crosslink on type I collagen gel properties, (*i.e.*, the fiber network properties).

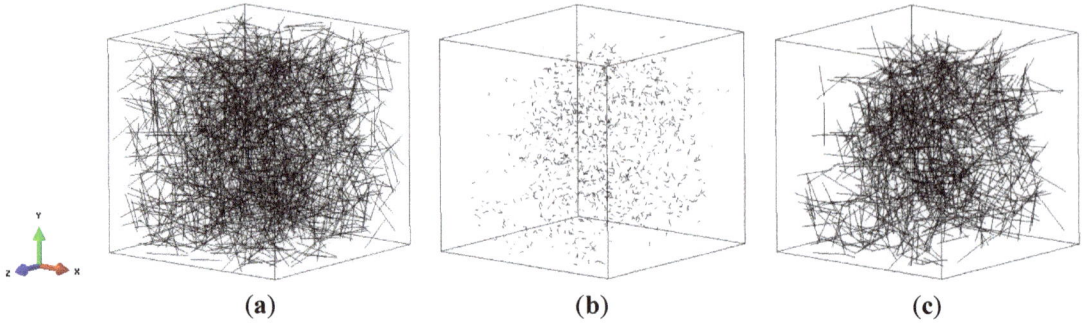

Figure 1. Representative volume element (RVE) with random distributed fibers (**a**) Before cross-linking; (**b**) Crosslinks; (**c**) Cross-linked fiber network.

The orientation of collagen fiber network was analyzed using the OrientationJ plugin [20] in ImageJ software (NIH, Bethesda, MD, USA). The non-affine deformation of fiber network is quantified as S [21]:

$$S = \sqrt{\frac{1}{N}\sum_{i=1}^{N}(\frac{d_i - x_i\varepsilon}{x_i\varepsilon})^2} \tag{1}$$

where d_i is the displacement for the ith node located at x_i from previous step when the network has a macroscopic strain value of ε. A larger S indicates an increased non-affinie deformation of fiber networks, with 0 as an affine deformation.

3. Results

3.1. Model Validation

The experimental work by Roeder *et al.* [6] was simulated using our RVE model subjected to 40% strain along x-direction as shown in Figure 2a. The fiber diameter was measured as 435 nm with the Young's modulus of 79 MPa, and the crosslink threshold was assumed as 450 nm. The fiber network strain was estimated from the relative edge displacement. The fiber network stress along x-direction was calculated by the edge reaction force divided by the total fiber cross-section area on the y–z plane. The stress-strain relationship of 3D collagen gel was depicted in Figure 2c, with comparison to the experimental measurements. It was clear that our RVE simulation agreed well with the experiments, especially at strain less than 20%. The discrepancy at larger strain could be explained by the actual heterogeneous fiber dimensions.

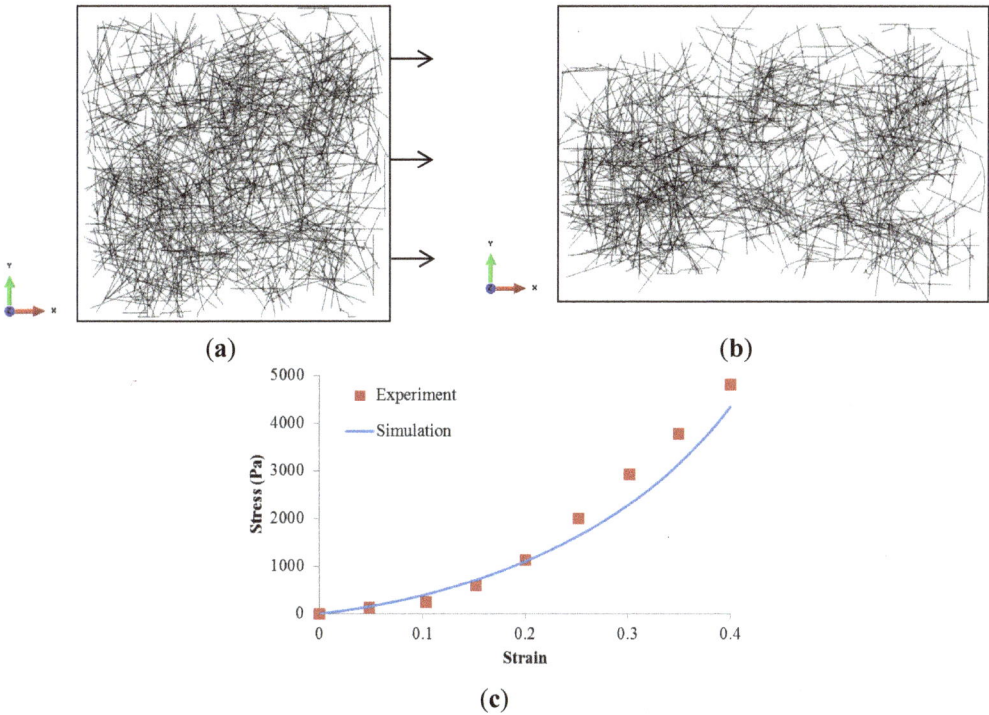

Figure 2. Cross-linked collagen fiber network (**a**) at zero loading; (**b**) at 40% strain along *x*-direction; and (**c**) stress–strain relationship.

3.2. Strain Stiffening Effect in the Baseline Model

Even though both collagen fibers and crosslinks were modeled as linear elastic materials, the fiber network exhibited obvious strain stiffening (Figure 3a). It was clear that the network stiffness was increased with strain, and its magnitude is much less than the stiffness of either fibers and crosslinks due to low fiber volume fraction. This could be explained by the fiber alignment. Therefore the orientation of collagen fibers as well as the non-affine motion property S was monitored. The collagen fiber orientation was quantified as the percentage distribution of collagen fibers within −15 to +15 degree angle relative to the loading axis *x*, which shows the occurrences of fiber alignment along the loading axis at a defined range as a precentage of the total number of fibers. It was shown in Figure 3b that fiber alignment along loading axis in both *xy* and *xz* planes increased with a larger strain. It was also found that the non-affine deformation parameter S decreased with the increased strain. The non-affine deformation was dominated at strains less than 10% which correspond to the reorganization of random distributed collagen fibers. As strain exceeded 10%, the fiber alignment along the loading axis continued to increase with the strain, however, the network deformation tended to be more affine.

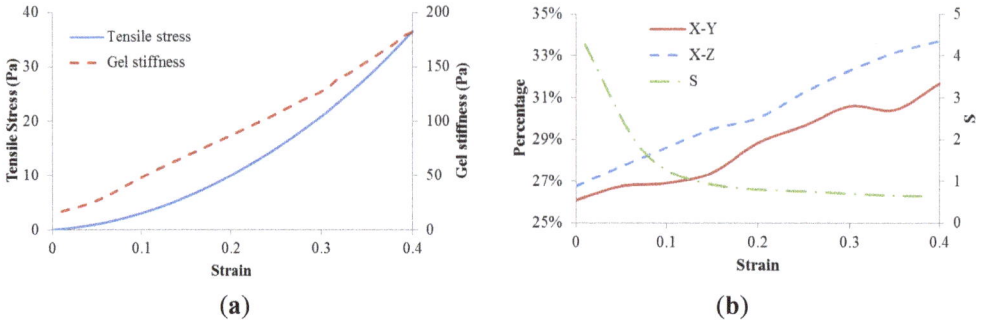

Figure 3. Dynamic results of baseline model. (**a**) Stress–strain relationship and gel stiffness along loading direction; (**b**) Percentage distribution of fibers in both x–y and x–z planes within −15 to +15 degree angle relative to the loading axis, as well as the non-affine parameter S.

3.3. Effect of Crosslink Density

In this work, crosslink density was regulated by the crosslink threshold as listed in seven RVE models in Table 1. As the crosslink threshold, (*i.e.*, maximum crosslink distance), increased from 800 nm in the baseline model to 1600 nm, the number of crosslinks surged from 2360 to 8572, however the number of cross-linked fibers only increased from 1130 to 1931, leading to crosslink density varying from 2.09 to 4.439. The crosslink threshold regulated microscale fiber network configurations was also depicted in Figure 4a. The increase of crosslink threshold resulted in larger numbers of crosslinks and crosslink density as well. However, a plateau was clearly observed for the number of cross-linked fibers as the maximum crosslink distance exceeded 1200 nm. This indicated a fully cross-linked collagen fiber network. Figure 4b plotted the relationship between the crosslink density and collagen gel stiffness. It was clearly observed that the gel stiffness increased with crosslink density. Specifically, a bilinear relationship was obtained. The rate of stiffness growth increased almost four times when the crosslink density was larger than 3.44, corresponding to the crosslink threshold of 1200 nm.

Table 1. RVE models with different crosslink densities.

Cases	Base	1	2	3	4	5	6
Crosslink threshold (nm)	800	850	900	1000	1200	1400	1600
No. of Crosslinks	2360	2933	3594	4687	6467	7776	8572
No. of Cross-linked fibers	1130	1340	1550	1749	1878	1925	1931
Crosslink density	2.09	2.18	2.32	2.68	3.44	4.039	4.439
Gel stiffness (Pa)	30.02	40.823	154.27	545.3	1280.3	4379.6	5659.1

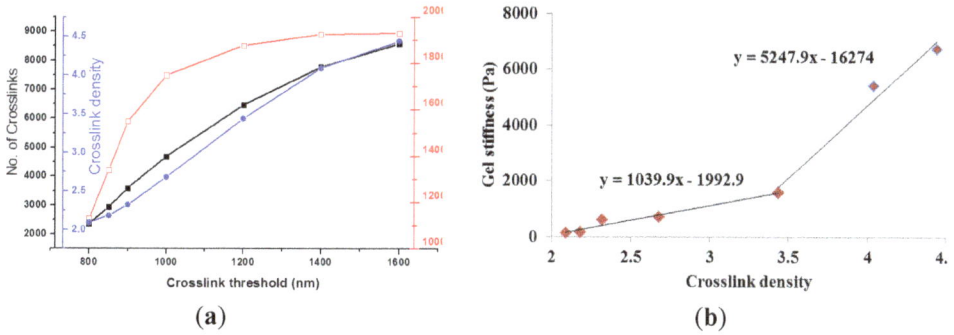

Figure 4. (**a**) Crosslink threshold regulated microstructure changes of collagen gel; (**b**) Correlation between crosslink density and gel stiffness.

3.4. Effect of Crosslink Stiffness

Crosslink stiffness was varied from 50 MPa in the baseline model to 25 MPa, 75 MPa and 100 MPa for studying its role on the type I collagen gel behavior. It was clear from Table 2 that stiffer crosslinks resulted in increased gel stiffness. This could be attributed to the increased load sharing capacity of crosslinks. It was shown from our models that the percentage of load shared by crosslinks was increased with the crosslink stiffness; however its share on the percentage of strain energy was reduced.

Table 2. Role of crosslink stiffness on Load sharing capacity of crosslinks.

Crosslink Stiffness	25 MPa	50 MPa	75 MPa	100 MPa
Gel stiffness	23.3 Pa	30.0 Pa	31.4 Pa	32.2 Pa
Percentage of total load shared by crosslinks	0.09%	1.55%	2.35%	2.92%
Percentage of strain energy shared by crosslinks	19.3%	11.9%	8.7%	6.9%

4. Discussion

Type I collagen was cross-linked under different conditions to formulate collagen gels for various tissue engineering applications [22]. Cross-linking plays an important role on the mechanical stability of gel. The structural properties of the gel will affect the motility of cells and the function of the regenerated tissue. In this study, a three-dimensional collagen fiber network equivalent to a concentration of 1 mg/mL was developed in a microscale RVE to investigate the role of crosslinks on mechanical responses of collagen gels. The model was validated against published experimental data. The obtained classical strain stiffening effect was elucidated by the role of fiber networking without considering the nonlinear elasticity of collagen fibers. By monitoring the change of fiber orientation angle, the strain stiffening effect could be visualized by continuous fiber alignments. This result is consistent with the experimental study by Vader et al. [2] who attributed the strain stiffening of collagen gels to the fiber alignment and densification. It also supports the theoretical hypothesis by Onck et al. [16] that strain stiffening in polymer gel was governed by the fiber rearrangement. We also observed the profound reduction of non-affine property S at lower strain, which corresponded to the reconfiguration of random distributed fibers, especially for lower crosslink density or threshold induced relative weak network (Figure 5). The non-affine deformation reduced with stretches corresponding to the gel stiffening. This agrees with the experimental observations by Wen *et al.* [21]. The results, taken together, suggest that the fiber network dominated the strain stiffening of collagen gel.

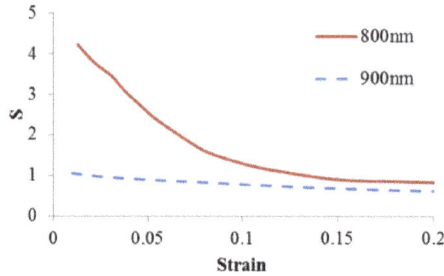

Figure 5. The impact of crosslink threshold on non-affine deformation.

The degree of cross-linking could be regulated by parameters including temperature, pH, collagen concentration and adopted polymerization agents [9,10]. In the current study, crosslink threshold, (*i.e.*, the inter-fiber nodal spacing), was used in our model to control the crosslink density. The adopted crosslink threshold from 800 to 1600 nm is based on the study by Lindstrom *et al.* [19]. Our results have demonstrated that the gel stiffness increased approximately 40 times by doubling the crosslink density with the same collagen concentration. A positive bilinear correlation was found between the crosslink density and gel stiffness. The turning point at the crosslink threshold of 1200 nm, *i.e.*, crosslink density of 3.44, was when all fibers were cross-linked. Further cross-linking treatment after this density point would dramatically increase the gel stiffness. This could be explained by the growth ratio of crosslinks/cross-linked fibers within the network. With a crosslink density larger than 3.44, the cross-linked fibers increased 2.8% and the number of crosslinks increased 32.5%, which resulted in a growth ratio of 11.6, compared to the ratio of 3.6 for the crosslink density less than 3.44 (Table 1). This indicated that crosslinks was mainly used to recruit more fibers into the network before the turning point, and then contributed to the reinforcement of fiber network with more crosslinks per node resulting in the pronounced increase in gel stiffness. Even a linear correlation between the crosslink density and engineered tissue stiffness was well accepted [23]. Sheu *et al.* observed a dramatically increase of collagen gel stiffness after fully cross-linking [9]. Gardel *et al.* also demonstrated a nonlinear correlation between the crosslink density and the stiffness of actin fiber network [24]. It should be noted that their crosslink density was calculated as the crosslink mass, which included all the fibers regardless of their crosslinking status. In our work, the crosslink density was based on the amount of cross-linked fibers. However, our adopted definition of crosslink density won't change the observed bilinear correlation between crosslink density and gel stiffness. Moreover, no experimental evidence has ever demonstrated this bilinear correlation. Our observations might shed lights on future testing and experiments.

The role of crosslink stiffness on type I collagen gel behaviors was also examined. Corresponding to different chemical structures of crosslinks [10], four crosslink stiffnesses with an increment of

25 MPa (e.g., 25 MPa, 50 MPa, 75 MPa, and 100 MPa), were studied. The gel stiffness increased 28% as the crosslink stiffness increased fourfold (Table 2). This observation aligned with our intuition, and we speculated that the increased gel stiffness was attributed to the increased load sharing capacity of crosslinks. It was calculated that percentage of load shared by crosslinks with stiffness of 25 MPa was 0.09%, compared to 2.92% for crosslink stiffness of 100 MPa. In addition, the percentage of strain energy shared by crosslinks was reduced from 19.3% to 6.9%, which indicated that stiffer crosslinks resulted in much less deformations, which mainly served as the load transmitter. This agrees with the observations by Gardel *et al*. [25] that softer F-actin filaments rather than the stiff crosslinks determined the mechanical response of the network. In addition, our baseline model could be used to illustrate the mechanical behaviors of interfibrillar entanglements for pepsin- and acid-solubilized collagen since crosslinks and entanglements could be considered the same in terms of structure configurations [7]. Compared to the impact of crosslink density, the effect of crosslink stiffness on gel stiffness was insignificant.

The non-fibrous matrix and the statistical estimates of gel stiffness was not considered in our models due to its minimal contribution to the gel mechanics [26]. The fiber curvature and its nonlinearity were also overlooked. In the future, we might consider the potential failure mode of crosslinks at large strains. Regardless of these simplifications, our study has demonstrated the importance of crosslink properties on the mechanical response of collagen gels. Specifically, the delicate microstructural changes in crosslink density and stiffness led to profound change in gel properties.

5. Conclusions

In this study, a RVE model of collagen fiber network was proposed to predict its tensile behavior under various crosslink density and crosslink stiffness. The required model input, such as the dimension, stiffness, or volume fractions of fibers, were adopted from published data. Our model prediction was validated by mimicking a tensile test by Roeder *et al*. [6]. Utilizing the inter-fiber spacing to regulate the crosslink density, we are able to provide some novel insights into the role of crosslink density and stiffness on the mechanical response of type I collagen gel, which can be summarized as:

- The strain stiffening effect of the collagen gel was dominated by the fiber alignment.
- The increased crosslink density has much more impact on the gel stiffening than the crosslink stiffness. A positive bilinear correlation between the crosslink density and gel stiffness was predicted.

These results could improve the understanding of the mechanics of collagen networks for designing and regulating clinical relevant biomaterials. This work could also be extended to study how cells respond to different micromechanical environments.

Acknowledgments

This work was supported by the National Science Foundation Faculty Early Career Development (CAREER) award (CBET-1254095), and the University of Nebraska-Lincoln

Research Council Interdisciplinary Grant. The authors also thank Christopher L. Ong for proofreading this manuscript.

Author Contributions

Shengmao Lin contributed to the development of computational models, interpretation of results and preparation of the manuscript. Linxia Gu supervised the student and the project with contributions to conceiving and designing the virtual experiments, data interpretation, and preparation of the manuscript.

Conflicts of Interest

The authors declare no conflict of interest.

References

1. Ingber, D.E. Mechanical and chemical determinants of tissue development. In *Principles of Tissue Engineering*; Academic Press, Inc.: San Diego, CA, USA, 2000; pp. 101–110.
2. Vader, D.; Kabla, A.; Weitz, D.; Mahadevan, L. Strain-induced alignment in collagen gels. *PLoS One* **2009**, *4*, doi:10.1371/journal.pone.0005902.
3. Lieleg, O.; Claessens, M.M.; Bausch, A.R. Structure and dynamics of cross-linked actin networks. *Soft Matter* **2010**, *6*, 218–225.
4. Bozec, L.; Horton, M. Topography and mechanical properties of single molecules of type I collagen using atomic force microscopy. *Biophys. J.* **2005**, *88*, 4223–4231.
5. Raub, C.B.; Suresh, V.; Krasieva, T.; Lyubovitsky, J.; Mih, J.D.; Putnam, A.J.; Tromberg, B.J.; George, S.C. Noninvasive assessment of collagen gel microstructure and mechanics using multiphoton microscopy. *Biophys. J.* **2007**, *92*, 2212–2222.
6. Roeder, B.A.; Kokini, K.; Sturgis, J.E.; Robinson, J.P.; Voytik-Harbin, S.L. Tensile mechanical properties of three-dimensional type I collagen extracellular matrices with varied microstructure. *J. Biomech. Eng.* **2002**, *124*, 214–222.
7. Motte, S.; Kaufman, L.J. Strain stiffening in collagen I networks. *Biopolymers* **2013**, *99*, 35–46.
8. Zeugolis, D.I.; Paul, G.R.; Attenburrow, G. Cross-linking of extruded collagen fibers—A biomimetic three-dimensional scaffold for tissue engineering applications. *J. Biomed. Mater. Res. A* **2009**, *89*, 895–908.
9. Sheu, M.-T.; Huang, J.C.; Yeh, G.C.; Ho, H.O. Characterization of collagen gel solutions and collagen matrices for cell culture. *Biomaterials* **2001**, *22*, 1713–1719.
10. Charulatha, V.; Rajaram, A. Influence of different crosslinking treatments on the physical properties of collagen membranes. *Biomaterials* **2003**, *24*, 759–767.
11. Stein, A.M.; Vader, D.A.; Weitz, D.A.; Sander, L.M. The micromechanics of three-dimensional collagen-I gels. *Complexity* **2011**, *16*, 22–28.
12. Stylianopoulos, T.; Barocas, V.H. Volume-averaging theory for the study of the mechanics of collagen networks. *Comput. Methods Appl. Mechan. Eng.* **2007**, *196*, 2981–2990.

13. Head, D.A.; Levine, A.J.; MacKintosh, F.C. Distinct regimes of elastic response and deformation modes of cross-linked cytoskeletal and semiflexible polymer networks. *Phys. Rev. E* **2003**, *68*, doi:10.1103/PhysRevE.68.061907.

14. Heussinger, C.; Schaefer, B.; Frey, E. Nonaffine rubber elasticity for stiff polymer networks. *Phys. Rev. E* **2007**, *76*, doi:10.1103/PhysRevE.76.031906.

15. Heussinger, C.; Frey, E. Stiff polymers, foams, and fiber networks. *Phys. Rev. Lett.* **2006**, *96*, doi:10.1103/PhysRevLett.96.017802.

16. Onck, P.R.; Koeman, T.; van Dillen, T.; van der Giessen, E. Alternative explanation of stiffening in cross-linked semiflexible networks. *Phys. Rev. Lett.* **2005**, *95*, doi:10.1103/PhysRevLett.96.017802.

17. Hulmes, D.J.; Miller, A. Quasi-hexagonal molecular packing in collagen fibrils. *Nature* **1979**, *282*, 878–880.

18. Digimat, A. *Software for the Linear and Nonlinear Multi-Scale Modeling of Heterogeneous Materials*; e-Xstream Engineering: Louvain-la-Neuve, Belgium, 2011.

19. Lindström, S.B.; Vader, D.A.; Kulachenko, A.; Weitz, D.A. Biopolymer network geometries: Characterization, regeneration, and elastic properties. *Phys. Rev. E* **2010**, *82*, doi:10.1103/PhysRevE.82.051905.

20. Rezakhaniha, R.; Agianniotis, A.; Schrauwen, J.T.; Griffa, A.; Sage, D.; Bouten, C.V.; van de Vosse, F.N.; Unser, M.; Stergiopulos, N. Experimental investigation of collagen waviness and orientation in the arterial adventitia using confocal laser scanning microscopy. *Biomech. Model. Mechanobiol.* **2012**, *11*, 461–473.

21. Wen, Q.; Basu, A.; Winer, J.P.; Yodh, A.; Janmey, P.A. Local and global deformations in a strain-stiffening fibrin gel. *New J. Phys.* **2007**, *9*, doi:10.1088/1367-2630/9/11/428.

22. Parenteau-Bareil, R.; Gauvin, R.; Berthod, F. Collagen-based biomaterials for tissue engineering applications. *Materials* **2010**, *3*, 1863–1887.

23. Balguid, A.; Rubbens, M.P.; Mol, A.; Bank, R.A.; Bogers, A.J.; van Kats, J.P.; de Mol, B.A.; Baaijens, F.P.; Bouten, C.V. The role of collagen cross-links in biomechanical behavior of human aortic heart valve leaflets-relevance for tissue engineering. *Tissue Eng.* **2007**, *13*, 1501–1511.

24. Gardel, M.L.; Shin, J.H.; MacKintosh, F.C.; Mahadevan, L.; Matsudaira, P.; Weitz, D.A. Elastic behavior of cross-linked and bundled actin networks. *Science* **2004**, *304*, 1301–1305.

25. Gardel, M.L.; Kasza, K.E.; Brangwynne, C.P.; Liu, J.; Weitz, D.A. Mechanical response of cytoskeletal networks. *Methods Cell Biol.* **2008**, *89*, 487–519.

26. Lake, S.P.; Hadi, M.F.; Lai, V.K.; Barocas, V.H. Mechanics of a fiber network within a non-fibrillar matrix: Model and comparison with collagen-agarose co-gels. *Ann. Biomed. Eng.* **2012**, *40*, 2111–2121.

Hybrid Membranes of PLLA/Collagen for Bone Tissue Engineering: A Comparative Study of Scaffold Production Techniques for Optimal Mechanical Properties and Osteoinduction Ability

Flávia Gonçalves, Ricardo Bentini, Mariana C. Burrows, Ana C. O. Carreira, Patricia M. Kossugue, Mari C. Sogayar and Luiz H. Catalani

Abstract: Synthetic and natural polymer association is a promising tool in tissue engineering. The aim of this study was to compare five methodologies for producing hybrid scaffolds for cell culture using poly-L-lactide (PLLA) and collagen: functionalization of PLLA electrospun by (1) dialkylamine and collagen immobilization with glutaraldehyde and by (2) hydrolysis and collagen immobilization with carbodiimide chemistry; (3) co-electrospinning of PLLA/chloroform and collagen/hexafluoropropanol (HFP) solutions; (4) co-electrospinning of PLLA/chloroform and collagen/acetic acid solutions and (5) electrospinning of a co-solution of PLLA and collagen using HFP. These materials were evaluated based on their morphology, mechanical properties, ability to induce cell proliferation and alkaline phosphatase activity upon submission of mesenchymal stem cells to basal or osteoblastic differentiation medium (ODM). Methods (1) and (2) resulted in a decrease in mechanical properties, whereas methods (3), (4) and (5) resulted in materials of higher tensile strength and osteogenic differentiation. Materials yielded by methods (2), (3) and (5) promoted osteoinduction even in the absence of ODM. The results indicate that the scaffold based on the PLLA/collagen blend exhibited optimal mechanical properties and the highest capacity for osteodifferentiation and was the best choice for collagen incorporation into PLLA in bone repair applications.

Reprinted from *Materials*. Cite as: Gonçalves, F.; Bentini, R.; Burrows, M.C.; Carreira, A.C.O.; Kossugue, P.M.; Sogayar, M.C.; Catalani, L.H. Hybrid Membranes of PLLA/Collagen for Bone Tissue Engineering: A Comparative Study of Scaffold Production Techniques for Optimal Mechanical Properties and Osteoinduction Ability. *Materials* **2015**, *8*, 408-423.

1. Introduction

The electrospinning process was patented in 1934 [1], and its use for cell scaffold design is widely accepted and explored in various scientific fields [2,3]. The process for obtaining electrospun materials consists of establishing a potential difference between the polymeric solution or melted polymer and the collector, creating an electric field between them and ejecting the electrically charged polymeric solution onto the collector. The electric field stretches the polymer chains, and the polymer is randomly deposited on the collector [2].

Electrospun materials can be produced from synthetic or natural polymers [3]. Among the synthetic polymers that can be used, aliphatic polyester poly-L-lactide (PLLA) excels as a biomaterial because of its semi-crystalline form, which enables the production of high tensile strength [4], biocompatible, easily processed materials with satisfactory mechanical properties for bone regeneration [4]. However, PLLA does not have specific binding sites for cell adhesion and is highly hydrophobic, causing poor interaction with cells and precluding the flow of nutrients [5,6].

Among the natural polymers used as biomaterials, collagen is prominent. As a protein formed mainly of glycine, proline and hydroxyproline, collagen exhibits a highly organized structure [7]. Collagen is the primary extracellular matrix protein of several tissues, such as bone and conjunctive tissue, exhibiting specific sites for cell adhesion and good compatibility [8]. However, the use of collagen presents some limitations, such as a high degradation rate, low processability and inadequate mechanical properties for bone-related applications [9,10]. Hence, the concomitant use of synthetic and natural materials, such as PLLA and collagen, has proven to be an excellent alternative for combining the positive aspects of both materials [5,11].

Depending on the conditions used, the electrospinning of collagen solutions may result in fiber mats with a diameter near that of natural collagen fibrils [12]. However, the main solvents used to dissolve collagen, such as fluorinated alcohols 1,1,1,3,3,3-hexafluoro-2-propanol (HFP) or 2,2,2-trifluoroethanol (TFE), also destabilize the hydrogen bonding of collagen's triple-helix structure, leading to denaturation [13]. A circular dichroism study reported the denaturation of 45% of the triple-helix structure in an 8% wt/vol collagen-HFP solution [12]. Using gel electrophoresis, analyses of gelatin and collagen bands when solubilized in HFP indicated a 58% loss of collagen. After electrospinning, the total loss increased to 68%. These values increased to 93% and 99.5%, respectively, when submitted to pepsin degradation. Pepsin is not known to degrade intact structures of collagen's triple-helix backbone, but it is known to be active on modified structures, providing evidence of collagen denaturation [13]. Other authors have suggested that the use of acetic acid as a solvent minimizes the degradation of native collagen structures. However, more studies are required to understand the exact effect of acetic acid as a collagen solvent because only a small advantage is observed with respect to HFP use [14].

One approach to avoid electrospinning-related problems is through the immobilization of collagen onto a polymer matrix surface, e.g., PLLA, which has been demonstrated to be an excellent alternative for increasing hydrophilicity and thus cell adhesion and proliferation while maintaining cell functionality [5,11]. Immobilization can be achieved by anchoring amine or carboxyl groups onto polymer surfaces previously generated through, for example, aminolysis or hydrolysis [15,16]. Although these methodologies have been successfully employed and intensively studied in films and sintered three-dimensional (3D) matrices [17], their use in electrospinning has rarely been reported [18,19]. Electrospun PLLA fiber aminolysis, followed by chitosan immobilization, resulted in increased fibroblast proliferation while maintaining the mechanical properties of the fibers [18]. Accordingly, the attachment of fibronectin to PLLA fibers through aminolysis led to increased epithelial cell proliferation and collagen type IV expression [19].

The aim of the present study was to evaluate and compare different types of collagen associations to PLLA in electrospun fibers. The methodologies used were as follows: (i) electrospinning of co-solutions, resulting in blends; (ii) electrospinning of PLLA followed by collagen immobilization onto the surfaces of fibers; and (iii) co-electrospinning of PLLA and collagen isolated solutions, resulting in hybrid mats. Various solvents and immobilization techniques were quantitatively and qualitatively evaluated and compared.

2. Results and Discussion

2.1. Generation and Characterization of Scaffolds

Aggregating biological components and synthetic polymers to mimic extracellular matrix (ECM) properties is a widespread technique [5,20]. The use of collagen as the bio-component has resulted in numerous successful examples of such mimicry, most of which have been constructed from collagen-decorated surfaces, in which the amount of collagen is usually scant [20]. Alternatively, blending both components through co-dissolution, followed by film or fiber processing, can yield materials with higher proportions of collagen. Depending on the polymer-to-collagen ratio, the electrospinning of a co-solution may render hybrid materials with fiber populations of different compositions. Co-electrospinning represents the extreme case of this situation, in which two opposed jets of distinct solutions from isolated components are aimed at the same rotating target.

Herein, five different approaches for incorporating a large or reduced amount of collagen into PLLA electrospun scaffolds are compared. Two of these methods are based on collagen immobilization on electrospun PLLA surfaces after modification through aminolysis or hydrolysis. The two other methods involve the co-electrospinning of PLLA/chloroform/DMF (9:1) and collagen/HFP or collagen/acetic acid solutions. Finally, a PLLA/collagen blend was generated from a co-solution in HFP.

An advantage of collagen immobilization onto PLLA is lower degradation and denaturation because collagen is solubilized in an acetic acid solution (4%) [21]. However, as demonstrated below, disadvantages of this method are the following: (i) only small amounts of collagen can be incorporated into the matrix; (ii) fibers must be functionalized prior to processing; and (iii) the scaffold mechanical properties decrease.

Aminolysis (using alkanediamines) and hydrolysis (using basic aqueous solutions) generate respective amino ($-NH_2$) and carboxyl ($-COOH$) groups on the polymer surface. These reactions are dependent on several parameters, such as (i) solvent; (ii) amine type and concentration; (iii) temperature; and (iv) total reaction time [16]. In this study, milder conditions were selected and compared with those used for films and 3D scaffolds [16,17,22]. In all cases, a decrease or complete loss of polymer crystallinity during electrospinning has been observed [23], which could directly affect the material properties and reactivity, thus affecting aminolysis and hydrolysis reaction rates. Additionally, the electrospinning technique is susceptible to the effects of several variables (e.g., polymer concentration, solvent type, flow rate, voltage and needle-collector plate distance) [3], which might govern the diameter of the fibers and indirectly affect the scaffolds' response to aminolysis and hydrolysis conditions.

Although Chen *et al.* [15] reported the differences in electrospun PLLA fiber hydrolysis using a 0.5 M NaOH solution for 5, 10 and 30 min, we determined that the limiting condition before extensive scaffold degradation was 0.1 M NaOH solution for 3 min. The statistically identical values of $-COOH$ concentration in Table 1 for reaction times of >3 min indicate that hydrolysis reaches a maximum at the surface while continuously leaching to the inner volume of the fibers, compromising the mechanical properties of the fibers instead of creating more reactive surface groups.

Table 1. Total carboxyl group concentration (mol/g) inserted into PLLA scaffolds as a function of the reaction time (min).

Material	Hydrolysis time (s) [#]	Carboxyl group concentration (mol/g)
	0	$(3.7 \pm 0.5) \times 10^{-5}$
	20	$(5.6 \pm 0.9) \times 10^{-5}$
PLLA	45	$(5.5 \pm 1.0) \times 10^{-5}$
	60	$(5.8 \pm 0.7) \times 10^{-5}$ *
	180	$(5.9 \pm 1.0) \times 10^{-5}$ *

[#] Reaction with 0.1 M NaOH; * Significant difference when compared to PLLA scaffolds without hydrolysis (0 s); $p < 0.05$.

We also observed severe limitations for aminolysis, in which total degradation of the fibers was observed when using conditions reported in literature [24,25]. As shown in Table 2, we observed a reaction time limit of 5 min using diamine concentrations that were 3–8 times lower than previously reported, whereas the amine concentration on the surface varied by only 2.5-fold.

Table 2. Total amine group concentration (mol/g) inserted into PLLA scaffolds as a function of reaction time (min).

Material	Aminolysis Time (min) [#]	Amine Group Concentration (mol/g)
	0.5	$(1.2 \pm 0.2) \times 10^{-5}$
	1	$(1.6 \pm 0.2) \times 10^{-5}$
PLLA	3	$(1.9 \pm 0.2) \times 10^{-5}$ **
	5	$(3.2 \pm 0.2) \times 10^{-5}$ ***

[#] Reaction at 0.008 g/mol of HAD; ** Significant difference when compared to PLLA scaffold with 0.5 s of aminolysis at $p < 0.01$ and *** $p < 0.001$.

After aminolysis or hydrolysis, dialdehyde (GTA) and carbodiimide (EDC) chemistry can be respectively employed to immobilize collagen onto the polymer surface [11,20], although some cytotoxicity is attributed to GTA [24].

The amount of collagen incorporated onto the scaffolds' surface was measured with the ninhydrin test and elemental analyses (Table 3). Using both methods, approximately 1.5% collagen could be inserted via hydrolysis/EDC methodology. However, for the aminolysis/GTA method, the amount of collagen detected by elemental analysis was 5-fold higher than the values determined using the ninhydrin test. Although similar values were observed among hydrolysis/EDC, elemental analyses exhibited high standard deviation and coefficient of variance values. Because the collagen was not well distributed in these materials, the high variability can be attributed to the small amount of material used for analyses. However, the ninhydrin test quantification results, which required more material for analyses, exhibited lower variance; thus, the ninhydrin test was a more efficient quantification methodology for these materials. This observation is in agreement with the higher concentration of functional groups obtained after surface modification, establishing functionalization as the limiting step of this approach.

Table 3. Means and standard deviations of collagen (wt%) inserted into the scaffolds obtained from elemental analyses and the ninhydrin test.

Material	Collagen Concentration (wt%)	
	Elemental analysis	Ninhydrin test
PLLA	0	-
PC_hydrolysis	1.4 ± 0.5	1.5 ± 0.2
PC_aminolysis	1.5 ± 0.6	0.30 ± 0.05
PC_blend	47 ± 1	-
PC_cf_HFP	71 ± 10	-
PC_cf AA	15 ± 6	-

Co-electrospinning is a technique that offers the advantage of incorporating large amounts of collagen. In addition, because this technique creates segregated fibers of each of the compounds from different solutions, alternative solvents can be selected, allowing a lower extent of collagen denaturation during electrospinning [13,14]. In this study, acetic acid and HFP were compared as solvents that have different effects on collagen denaturation. Collagen is highly stable in these solvents, however, resulting in a high threshold concentration for achieving the viscosity conditions necessary for electrospinning to occur without bead formation.

The co-electrospinning of PLLA and collagen was optimum when using concentrations of 50 wt% polymers. Although initial solutions contained the same concentrations of PLLA and collagen, their distinct properties, such as viscosity, surface tension and charge density, affected their ability to be electrospun [25], hence yielding hybrid mats of varied composition. Elemental analyses (CHN) were performed to determine the final amounts of collagen in the materials (Table 3). The collagen solution formed using HFP was able to incorporate more collagen than the acetic acid solution: the former yielded mats containing 71% collagen, whereas the latter yielded mats containing only 15% collagen.

Finally, PLLA and collagen blends were produced using a co-solvent that equally dissolved both polymers. In this case, HFP was observed as the only solvent in which the two polymers were miscible, despite the loss of collagen's most important structural features, such as its triple helices assembly [13]. To create a standard for comparison, co-solutions containing 50 wt% of both polymers were electrospun. The resulting blended material exhibited collagen incorporation closest to its initial composition (47% collagen, Table 3).

Figure 1 presents SEM images of the different scaffolds evaluated in this study. When collagen was immobilized on the surface of PLLA through hydrolysis (Figure 1B) or aminolysis (Figure 1C), the fibers became thicker, while a fibrous structure was maintained. For comparison, Figure 1A shows plain PLLA fibers. The scaffold obtained from PLLA/collagen blend electrospinning (Figure 1D) contained regular but thinner fibers, whereas the materials obtained by co-electrospinning clearly consisted of molten fibers, a typical result observed when pure collagen is spun. Nonetheless, the fibers obtained from the co-spinning of the acetic acid/collagen solution were more regular than those obtained using HFP.

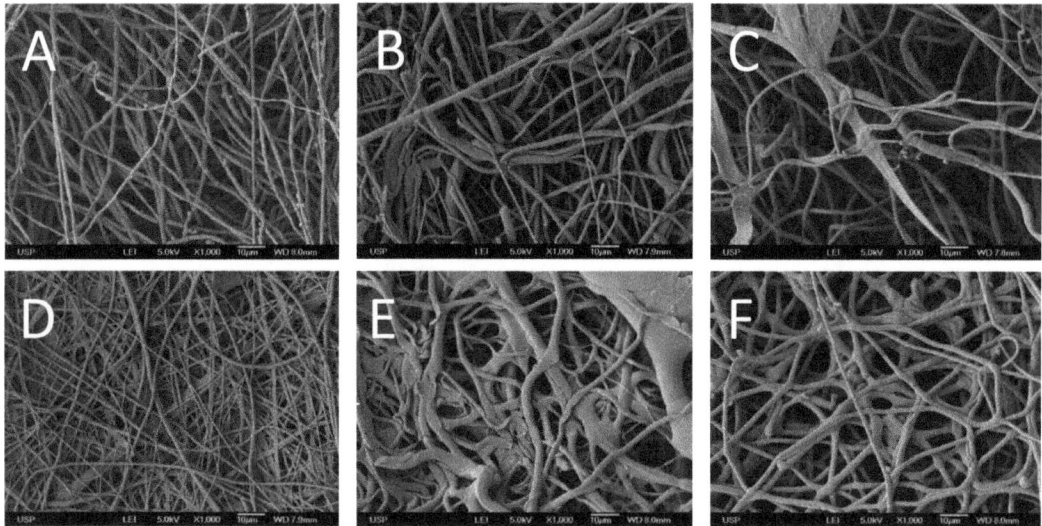

Figure 1. SEM images (×1000 magnification) of electrospun scaffolds: (**A**) PLLA; (**B**) PLLA after hydrolysis and collagen incorporation with EDC; (**C**) PLLA after aminolysis and collagen incorporation with GTA; (**D**) PLLA/collagen electrospun by blending; (**E**) PLLA/collagen electrospun by co-electrospinning and using HFP as the collagen solvent; and (**F**) PLLA/collagen electrospun by co-electrospinning and using acetic acid solution as the collagen solvent.

Table 4 presents the mechanical scaffold characterization results, a property of utmost importance when evaluating the possible applications of these materials. The materials subjected to hydrolysis and aminolysis, followed by collagen immobilization, exhibited the lowest Young's modulus and tensile strength values. These findings are somewhat expected because diamines are known to percolate through the polymer matrix, followed by lysis reactions and disruptions of fibers, decreasing the strength and elasticity of the material [26]. Kim and Park [26] reported that PLLA nanofibers subjected to aminolysis developed stacked lamellae through transverse oriented degradation, resulting in lower mechanical properties.

However, basic hydrolysis is expected to be solely a surface reaction. Sun *et al.* [27] demonstrated that alkali etching of PLLA yarns promotes time-dependent mass loss, concluding that the process is surface-limited. This surface peeling causes a reduction in fiber diameter, which is also dependent on alkali concentration and temperature. However, in this study, the electrospun fiber diameters were 2 orders of magnitude lower than those of the yarns, and the same peeling was expected to break down most of the low-diameter fibers, explaining the results reported in Table 4.

Table 4. Mechanical properties of PLLA/collagen electrospun materials: means and standard deviations of Young's modulus [1], tensile strength and elongation at break values of the six scaffolds tested.

Material	Elongation at Break (%)	Tensile Strength (MPa)	Young's Modulus [1] (GPa)
PLLA	39 ± 8	18 ± 3	0.19 ± 0.03
PC_hydrolysis	14 ± 2 ***	4.8 ± 0.8 *	0.07 ± 0.01
PC_aminolysis	5.8 ± 0.3 ***	7 ± 1	0.10 ± 0.02
PC_blend	$(5 \pm 1) \times 10^{1}$	$(8 \pm 1) \times 10^{1}$ ***	1.0 ± 0.1 ***
PC_cs_HFP	6 ± 1 ***	38 ± 6 ***	1.4 ± 0.2 ***
PC_cs_AA	14 ± 2 ***	24 ± 4	0.7 ± 0.1 ***

[1] Young's modulus is calculated from the linear region of the stress-strain curves; * Significant difference when compared to PLLA scaffold at $p < 0.05$ and *** $p < 0.001$.

Although hydrolysis results in higher scaffold degradation and consequently, lower mechanical properties, the higher surface enables the incorporation of higher amounts of collagen, when compared to aminolysis, as previously discussed.

Regarding DMA analyses of materials with a large amount of collagen obtained by co-electrospinning, the lower elongation of these materials at break is believed to be a function of neat-collagen crosslinked fiber friability. While electrospun collagen fiber elongation has been observed to reach values of 26% and 33% in literature [28,29], these values refer to a scaffold not submitted to any crosslinking treatment; in this present study, materials were analyzed after being crosslinked. Although collagen scaffolds could be obtained for electrospinning, they could not be tested due to their friability after crosslinking treatment. Co-electrospun scaffolds also exhibited higher rigidity, as indicated by their high elastic moduli (Table 4). Previous studies [21,29] have reported that the Young's moduli of electrospun collagen fibers are related to fiber diameter. Yang *et al.* demonstrated that the Young's modulus ranged from 1.0 to 3.9 GPa in scaffolds with mean fiber diameters of 187 and 305 nm, respectively. In addition, no significant difference was observed in the Young's moduli between crosslinked and uncrosslinked collagen fibers. These values are at least 5 times higher compared with those of PLLA examined in this present study, explaining the higher Young's modulus values obtained in co-electrospun and blended materials.

In the blended materials, the PLLA/collagen scaffold was able to maintain high elongation via PLLA and high tensile strength via collagen. Consequently, this material exhibited better mechanical properties than the other scaffolds evaluated in this study. These results agree with those of Ngiam *et al.* [30] who reported that 1:1 PLLA/collagen scaffold blend exhibited an elasticity modulus that was approximately 3.5 times higher than that of pure PLLA after hydroxyapatite incorporation.

2.2. Cell Culture

Fluorescence images of labeled cell nuclei obtained by confocal microscopy are presented in Figure 2; the cells were homogeneously distributed on top of the majority of the materials, except on the PC_hydrolysis scaffold and to a lower extent on the PC_aminolysis scaffold, which contained regions of concentrated cells (Figure 2B,C). Because small amounts of collagen were bound to the scaffold in these methods, collagen only partially covered the fibers, creating preferential sites for cell adhesion and proliferation. In fact, the collagen concentration increased the number of sites available for integrin and other transmembrane collagen receptor attachment, responsible for cell adhesion and proliferation [31,32]. Likewise, studies have demonstrated that increasing the concentration of immobilized collagen on PLLA/collagen membrane surfaces is correlated with increasing cell density [20,33].

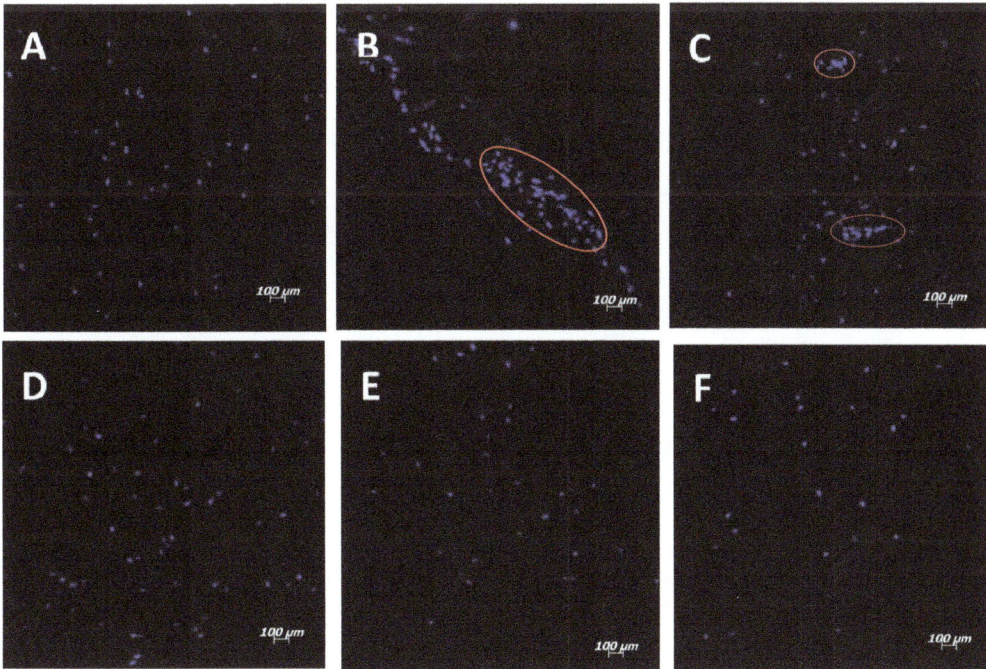

Figure 2. Fluorescence confocal images (100× magnification) of cells on electrospun scaffolds: (**A**) PLLA; (**B**) PLLA after hydrolysis and collagen incorporation with EDC; (**C**) PLLA after aminolysis and collagen incorporation with GTA; (**D**) PLLA/collagen electrospun by blending; (**E**) PLLA/collagen electrospun by co-electrospinning and using HFP as the collagen solvent; and (**F**) PLLA/collagen electrospun by co-electrospinning and using acetic acid solution as the collagen solvent.

Cell proliferation was measured by ^{3}H-thymidine incorporation into DNA (Figure 3), and cell differentiation (osteogenesis) was assessed by alkaline phosphatase (ALP) activity (Table 5). The results show a different growth rate for each type of scaffold upon culturing human mesenchymal stem-cells from exfoliated teeth dental pulp (SHEDs) for up to 14 days. Initially, up to day 3, PC_hydrolysis exhibited higher adhesion and proliferation than all of the other materials. One possible explanation for this result is that the collagen attached to this functionalized surface was less degraded than electrospun collagen [21], facilitating cell adhesion [5,16,20] through a highly specialized mechanism of interaction with the collagen triple-helix [32]. Furthermore, hydrolysis increases polymer hydrophilicity, favoring cell attachment and growth [34]. The variation observed between PC_hydrolysis and PC_aminolysis can be attributed to the 5-fold lower content of collagen and the presence of half as many functional groups during PC_aminolysis, limiting the effect of this process.

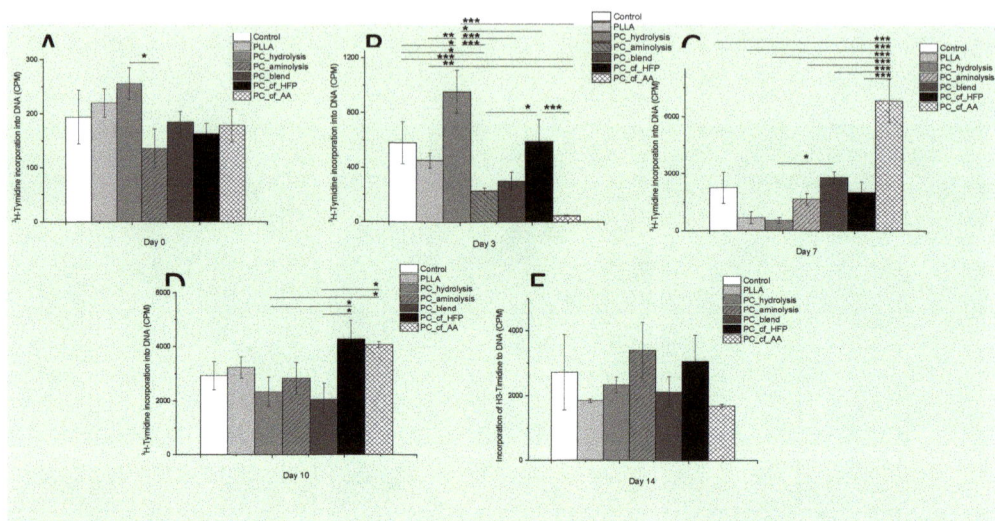

Figure 3. Incorporation of ^{3}H-thymidine (CPM) into the DNA of SHEDs after (**A**) 0; (**B**) 3; (**C**) 7; (**D**) 10; and (**E**) 14 days of culture on different PLLA and collagen scaffolds. Statistically significant difference at * $p < 0.05$, ** $p < 00.01$ and *** $p < 00.001$.

Table 5. ALP activity (U/L) means and standard deviations in SHEDs cultured on scaffolds under ODM and DMEM conditions.

Material	ALP (U/L)	
	ODM	DMEM
PLLA	0.16 ± 0.05	Not detected
PC_hydrolysis	0.60 ± 0.08	0.15 ± 0.04
PC_aminolysis	0.5 ± 0.1	Not detected
PC_blend	1.2 ± 0.1 ***	0.8 ± 0.1
PC_Cf_HFP	1.0 ± 0.2 ***	0.34 ± 0.08 **
PC_cf_AA	1.6 ± 0.3 ***	Not detected
Control	1.3 ± 0.1 ***	Not detected

*** Significant difference of ALP in ODM medium when compared to the PLLA scaffold at $p < 0.001$;
** Significant difference of ALP in DMEM medium when compared to the PLLA/collagen blended material at $p < 0.01$.

From day 7 to 10, higher cell proliferation rates on the co-electrospun materials were observed. These results can be attributed to two main factors, namely, the amount of collagen and cells adhering to cryptic binding sites in collagen, which are only exposed after collagen denaturation [31]. Finally at day 14, advanced cell maturation leads to decreased cell metabolism and equalized growth rates for all groups. These results agree with studies that have demonstrated higher early cell adhesion when collagen was added to scaffolds of various compositions [20,33].

ALP is an enzyme that is used as a biochemical marker of osteoblastic activity. ALP activity was used to assess SHED cell differentiation when plated onto scaffolds and cultured with ODM or DMEM. Osteodifferentiation should be observed on the materials cultured in the presence of ODM, whereas cells that are cultivated in basal medium (DMEM) should differentiate as a function of the scaffold material. Table 5 shows that the materials containing higher amounts of collagen, as obtained by blending or co-electrospinning, exhibited higher expression of ALP in ODM as well as in pure DMEM. These findings can be attributed to the larger number of sites for integrin adhesion, which promote better cell dispersion and improve osteoblast differentiation [35,36]. Integrins are a class of transmembrane proteins known for mediating cell adhesion to the extracellular matrix and consequently, providing an intracellular signaling for cell proliferation, functionality and/or differentiation [35]. Studies performed by Mizuno *et al.* [35] and Schneider *et al.* [36] reported that the Asp-Gly-Glu-Ala domain of type I collagen interacts with α2β1 integrin, increasing gene expression of bone markers and matrix mineralization [35,36].

However, not all materials were able to promote the differentiation of mesenchymal stem cells (SHEDs) into osteoblasts without ODM. This osteodifferentiation ability was observed on only three materials, namely, the scaffolds obtained by co-electrospinning using HFP, by collagen immobilization after hydrolysis and by polymer blending, with the latter exhibiting the highest extent of osteogenic induction. The main but not exclusive factor responsible for osteoblastic differentiation was the presence of collagen on the scaffold because the ALP activity was very low or non-detectable in both pure PLLA or in the control without collagen. These findings agree with those of a previous study that demonstrated gene expression of bone markers in collagen but not in PLLA scaffolds under basal conditions [37].

However, the presence of collagen does not ensure differentiation because the MSCs grown on the PC_aminolysis and PC_cs_AA scaffolds did not exhibit ALP activity in basal medium. Other factors operate in conjunction to determine the osteodifferentiation capacities of a material. Comparing a non-osteoinductive material (PC_aminolysis) with an osteoinductive material (PC_hydrolysis) reveals that the main difference between them is collagen content, indicating that a minimum amount of collagen in the scaffold is required to promote cell differentiation.

When comparing the materials obtained by co-electrospinning, both were observed to contain independent fibers of PLLA and collagen, with the main difference between them being the solvent used to obtain the collagen solution. According to Table 5, collagen dissolved in HFP is able to promote cell differentiation in DMEM, whereas in acetic acid, cell differentiation cannot be promoted. Dong and co-workers [38] demonstrated that HFP solubilizes collagen not only by breaking the hydrophobic interactions via the trifluoromethyl groups but also by breaking hydrogen bonds via the mildly acidic secondary alcohol hydroxyl. Acetic acid would promote solubilization only via hydrogen bond rupture, which may produce differences in the collagen fibers obtained using these methodologies. A study by Liu *et al.* [14] using circular dichroism analysis demonstrated that electrospun scaffolds of collagen obtained with HFP and acetic acid caused collagen denaturation due to the loss of the triple-helix conformation. Moreover, when using acetic acid, the extent of collagen degradation was reduced compared with that observed using HFP [14]. The higher degradation afforded by HFP would easily expose cryptic binding sites to integrin receptors, as previously described [31]. However, partial degradation by acetic acid would induce sterically hindered links to these sites. Therefore, higher bone differentiation could be obtained using HFP.

The highest extent of osteoinduction was observed on the blended material. Several factors that are unique to these materials' design can interact to induce and improve bone differentiation, such as reduced fiber size, fibers composed of both collagen and PLLA and highest mechanical properties [37,39].

The data obtained in this study allow us to conclude that the material obtained by electrospinning a co-solution of PLLA/collagen yielded the best performance for bone tissue engineering applications because this material demonstrated the highest mechanical properties and high values of ALP expression in osteoinductive and basal media, exhibiting osteoconduction and osteoinduction properties.

3. Experimental Section

Five methodologies were designed to compare the mechanisms by which collagen is inserted into scaffolds, namely: (1) immobilization of collagen on the surface of electrospun PLLA subjected to hydrolysis using carbodiimide as the crosslinking agent (PC_hydrolysis); (2) immobilization of collagen on the surface of electrospun PLLA subjected to aminolysis and using glutaraldehyde as the crosslinking agent (PC_aminolysis); (3) electrospinning of a co-solution of PLLA and collagen (1:1) using HFP as co-solvent (PC_blend); (4) co-electrospinning (concomitant spinning of isolated solutions) of PLLA/chloroform and collagen/HFP solutions (1:1) (PC_cs_HFP); (5) co-electrospinning of PLLA/chloroform and collagen/acetic acid solutions (PC_cs_AA).

3.1. Electrospinning of PLLA, Collagen and PLLA/Collagen Solutions

A PLLA/collagen blend scaffold in a 1:1 ratio was obtained by the co-dissolution of both polymers in HFP, producing a 5 wt% solution, followed by electrospinning in one syringe.

PLLA/collagen hybrid scaffolds formed in a 1:1 ratio using different solutions were prepared through the simultaneous co-electrospinning of a PLLA solution (5 wt%) in chloroform/DMF (9:1) and a collagen solution (25 wt%) in acetic acid (40 vol%) or a collagen solution (5 wt%) in HFP. Solutions were electrospun using different syringes coupled along the same axis but in different directions and perpendicular to the collector.

Pure PLLA solution (5 wt%) in chloroform/DMF (9:1) was electrospun for further collagen immobilization. All solutions were electrospun using the following parameters: flow rate of 2 mL/h, distance of 9 cm between the needle and collector and applied voltage of 12.5 kV. The collector was a rotor operated at low speed.

3.2. Scaffold Crosslinking

Because collagen fibers are partially soluble in aqueous media after electrospinning, the scaffolds obtained using co-spun or blended fibers were stored in a glutaraldehyde (GTA) atmosphere for 24 h to crosslink the scaffold fibers. After crosslinking treatment, the scaffolds were washed four times with a 0.02 M glycine solution for 20 min and once with deionized water to remove and neutralize the remaining GTA.

3.3. PLLA Functionalization and Collagen Immobilization on the Surface

PLLA electrospun scaffolds were submitted to aminolysis with 1,6-hexanediamine (HDA) or to hydrolysis to create surface amino and carboxylic groups, respectively.

For aminolysis, the scaffolds were immersed in a propanol solution containing 8 mg/mL of HDA at 50 °C for 5 min. Collagen was immobilized on the scaffold surfaces by immersion for 3 h at room temperature in a solution containing 1 wt% GTA in PBS (pH 7.4), rinsed with deionized water for 2 h and immersed in a collagen solution at 4 °C for 24 h [20].

For hydrolysis, PLLA scaffolds were immersed in an aqueous solution of 0.1 M NaOH at 37 °C for 3 min. A 2 mg/mL collagen solution was prepared in aqueous acetic acid (3 vol%). These scaffolds were then immersed in a 48 mM 1-ethyl-3-(3-dimethylaminopropyl)carbodiimide chloride (EDC) solution, 6 mM N-hydroxysuccinimide and 50 mM MES (2-(N-morpholino)ethanesulfonic acid) buffer (pH 5.0) for 24 h at 4 °C, rinsed with deionized water for 2 h and immersed in a collagen solution for 24 h at 4 °C [11].

3.4. Aminolysis, Hydrolysis and Collagen Quantification

The number of amine groups inserted into the scaffolds was measured using the ninhydrin test. Briefly, scaffolds (r = 0.6 mm, approximately 5 mg per disc) were immersed for 45 s in an ethanol solution containing 0.01 M ninhydrin, transferred to a clean glass tube and heated to 70 °C for 10 min. Scaffolds were solubilized in a solution containing dichloromethane and isopropanol (1:1). The absorbance of ninhydrin-amine group complexation was measured at 540 nm and compared with a calibration curve to determine the amine concentration in the scaffolds.

The number of carboxyl groups inserted into the scaffolds was measured using rhodamine-carboxylic acid interaction. Briefly, rhodamine-6G hydrochloride was dissolved in a phosphate buffer (pH 12), followed by toluene extraction. PLLA scaffold discs (r = 0.6 mm; approximately 5 mg per disc) were solubilized in 1 mL of dichloromethane, followed by the addition of 1 mL of a neutral rhodamine-6G toluene solution. The solution was incubated in the dark for 1 h. The absorbance of rhodamine-carboxylic acid complexation at 535 nm was measured and compared with a calibration curve to determine the carboxyl concentration in the scaffolds.

The amount of immobilized collagen was measured using the ninhydrin test. Scaffold discs (r = 0.6 mm, approximately 5 mg per disc) were hydrolyzed with a 6 M HCl aqueous solution for 24 h at 120 °C under a nitrogen atmosphere. The solution was dried, and a combination of a ninhydrin solution (0.04 M) and 0.1 M citric acid buffer (pH 5.5) was added. The final solution was heated to 70 °C for 10 min and cooled to 4 °C for 5 min. The absorbance of ninhydrin-amine group complexation (amino acid groups from hydrolyzed collagen) was measured at 560 nm and compared with a calibration curve to determine the collagen concentration in the scaffolds.

3.5. Scanning Electron Microscopy

To analyze and compare scaffold morphologies, SEM images were obtained using a JEOL FEG 741F field-emission electron microscope (Tokyo, Japan).

3.6. Elemental Analyses

CHN analyses on the scaffolds were performed using a Perkin-Elmer Elemental Analyzer model 2400 Series II. Because only collagen molecules contained nitrogen, pure collagen mats were used to calibrate the amount of collagen inserted into the other scaffolds based on the proportion of nitrogen detected in each material.

3.7. Dynamic Mechanical Analysis (DMA)

Scaffolds were cut into specimens measuring 30 mm × 7 mm. Scaffold thickness was measured using a micrometer (Mitutoyo, Japan), and the specimens were subjected to a stress-strain test in a TA Instruments Q800 DMA tester (New Castle, PA, USA). Briefly, the specimens were fixed between two clamps, and a pre-load of 0.01 N was applied for 5 min. The temperature was maintained at 30 °C, and a force ramp of 2 N/min was applied. The tensile strength, Young's modulus and maximum elongation at break were obtained for each material.

3.8. Cell Proliferation

A cell suspension (2.0×10^4 cells/well) of human mesenchymal stem-cells from exfoliated teeth dental pulp (SHEDs) was added to a 24-well culture tray coated with scaffold and cultured in osteoblastic differentiation media (ODM), composed of Dulbecco's modified Eagle's medium (DMEM), 10% fetal bovine serum, 10 mM β-glycerophosphate disodium salt hydrate, 50 μg/mL L-ascorbic acid and 1% penicillin/streptomycin (10,000 U/mL/10,000 μg/mL). Cell proliferation was determined using ^{3}H-thymidine uptake into DNA. Cultures were maintained for 0, 3, 7, 10 and 14 days. At each time point, cells were labeled with ^{3}H-thymidine (0.037 MBq/well or 0.5 μCi/well) for 18 h before harvesting. Cultures were washed with PBS twice before the addition of 500 μL of 5% TCA (twice) to remove unincorporated label, and the cells were then lysed in 0.1 N NaOH and 0.1% SDS for 2 h and harvested onto glass fiber filters. The filters containing ^{3}H-thymidine-labeled DNA were counted using a Perkin-Elmer liquid scintillation counter. The results are expressed in counts per minute (CPM).

3.9. Alkaline Phosphatase Assay (ALP)

Human MSCs from the dental pulp of exfoliated teeth (SHEDs) were cultured on the top of the scaffolds for 14 days. Half of the scaffolds and cells were maintained in ODM, and the other half were cultured in DMEM only. For the ALP assay, cells were lysed in lysis buffer composed of 1% Triton X-100, 0.9% NaCl and 0.5 M Tris (pH 9.0) under agitation for 30 min at 4 °C. Cell suspensions were then maintained in an ultrasonic bath for 10 min and centrifuged at 1500 rpm for 15 min before supernatant collection. ALP activity assays were performed according to the manufacturer's protocol (Labtest, Monte Claros, Brasil), and the absorbance at 590 nm was measured using a spectrophotometer (Tecan, Infinite 200 PRO, Mannedorf, Switzerland).

3.10. Cell Distribution on Scaffolds

The cell distributions on the scaffolds were determined by fluorescent staining with nucleus marker 4',6-diamidino-2-phenylindole (DAPI) at an excitation wavelength of 358 nm and an emission wavelength of 461 nm. After 14 days of culture in ODM, the scaffolds were rinsed with PBS and immersed in a 14.3 mM DAPI solution for 5 min. The scaffolds were rinsed three times with PBS and mounted in glass slides using Vectashield as the mounting medium. The scaffolds were analyzed using a confocal microscope (Zeiss, LMS 510 META, Jena, Germany).

3.11. Statistical Analyses

For each test, data were subjected to one-way ANOVA and Tukey's test ($\alpha = 0.05$) while considering homoscedasticity and normality.

Acknowledgments

The authors would like to thank FAPESP (2010/17698-2 and 2011/21442-6) and CNPq for funding this project at Catalani's laboratory and BNDES, CNPq, FAPESP, FINEP, MCTI and MS-DECIT for funding Sogayar lab.

Author Contributions

Flávia Gonçalves, Luiz H. Catalani and Mari C. Sogayar conceived and designed the experiments; Ricardo Bentini (mechanical properties), Ana C. O. Carreira (cell proliferation assay), Patricia Kossugue (SHEDs isolation and culturing) and Flávia Gonçalves (other experiments) performed the experiments; Luiz H. Catalani, Flávia Gonçalves and Mariana C. Burrows analyzed the data; Mari C. Sogayar and Luiz H. Catalani contributed reagents and materials; Flávia Gonçalves, Luiz H. Catalani and Mari C. Sogayar wrote and revised the manuscript. Ana C. O. Carreira and Mari C. Sogayar discussed the experiments and the manuscript.

Conflicts of Interest

The authors declare no conflict of interest.

References

1. Formhals, A. Process and Apparatus for Preparing Artificial Threads. U.S. Patent 1,975,504, 2 October 1934.
2. Bhardwaj, N.; Kundu, S.C. Electrospinning: A fascinating fiber fabrication technique. *Biotechnol. Adv.* **2010**, *28*, 325–347.
3. Hong, J.K.; Madihally, S.V. Next generation of electrosprayed fibers for tissue regeneration. *Tissue Eng. Part B Rev.* **2011**, *17*, 125–142.

4. Sabir, M.I.; Xu, X.X.; Li, L. A review on biodegradable polymeric materials for bone tissue engineering applications. *J. Mater. Sci.* **2009**, *44*, 5713–5724.

5. Ma, Z.; Gao, C.; Gong, Y.; Ji, J.; Shen, J. Immobilization of natural macromolecules on poly-L-lactic acid membrane surface in order to improve its cytocompatibility. *J. Biomed. Mater. Res.* **2002**, *63*, 838–847.

6. Van Wachem, B.; Beugeling, T.; Feijen, J.; Bantjes, K.; Detmers, J.P.; van Aken, W.G. Interaction of cultured human endothelial cells with polymeric surfaces of different wettabilities. *Biomaterials* **1985**, *6*, 403–408.

7. Orgel, J.P.; San Antonio, J.D.; Antipova, O. Molecular and structural mapping of collagen fibril interactions. *Connect. Tissue Res.* **2011**, *52*, 2–17.

8. Cen, L.; Liu, W.; Cui, L.; Zhang, W.; Cao, Y. Collagen tissue engineering: Development of novel biomaterials and applications. *Pediatr. Res.* **2008**, *63*, 492–496.

9. Bottino, M.C.; Thomas, V.; Schmidt, G.; Vohra, Y.K.; Chu, T.M.; Kowolik, M.J.; Janowski, G.M. Recent advances in the development of GTR/GBR membranes for periodontal regeneration—A materials perspective. *Dent. Mater.* **2012**, *28*, 703–721.

10. Glowacki, J.; Mizuno, S. Collagen scaffolds for tissue engineering. *Biopolymers* **2008**, *89*, 338–344.

11. Zhu, Y.; Gao, C.; Liu, Y.; Shen, J. Endothelial cell functions *in vitro* cultured on poly(L-lactic acid) membranes modified with different methods. *J. Biomed. Mater. Res. A* **2004**, *69*, 436–443.

12. Yan, S.; Li, X.Q.; Liu, S.P.; Wang, H.S.; He, C.L. Fabrication and properties of PLLA-gelatin nanofibers by electrospinning. *J. Appl. Polym. Sci.* **2010**, *114*, 542–547.

13. Zeugolis, D.I.; Khew, S.T.; Yew, E.S.; Ekaputra, A.K.; Tong, Y.W.; Yung, L.Y.; Hutmacher, D.W.; Sheppard, C.; Raghunath, M. Electro-spinning of pure collagen nano-fibres—Just an expensive way to make gelatin? *Biomaterials* **2008**, *29*, 2293–2305.

14. Liu, T.; Teng, W.K.; Chan, B.P.; Chew, S.Y. Photochemical crosslinked electrospun collagen nanofibers: Synthesis, characterization and neural stem cell interactions. *J. Biomed. Mater. Res. A* **2010**, *95*, 276–282.

15. Chen, J.; Chu, B.; Hsiao, B.S. Mineralization of hydroxyapatite in electrospun nanofibrous poly(L-lactic acid) scaffolds. *J. Biomed. Mater. Res. A* **2006**, *79*, 307–317.

16. Zhu, Y.; Gao, C.; Liu, X.; He, T.; Shen, J. Immobilization of biomacromolecules onto aminolyzed poly(L-lactic acid) toward acceleration of endothelium regeneration. *Tissue Eng.* **2004**, *10*, 53–61.

17. Zhang, H.; Lin, C.Y.; Hollister, S.J. The interaction between bone marrow stromal cells and RGD-modified three-dimensional porous polycaprolactone scaffolds. *Biomaterials* **2009**, *30*, 4063–4069.

18. Cui, W.; Cheng, L.; Li, H.; Zhou, Y.; Zhang, Y.; Chang, J. Preparation of hydrophilic poly(L-lactide) electrospun fibrous scaffolds modified with chitosan for enhanced cell biocompatibility. *Polymer* **2012**, *53*, 2298–2305.

19. Zhu, Y.; Leong, M.F.; Ong, W.F.; Chan-Park, M.B.; Chian, K.S. Esophageal epithelium regeneration on fibronectin grafted poly(L-lactide-co-caprolactone) (PLLC) nanofiber scaffold. *Biomaterials* **2007**, *28*, 861–868.

20. Tan, H.; Wan, L.; Wu, J.; Gao, C. Microscale control over collagen gradient on poly(L-lactide) membrane surface for manipulating chondrocyte distribution. *Colloids Surf. B Biointerfaces* **2008**, *67*, 210–215.

21. Yang, L.; Fitie, C.F.; van der Werf, K.O.; Bennink, M.L.; Dijkstra, P.J.; Feijen, J. Mechanical properties of single electrospun collagen type I fibers. *Biomaterials* **2008**, *29*, 955–962.

22. Bramfeldt, H.; Vermette, P. Enhanced smooth muscle cell adhesion and proliferation on protein-modified polycaprolactone-based copolymers. *J. Biomed. Mater. Res. A* **2009**, *88*, 520–530.

23. Zeng, J.; Chen, X.; Liang, Q.; Xu, X.; Jing, X. Enzymatic degradation of poly(L-lactide) and poly(epsilon-caprolactone) electrospun fibers. *Macromol. Biosci.* **2004**, *4*, 1118–1125.

24. Khor, E. Methods for the treatment of collagenous tissues for bioprostheses. *Biomaterials* **1997**, *18*, 95–105.

25. Kriegel, C.; Arecchi, A.; Kit, K.; McClements, D.J.; Weiss, J. Fabrication, functionalization, and application of electrospun biopolymer nanofibers. *Crit. Rev. Food Sci. Nutr.* **2008**, *48*, 775–797.

26. Kim, T.G.; Park, T.G. Biodegradable polymer nanocylinders fabricated by transverse fragmentation of electrospun nanofibers through aminolysis. *Macromol. Rapid Commun.* **2008**, *29*, 1231–1236.

27. Sun, M.; Downes, S. Physicochemical characterisation of novel ultra-thin biodegradable scaffolds for peripheral nerve repair. *J. Mater. Sci. Mater. Med.* **2009**, *20*, 1181–1192.

28. Carlisle, C.R.; Coulais, C.; Guthold, M. The mechanical stress-strain properties of single electrospun collagen type I nanofibers. *Acta Biomater.* **2010**, *6*, 2997–3003.

29. Ji, J.; Bar-On, B.; Wagner, H.D. Mechanics of electrospun collagen and hydroxyapatite/collagen nanofibers. *J. Mech. Behav. Biomed. Mater.* **2012**, *13*, 185–193.

30. Ngiam, M.; Liao, S.; Patil, A.J.; Cheng, Z.; Yang, F.; Gubler, M.J.; Ramakrishna, S.; Chan, C.K. Fabrication of mineralized polymeric nanofibrous composites for bone graft materials. *Tissue Eng. Part A* **2009**, *15*, 535–546.

31. Heino, J. The collagen family members as cell adhesion proteins. *Bioessays* **2007**, *29*, 1001–1010.

32. Leitinger, B. Transmembrane collagen receptors. *Annu. Rev. Cell Dev. Biol.* **2011**, *27*, 265–290.

33. Cai, K.; Kong, T.; Wang, L.; Liu, P.; Yang, W.; Chen, C. Regulation of endothelial cells migration on poly(D, L-lactic acid) films immobilized with collagen gradients. *Colloids Surf. B Biointerfaces* **2010**, *79*, 291–297.

34. Chua, K.N.; Chai, C.; Lee, P.C.; Tang, Y.N.; Ramakrishna, S.; Leong, K.W.; Mao, H.Q. Surface-aminated electrospun nanofibers enhance adhesion and expansion of human umbilical cord blood hematopoietic stem/progenitor cells. *Biomaterials* **2006**, *27*, 6043–6051.

35. Mizuno, M.; Kuboki, Y. Osteoblast-related gene expression of bone marrow cells during the osteoblastic differentiation induced by type I collagen. *J. Biochem.* **2001**, *129*, 133–138.

36. Schneider, G.B.; Zaharias, R.; Stanford, C. Osteoblast integrin adhesion and signaling regulate mineralization. *J. Dent. Res.* **2001**, *80*, 1540–1544.

37. Schofer, M.D.; Boudriot, U.; Wack, C.; Leifeld, I.; Grabedunkel, C.; Dersch, R.; Rudisile, M.; Wendorff, J.H.; Greiner, A.; Paletta, J.R.; *et al.* Influence of nanofibers on the growth and osteogenic differentiation of stem cells: A comparison of biological collagen nanofibers and synthetic plla fibers. *J. Mater. Sci. Mater. Med.* **2009**, *20*, 767–774.

38. Dong, B.; Arnoult, O.; Smith, M.E.; Wnek, G.E. Electrospinning of collagen nanofiber scaffolds from benign solvents. *Macromol. Rapid Commun.* **2009**, *30*, 539–542.

39. Sisson, K.; Zhang, C.; Farach-Carson, M.C.; Chase, D.B.; Rabolt, J.F. Fiber diameters control osteoblastic cell migration and differentiation in electrospun gelatin. *J. Biomed. Mater. Res. A* **2010**, *94*, 1312–1320.

MDPI AG
Klybeckstrasse 64
4057 Basel, Switzerland
Tel. +41 61 683 77 34
Fax +41 61 302 89 18
http://www.mdpi.com/

Materials Editorial Office
E-mail: materials@mdpi.com
http://www.mdpi.com/journal/materials

www.ingramcontent.com/pod-product-compliance
Lightning Source LLC
Chambersburg PA
CBHW051923190326

41458CB00026B/6384